ADVANCED TOPICS IN SCIENCE AND TECHNOLOGY IN CHINA

国家科学技术学术著作出版基金资助出版

王　平　刘清君
吴春生　Chung-Chiun Liu　编著

生物医学传感与检测
Biomedical Sensors and Measurement

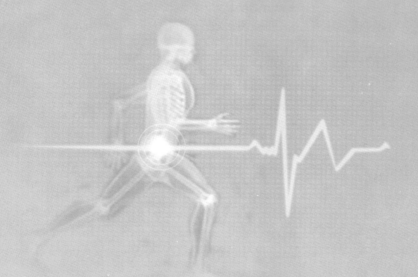

浙江大学出版社

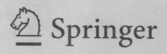
Springer

图书在版编目(CIP)数据

生物医学传感与检测 / 王平等编著. —杭州：浙江大学出版社，2012.10

ISBN 978-7-308-10069-4

Ⅰ.①生… Ⅱ.①王… Ⅲ.①生物传感器—检测 Ⅳ.①TP212.3

中国版本图书馆CIP数据核字(2012)第120335号

Not for sale outside Mainland of China
此书仅限中国大陆地区销售

生物医学传感与检测

王 平　刘清君　吴春生　C. C. Liu(美)　编著

责任编辑	张　琛 (zerozc@zju.edu.cn)
封面设计	俞亚彤
出版发行	浙江大学出版社
	(杭州市天目山路148号　邮政编码310007)
	(网址：http://www.zjupress.com)
排　版	杭州中大图文设计有限公司
印　刷	杭州杭新印务有限公司
开　本	787 mm × 1092 mm　1/16
印　张	18.75
字　数	680千
版 印 次	2012年10月第1版　2012年10月第1次印刷
书　号	ISBN 978-7-308-10069-4
定　价	48.00元

版权所有　翻印必究　　印装差错　负责调换
浙江大学出版社发行部邮购电话　(0571)88925591

Forward

This book introduces the basic fundamentals of biomedical sensors and the measurement technology as well as the recent advancements in this field in recent years. There are five chapters in this book which can be subdivided into two major parts. The first part places emphasis on the fundamentals and the development of modern biomedical sensors and measurement technology including their basic features and special requirement in application. This part also provides essential information on the basic sensitive reaction mechanisms, characteristics and processing approaches of the biomedical sensors. The second part introduces the typical sensors including the physical, chemical and biological sensors as well as the discussion of their measurement techniques. The practical applications of each of these sensors are also described in detail. There are two unique features in this book: (1) The combination of the discussion on the biomedical sensor technologies and their required measurement techniques which include the fundamentals and the practical applications of the biomedical sensors; (2) The rationale and the needs of the integration of discrete sensing elements into a meaningful and practical sensor array which can become an intelligent sensing system. The authors have given a very persuasive and sound approach in this important scientific and practical endeavor. It also should be acknowledged that the authors have systematically provided a clear roadmap for the development of various sensors by first introducing macro-size sensors for the detection of physical properties and then leading to the advancement of micro-size chemical and biological sensing systems.

    The advancement of the micro-size for chemical, biological and biomedical sensors or sensor micro-systems requires multi-disciplinary skills and expertise. This includes the understanding and expertise in microfabrication and micromachining processing, electronic and ionic conductive materials, sensor operational principles, electronic transduction interface technologies and many others. In this book, the authors have logically and systematically discussed and analyzed the interwoven relationship of these techniques and their applications to the development of scientifically and commercially sound practical chemical, biological and biomedical sensors.

    The authors have been engaging, for many years, in the research and teaching of sensor technology, particularly in the field of biomedical sensors. They have been involved in this research effort more than a decade and, their appreciation of the multi-disciplinary nature of the sensor research and the unique requirements for the advancement of the biomedical sensors and their measurement techniques can be well recognized throughout this book. As mentioned, this book is derived from the research work on biomedical sensors and measurement, in recent years, at Zhejiang University, Hangzhou, China. While it also contained many references to the teaching materials from Case Western Reserve University, Clevland, USA. Thus, this book will serve as an excellent reference source for researchers in biomedical sensor and measurement. It is intended for scientists, engineers, and manufacturers involved in the development, design, and application of biomedical sensors and measurement. The reference list given in each chapter is very thorough and relevant. This book will also be a very good book for the senior undergraduate and graduate students who wish to pursue a professional career in this field.

Forward

Biomedical sensors and the measurement are scientifically and commercially important in numerous applications at this juncture, and this book will be most welcome for researchers and students who wish to understand this field further and to make a meaningful contribution to this important endeavor.

Chung-Chiun Liu
Wallace R. Persons Professor of Sensor Technology and Control
Professor of Chemical Engineering
Director, Electronics Design Center
Case Western Reserve University
Cleveland, Ohio
U.S.A.
August, 2012

Preface

In the 21st century, technological innovation has had such a rapid development that science and technology permeates all aspects of our lives, especially in the biomedical field which attracts a large number of professional scientists and engineers. Biomedical engineering is a combination of two developing areas: biomedicine and information technology. It promotes the biomedical sensor design and development, as well as applications in clinical diagnosis and treatment technology. Biomedical engineering covers many research areas: bio-mechanics, bio-materials, physiological modeling and sensing technology, detection technology, signal and image processing. One important research field is biomedical sensor and measurement technology, which obtains original information from primitive organisms (especially human body), one of the most crucial procedures.

In the 1960s, scientists and engineers paid more attention to sensors for they met many of the practical requirements. The development of chemical and biological sensors creates selective sensors, which makes the direct detection of a variety of ions and molecules possible. Micro-sensors and micro-electrodes quickly replaced traditional large-size sensors and were applied to biological and medical fields.

At present, the quick body digital thermometer, blood pressure monitors, and wearable home-used blood glucose meter have been widely used. CT (computed tomography scanning) and ultrasound technology are recognized as common advanced diagnostic tools. However, many have omitted that sophisticated sensors play an important role in these instruments. Sensors have brought about revolutionary changes in the field of biomedical diagnostics and application of medical instrumentation, and it will have a positive effect on human life quality in the 21st century. It has the following applications:

- Digital medical image tools like CT, ultrasound, etc.;
- For the traditional image tools such as X-ray machines, it improves and gets more information and reduces the amount of radiation;
- Portable clinical multi-parameter monitoring equipment;
- Portable home-use monitoring and diagnostic equipment;
- Implantable, self-calibration equipment which will be widely used in the future;
- Intelligent systems of sensors that can replace our sense system, such as sight, hearing, touch, smell and taste, etc.;
- Rapid diagnosis tools based on immunization and DNA-chip technology.

Although biomedical sensors are being applied more and more, in many cases, the theory is not entirely clear. It's controversial in the expression of stimulus signal theory, signal extraction and measurement. The development of new biomedical sensors indicates a great fundamental research work, which is the key part at present.

Biomedical sensors convert biomedical signals into easy-to-measure electrical or optical signals. It is the interface between organisms and electronic systems. Meanwhile, effective detection technology, including low-noise and anti-jamming circuits and data processing techniques are essential during the conversion from biomedical to electronic signals, as well as for further processing. This book adds the measurement technology with sensing technology according to the actual teaching requirement, so that students and other researchers may systematically learn and comprehend the relevant knowledge in this field.

This book can be used as a reference book for researchers and senior undergraduates and graduate students. This book combines measurement technology with sensor technology and strengthens the links between them. In addition, the authors have added an introduction of regular physical sensors and chemical sensors, reorganized and reviewed the latest international development trends of chemical sensors, biological sensors and their intelligent systems, such as electronic nose, electronic tongue, microfluidic chips and micro-nano biosensors, and their applications.

Biomedical sensors and measurement techniques require synthesizing the interdiscipline of physics, electronics, materials, chemistry, biology, and medicine, etc. The authors are trying to meet this requirement through a detailed description of working principles, sensing technology, detection circuit, and identification system theory of sensors or devices. We believe that this book will be of great value for those academics, engineers, graduates and senior undergraduates in the biomedical and relevant fields.

The book is composed of five chapters. Chapter 1 introduces the development of biomedical sensors and measurement technology; Chapter 2 describes fundamental knowledge of modern sensors and measurement technologies; Chapter 3 describes the physical sensors and measurement technology; Chapter 4 describes chemical sensors and measurement techniques; and Chapter 5 describes the biosensors and measurement technology. Some content in this book belongs to the international research frontier. Biomedical sensors and measurement technology promotes the reorganization of biomedical information transmission, processing and perception, as well as the development of biomedical engineering and the interdisciplinary field.

The book is the result of many years of study, research and development of the faculties, PhD candidates and many others affiliated to the Biosensor National Special Laboratory of Zhejiang University. We would like to give particular thanks to Jun Wang, Wei Cai, Qi Dong, Gong Cheng, Di Wang, Jun Zhou, Cong Zhao, Lin Wang, Liang Hu, Kai Yu, Wen Zhang, Huixin Zhao, Liping Du, Ning Hu, Yishan Wang, Yingchang Zou, Liujing Zhuang, Ning Xu, Qian Zhang, and Xuanlang Zhang. We sincerely thank them all for their contributions.

And we are deeply grateful to the financial supporting for us overall systematization, teaching, and research work on biomedical topics through over the past ten years: by National Natural Science Foundation and National Distinguished Young Scholars Fund of China (Grant Nos. 30627002, 60725102, 30700167, 30970765, 81027003, 81071226, 31000448), Zhejiang Provincial Natural Science Foundation of China (Grant Nos. 2006CB021, 2010C14006, Y2080673), National Basic Research Program of China (973 Program, Grant No. 2009CB320303) and National High Technology Research Program of China (863 Program, Grant No. 2007AA09210106), State Key Laboratory of Transducer Technology of China (Grant Nos. SKT0702, SKT1101), etc.

As biomedical sensors and measurement involved in wide interdisciplinary areas, and in consideration of the limit of authors' knowledge and experiences, errors of judgment are, of course, inevitable, therefore, comments and suggestions will, be very appreciated.

Ping Wang
Hangzhou, China
August, 2012

Contents

Chapter 1 Introduction 1
  1.1 Definition and Classification of Biomedical Sensors 1
    1.1.1 Basic Concept of Sensors 1
    1.1.2 Classification of Biomedical Sensors 2
  1.2 Biomedical Measurement Technology 2
    1.2.1 Bioelectrical Signal Detection 3
    1.2.2 Biomagnetic Signal Detection 3
    1.2.3 Other Physiological and Biochemical Parameter Detection 4
  1.3 Characteristics of Biomedical Sensors and Measurement 4
    1.3.1 Features of Biomedical Sensors and Measurement 5
    1.3.2 Special Requirements of Biomedical Sensors and Measurement 5
  1.4 Development of Biomedical Sensors and Measurement 6
    1.4.1 Invasive and Non-Invasive Detection 7
    1.4.2 Multi-Parameter Detection 7
    1.4.3 *In vitro* and *in vivo* Detection 8
    1.4.4 Intelligent Artificial Viscera 9
    1.4.5 Micro-Nano Systems 10
    1.4.6 Biochips and Microfluidics 11
    1.4.7 Biomimetic Sensors 11
  References 12

Chapter 2 Basics of Sensors and Measurement 13
  2.1 Introduction 13
  2.2 Sensor Characteristics and Terminology 14
    2.2.1 Static Characteristics 14
    2.2.2 Dynamic Characteristics 16
  2.3 Sensor Measurement Technology 19
    2.3.1 Measurement Methods 19
    2.3.2 Sensor Measurement System 21
    2.3.3 Signal Modulation and Demodulation 23
    2.3.4 Improvement of Sensor Measurement System 30
  2.4 Biocompatibility Design of Sensors 31
    2.4.1 Concept and Principle of Biocompatibility 31
    2.4.2 Biocompatibility for Implantable Biomedical Sensors 34
    2.4.3 Biocompatibility for *in vitro* Biomedical Sensors 36

Contents

  2.5   Microfabrication of Biomedical Sensors ⋯ 39
       2.5.1   Lithography ⋯ 39
       2.5.2   Film Formation ⋯ 39
       2.5.3   Etching ⋯ 40
       2.5.4   Design of the Biomedical Sensors ⋯ 42
  References ⋯ 44

Chapter 3   Physical Sensors and Measurement ⋯ 47

  3.1   Introduction ⋯ 47
  3.2   Resistance Sensors and Measurement ⋯ 48
       3.2.1   Resistance Strain Sensors ⋯ 48
       3.2.2   Piezoresistive Sensors ⋯ 59
  3.3   Inductive Sensors and Measurement ⋯ 62
       3.3.1   Basics ⋯ 62
       3.3.2   Applications in Biomedicine ⋯ 80
  3.4   Capacitive Sensors and Measurement ⋯ 81
       3.4.1   The Basic Theory and Configuration of Capacitive Sensors ⋯ 81
       3.4.2   Measurement Circuits ⋯ 87
       3.4.3   Biomedical Applications ⋯ 90
  3.5   Piezoelectric Sensors and Measurement ⋯ 96
       3.5.1   Piezoelectric Effect ⋯ 96
       3.5.2   Piezoelectric Materials ⋯ 98
       3.5.3   Measurement Circuits ⋯ 99
       3.5.4   Biomedical Applications ⋯ 102
  3.6   Magnetoelectric Sensors and Measurement ⋯ 106
       3.6.1   Magnetoelectric Induction Sensors ⋯ 106
       3.6.2   Hall Magnetic Sensors ⋯ 110
  3.7   Photoelectric Sensors ⋯ 115
       3.7.1   Photoelectric Element ⋯ 115
       3.7.2   Fiber Optic Sensors ⋯ 123
       3.7.3   Applications of Photoelectric Sensors ⋯ 124
  3.8   Thermoelectric Sensors and Measurement ⋯ 126
       3.8.1   Thermosensitive Elements ⋯ 126
       3.8.2   Thermocouple Sensors ⋯ 128
       3.8.3   Integrated Temperature Sensors ⋯ 130
       3.8.4   Biomedical Applications ⋯ 132
  References ⋯ 134

Chapter 4   Chemical Sensors and Measurement ⋯ 137

  4.1   Introduction ⋯ 137
       4.1.1   History ⋯ 137
       4.1.2   Definition and Principle ⋯ 139
       4.1.3   Classification and Characteristics ⋯ 139

4.2	Electrochemical Fundamental		140
	4.2.1	Measurement System	140
	4.2.2	Basic Conception	142
	4.2.3	Classification of Electrodes	147
4.3	Ion Sensors		149
	4.3.1	Ion-Selective Electrodes	149
	4.3.2	Ion-Selective Field-Effect Transistors	155
	4.3.3	Light Addressable Potentiometric Sensors	159
	4.3.4	Microelectrode Array	163
4.4	Gas Sensors		167
	4.4.1	Electrochemical Gas Sensors	167
	4.4.2	Semiconductor Gas Sensors	170
	4.4.3	Solid Electrolyte Gas Sensors	177
	4.4.4	Surface Acoustic Wave Sensors	180
4.5	Humidity Sensors		183
	4.5.1	Capacitive Humidity Sensors	184
	4.5.2	Resistive Humidity Sensors	185
	4.5.3	Thermal Conductivity Humidity Sensors	186
	4.5.4	Application	187
4.6	Intelligent Chemical Sensor Arrays		189
	4.6.1	e-Nose	189
	4.6.2	e-Tongue	196
4.7	Micro Total Analysis System		201
	4.7.1	Design and Fabrication	201
	4.7.2	Applications	209
4.8	Sensor Networks		209
	4.8.1	History of Sensor Networks	210
	4.8.2	Essential Factors of Sensor Networks	210
	4.8.3	Buses of Sensor Networks	211
	4.8.4	Wireless Sensor Network	214
References			219

Chapter 5 Biosensors and Measurement ··· 223

5.1	Introduction		223
	5.1.1	History and Concept of Biosensors	223
	5.1.2	Components of Biosensor	224
	5.1.3	Properties of Biosensors	226
	5.1.4	Common Bioreceptor Components	226
5.2	Catalytic Biosensors		227
	5.2.1	Enzyme Biosensors	227
	5.2.2	Microorganism Biosensors	231
5.3	Affinity Biosensors		233
	5.3.1	Antibody and Antigen Biosensors	233

      5.3.2 Nucleic Acid Biosensors ··· 240
      5.3.3 Receptor and Ion Channel Biosensors ································· 244
5.4 Cell and Tissue Biosensors ··· 247
      5.4.1 Cellular Metabolism Biosensors ······································· 248
      5.4.2 Cellular Impedance Biosensors ······································· 251
      5.4.3 Extracellular Potential Biosensors ···································· 254
5.5 Biochips ·· 258
      5.5.1 Chips of Microarray ·· 259
      5.5.2 Gene and Protein Chips ·· 259
      5.5.3 Tissue and Cell Chips ·· 264
      5.5.4 Lab-on-a-Chip ·· 266
5.6 Nano-Biosensors ··· 268
      5.6.1 Nanomaterials for Biosensors ·· 269
      5.6.2 Nanoparticles and Nanopores Biosensors ·························· 270
      5.6.3 Nanotubes and Nanowires Biosensors ······························ 278
References ·· 284

Index ··· **289**

Chapter 1

Introduction

1.1 Definition and Classification of Biomedical Sensors

Sensors are devices that can be used to transform non-electrical signals into electrical signals. The biomedical sensors are very important kinds of sensors. First we will introduce some basic knowledge about biomedical sensors including the definition and classification.

1.1.1 Basic Concept of Sensors

Sensors or transducers are devices which can respond to measured quantities and transform the quantities into signals which can be detected. A sensor is usually composed of a sensitive component which directly responds to a measured quantity, a conversion component and related electronic circuits. Sensors often provide information about the physical, chemical or biological state of a system.

Measurement is defined as operations that aim to obtain the measured value of the quantity. Sensor measurement technology uses sensors to transform other measured quantities into physical quantities which are easier for communication and processing; then we can go on with display, recording and analysis. Along with the development of modern electronic technology, micro-electronic technology and communication technology, electrical signals are most convenient for processing, transportation, display, and recording; which makes electrical signals represent some of the various types of useful signals. Therefore, sensors are also narrowly defined as devices that can transform non-electrical signals into electrical signals.

Biomedical sensors are special electronic devices which can transfer the various non-electrical quantities in biomedical fields into easily detected electrical quantities. They expand the sensing function of the human sense organ. For this reason, they are the key parts of various diagnostic medical analysis instruments and equipment and health care analysis. Biomedical sensing technology is the key to collecting human physiological and pathological information and is an important disciplinary branch of biomedical engineering.

1.1.2 Classification of Biomedical Sensors

Biomedical sensors can be classified into the following categories according to their working principle, including physical sensors, chemical sensors, and biological sensors.

Physical sensors: It refers to the sensor made according to physical nature and effect. This kind of sensor is mostly represented by sensors such as metal resistance strain sensors, semiconductor piezoresistive sensors, piezoelectric sensors, photoelectric sensors.

Chemical sensors: It refers to the sensor made according to chemical nature and effect. This kind of sensors usually uses ion-selective sensitive film to transform non-electrical quantities such as a chemical component, content, density, to related electrical quantities, such as various ion sensitive electrodes, ion sensitive tubes, humidity sensors.

Biological sensors or biosensors: It refers to the sensors using biological active material as a molecule recognition system. This kind of sensors usually uses enzyme to catalyze some biochemical reactions or detect the types and contents of large organic molecules through some specific combination. It is a kind of newly developed sensors in the second half of the century, and examples include enzyme sensors, microorganism sensors, immunity sensors, tissue sensors, DNA sensors, etc.

Classified by detection type, there are displacement sensors, flow sensors, temperature sensors, speed sensors, pressure sensors, etc. As for pressure sensors, there are metal strain foil pressure sensors, semiconductor pressure sensors, capacity pressure sensors and other sensors that can detect pressure. As for temperature sensors, it includes thermal resistance sensors, thermocouple sensors, PN junction temperature sensors and other sensors that can detect temperature.

There is another method that classifies sensors according to the human sense organ that the sensor can replace, such as vision sensors, including various optical sensors and other sensors that can replace the visual function; hearing sensors, including sound pick-up sensors, piezoelectric sensors, capacity sensors and other sensors that can replace the hearing function; olfaction sensors, including various gas sensors and other sensors that can replace the smelling function (Harsányi, 2000). This kind of classification is good for the development of simulation sensors.

In many situations, these classification methods mentioned above are used together. For example, the strain gauge pressure sensor, conductance cardiac sounds sensor, thermoelectric glucose sensor and so on. The classification has met problems as a result of the diverse development of sensing technology. Therefore the classification methods have their advantages and disadvantages. Any standard classification method hasn't existed so far.

1.2 Biomedical Measurement Technology

Biomedical signals are commonly weak, highly random with strong noise and interference, allowing dynamic changes, and exhibiting significant individual differences. Therefore, biomedical measurement technologies are more complex and rigid than common industrial detection technologies.

Biomedical measurement is a guiding technology in the acquisition and processing of biomedical information and is directly related to the research of biomedical sensing technology, biomedical measurement methods, electronics and measuring systems. Therefore, the innovative research and development in biomedical measurement has a direct effect on the design and application of sensors and medical instruments.

Biomedical measurement technology involves the detection of physical, chemical and biological signals in different levels of organisms. For example, Electrocardiograph (ECG), Electroencephalogram (EEG), and Electromyogram (EMG) are electrical physiological signals; while blood pressure, body temperature, breath, blood flow and pulse are non-electrical physiological signals; blood and urine are chemical or biological signals; enzymes, proteins, antibodies and antigens are biological signals. Similarly, the biomedical measurement systems demand particular reliability and security.

Nowadays the measurement of physical signals has been popularized and many measurements of chemical signals have practical applications. The measurement of biological signals is mostly at the laboratory research stage. With a greater combination of microelectronics, optoelectronics, quantum chemistry and molecular biology with traditional sensing technology, the measurement methods and systems for detecting complex organisms will enjoy a brighter future. Biomedical measurement technology will also develop into mini-type, multiple-parameter and practical applications. The advancements of electronics, IC technology, computer technology and advanced signal processing and intelligent algorithms will promote the application of biomedical measurement.

1.2.1 Bioelectrical Signal Detection

The detection of physiological quantities in the circulatory system and nervous system develops relatively early and rapidly, and its importance always leads to a large amount of research reports in this field. Take ECG as an example. Many researchers are still working on automatic extraction and discriminating arrhythmia information from ECG under strong interference. In addition, the detection of the P wave and the ST segment in ECG, the research on obtaining an ECG of a fetus from a mother's body surface and on high frequency ECG and on body surface, the real-time detection and the late potential detection have been improved to different extents. ECG detection is mainly applied in diagnosing heart diseases and preventing sudden cardiac death. Moreover, it could also aid in surgical investigational procedures (Tigaran et al., 2009). Although these research achievements are not mature enough to be put into clinical use, they improve the function of ECG diagnosis and monitoring devices.

1.2.2 Biomagnetic Signal Detection

The biomagnetic field comes from the human body with biological electrical activities, such as Magnetocardiacgram (MCG), Magnetoencephalogram (MEG), Magnetomyogram (MMG). In addition, it also includes the magnetic field caused by the magnetic medium in the tissue when affected by an external magnetic field. An invasive strong magnetic mass can also cause an internal biomagnetic field. At present we can detect these magnetic fields in the laboratory. However, commonly a biological magnetic field is very weak. For example, the intensity of MCG is about 10^{-10} T and the intensity of MEG is about 10^{-12} T. Therefore SQUID (superconducting quantum interference device) in the liquid nitrogen container is used to detect the biological magnetic field and the measurement system should be placed in a special shielding environment.

In contrast to the detection of bioelectricity, the detection of a biomagnetic field has many features. Take the measurement of MCG as an example. The detection system does not directly come in contact with the organism, which means that the detection uses a detecting coil rather than an electrode to pick up the biological signals. Therefore it receives no effect from the surface of the objects and does not cause an electrode artifact, which is electrically safe. Besides, the detecting signals come from a certain spot or place rather than the difference between two points. Therefore, a location measurement can take place. The magnetoconductivity in tissue is well-distributed which means that biomagnetic signals will not distort when spreading in the body. As a result, research on biomagnetic detecting methods has become one of the pioneering and hot topics and has good application prospects. With the development of room temperature superconductor technology, biomagnetic detection will reach the clinical application stage.

1.2.3 Other Physiological and Biochemical Parameter Detection

It has become a common practice to use sensors non-invasively to detect non-invasive blood pressure, blood flow, breath, pulse, body temperature and cardiac sounds, which leads to wide applications in clinical examinations and other monitoring techniques. The trend is to develop new non-invasive or slightly invasive detecting methods and use one sensor each time to detect multiple physiological parameters. For example, we use the photoelectric method to detect the pulse as well as other information such as the heart rate, blood pressure, oxygen saturation; use electromagnetic coupling or optical coupling to detect intracranial pressure, and pressure in the mouth. Non-contact and long-distance detection also lead current development trends.

Biochemical parameter detection usually uses blood and body fluid as samples to conduct the measurement. Therefore, most of the methods are invasive and cannot measure the changes of the parameters over a long-time and in real-time. At present, non-invasive and slightly invasive biochemical parameter detecting methods have received great attention. For example, researchers have detected phenacetin in the saliva and compared it with the results of blood plasma tests; researchers extract lixivium by exerting a small amount of negative pressure on the skin and then using ion field effect transistor sensors to detect blood sugar; dielectric spectroscopy (DS) has been applied to monitor changes in the glucose level by combining electromagnetic and optical sensors (Talary et al., 2007).

1.3 Characteristics of Biomedical Sensors and Measurement

Biomedical sensors and measurement have specificities when used for human signal detection such as interdisciplinarity, knowledge-intensity, biocompatibility, which is non-invasive, safe and reliable. The measurement of biomedical sensors has become an important research area in recent years. In this chapter, the features and special requirements of biomedical sensors and measurement are introduced. In addition, the most different aspect in designing biomedical sensors from other sensors, biocompatibility, is also discussed.

1.3.1 Features of Biomedical Sensors and Measurement

Interdisciplinary Research: Biomedical sensor technology, as an active discipline, combines electronic science with biomedicine. Biomedical sensor technology meets the requirement of early diagnosis, quick diagnosis, bedside monitoring, monitoring *in vivo* and more advanced health care, and provides indispensable support for gene probes, molecular recognition, monitoring of neurotransmitters and neuromodulators, and more advanced scientific study. The developing disciplines such as microelectronics technology, biological technology, molecular biology and photonics technology lay the foundation for biomedical sensors technology. With such a background, biomedical sensor technology has made significant and rapid progress.

Basic research and technological innovation: In the 1970s, sensors were involved in the technological and scientific fields and focused on new product development. The basic research paid more attention to the advanced and high-level product exploration process. The primary target was a description of the molecular recognition mechanism, which is the basis of improving SNR, the mastery of the interface process, and the key to shortened response time. To put results into products, all kinds of processing technology including precision machining, semiconductor technology, chemical etching and biotechnology should be applied to technological innovation.

Sensitive materials and film formation technology: The core components of a sensor-sensitive membrane consist of sensitive materials combined with the matrix material. As to popular film formation technology, semiconductor thin-film, thick-film and molecular beam extension are used in physical sensors, physical adsorption and embedded technology, chemical cross-linking and molecular assembly for chemical sensors, and multi-enzyme system membranes, monoclonal antibody films, conductor films and LB film for biochemistry sensitive membranes.

Knowledge-intensive: Many disciplines are involved in sensor design, production and utilization. Take a chemical sensor for example. A knowledge of quantum chemistry is necessary for sensitive materials design; the same as that super-molecular chemistry, host-slave chemistry and biotechnology for materials synthesis; interface chemistry, physical interface and molecular assembly technology for film formation technology; microelectronic technology, photonics technology and precision machining for transfer devices.

High reliability: For a biosensor to be in direct contact with the human body, it must have high reliability. Sensors should be controlled strictly by the FDA in America and put on the market only if proved to be safe for the human body in the long term and to provide reliable monitoring data. Sensors detecting body fluids should be corrosion resistant and be easy to clean; embedded or implanted sensors should withstand rejection by the human body.

Fine technology: Fine technology is necessary for high-precise sensors. A matrix sensor, in the operation of integration technology, needs special implantable technology to reject leaking or deformation when soaked for a long time; coupling of a sensitive membrane and fiber cross Section requires fine technology; although a glass microelectrode can be stretched by certain machines. Precision machining is the combination of machining and chemical technology. The sensor is not only a product but also a fine artwork.

1.3.2 Special Requirements of Biomedical Sensors and Measurement

For biomedical measurement, it has specificity when used for human signal detection: it is a non-invasive, safe and reliable measurement. It has become an important research project in recent years. Non-invasive detection, which causes no wound or a slight wound, is easily received by people. It helps to keep the physiological status of objects and long-time or real-time monitoring can take place. Therefore it is convenient for clinical examination, monitoring and recovering evaluation. Non-invasive detection has become an important part of biomedical measurement technology.

Biomedical measurement research fields involve some special measurement method, e.g., low-noise and anti-interference technology, picking up signals and analyzing and processing technology and measurement systems and analog-digital circuits and computer hardware and software and even BCI (brain-computer interface) technology, etc. It also depends on the development of life sciences (such as cytophysiology, neurophysiology, biochemistry, etc.). The diversity of research objects in biomedical detecting technology make the research projects dispersive in this field. However, any promotion of detection methods in physiological quantities and biochemical quantities will greatly compel the advancement of the whole life science as well as the invention of new diagnosis and treatment devices.

The most different aspect in designing biomedical sensors from other sensors is the consideration of biocompatibility. Because this type of sensors directly makes contact with tissue or blood, the sensor design should include hemocompatibility and histocompatibility.

The first and the most important issue in manufacturing sensors is the material selection. The metallic materials used in sensors should be inert metals such as stainless steels, titanium alloys. The polymers should be degradable materials, such as PMMA (polymethyl methacrylate), silicones. All the materials used for sensor structure should be strictly selected to avoid serious host response and should function normally after being inserted into the animal body. The rigidness and flexibility of materials should also meet the requirement since the implanted sensors need to adjust to anatomical structures of the measured objects.

Secondly, a series of experiments on animals and clinical trials should be carried out before clinical applications. Besides choosing inert and least harmful materials at present, we still have to do full sequence tests for biocompatibility because the implanted sensors are under a different physiological environment.

Finally, we apply biological methods to evaluate the host responses. The *in vivo* biocompatibility can be evaluated by analyzing the cell population present, measuring the mediator and metabolite cells excreted, and analyzing the morphologic characteristics of the tissue and the capsule thickness around the implant.

Besides, some biological samples such as enzymes, proteins, cells and tissues have to be analyzed externally. An appropriate immobilization on the sensor surface is required for maintaining biological viability and activity. Hence, the biocompatibility for *in vitro* biomedical sensors should also be taken into account in sensor design.

1.4 Development of Biomedical Sensors and Measurement

Biomedical sensors and measurements have been developing rapidly over the past 30 years, and the development is represented in various aspects. The development of the medical sensors has basically changed the traditional mode, forming the development trend of smart, micro, multi-parameter, remote-control and non-invasive, and achieved some technical breakthroughs. Other new types of sensors such as DNA sensors, fiber sensors, and biological tissue sensors are also being developed. The revolution of medical sensor technology will help promote the development of modern medicine.

1.4.1 Invasive and Non-Invasive Detection

Miniaturization of sensors makes direct and continuous monitoring of vascular parameters (such as blood pressure, temperature, and flow rate) possible, and these sensors have become new clinical diagnosis tools. Although commercial products have been put into practical use, their practical potential has not been fully developed. Chemical and biological sensor technology plays an important role in the field of public health by supporting those who focus on rapid detection, high sensitivity and expertise. Clinical doctors also need a way to monitor patients with key metabolite concentrations of various diseases. A lot of effort is being made in this respect, using chemical and biological sensors to expand into non-traditional clinical chemistry analysis.

In addition, non-invasive detection of body fluids is being developed. Traditional body fluids (blood, urine, myeloid fluid, saliva, sweat, ascites, semen, etc.) are required to be extracted from patients, and most are invasive or *in vitro* measurements. In order to be able to carry out continuous measurement, it is necessary to develop different types of non-invasive or minimally invasive detection of body fluids.

Non-invasive detection means no or nearly no invasion during detection. Non-invasive detection, which has little effect on the human body, is not only more acceptable by receivers, but also more reliable, easy-to-operate, and easy for sterilization and results in the possibility of less infection. Non-invasive detecting sensors have a higher sensibility, accuracy, anti-interference and signal-to-noise rate.

Nondestructive monitoring is the most receptive monitoring method for the patient and has received widespread attention. At present, progress has been made in percutaneous blood gas sensors which can monitor blood gas non-invasively (P_{O_2}, P_{CO_2}), and the use of non-blood measuring to monitor blood glucose, urea, etc.

International research on biomedical sensor technology is synchronous with advances in the development of biomedicine. A major issue is how to improve the clinical technology and develop biomedical research.

Biosensors are bioanalytical devices that use biological materials such as proteins, cells and tissues as sensitive elements to be integrated with various physicochemical transducers for sensing the desired signals. Continuous research on biomedicine, physics, chemistry and electronics, and the discovery of materials and inventions quickly lead to important applications in the area of biosensors, such as micro-structure and the integration of biomedical sensors, biochips, nanotechnology sensors.

1.4.2 Multi-Parameter Detection

In clinical medicine, physiological parameters and multiple sensors usually can help to support clinical operations. Multi-parameter detection sensors are detection systems with small dimensions and multiple functions, using a single sensor system to measure multiple parameters simultaneously to obtain the functions of multiple sensors. Multi-parameter sensors integrate various sensitive components on one chip. Since the working condition is the same, it's easy to compensate and correct the system deviations.

Compared with using multiple sensors, these sensors have higher accuracy, better stability, smaller dimension, less weight and lower cost.

Serial operation is hard to perform when discrete sensors are used for monitoring different parameters. The method also has low efficiency and cannot meet simultaneous requirements in terms of time and space. Integrated technology creates conditions for multi-parameter sensors. At the beginning of the 1980s, British researchers invented an integrated blood electrolyte sensor which could monitor 5 parameters (Na^+, K^+, Ca^{2+}, Cl^- and pH). At the end of the 1980s, some researchers designed LAPS (light addressable potentiometric sensor) for monitoring multiple biochemical parameters (Wu et al., 2001).

Nowadays, the improvement of living standards requires continued advancement in diagnosis and therapy methods. Multi-parameter detection will have more significant applications in the biomedical field, especially with the development of MEMS (micro-electro-mechanical system) technology for developing more precise and smaller sensors and novel measuring technology. It's apparent that biomedical sensors will become more multifunctionalized and more miniaturized, with higher precision and integration.

1.4.3 *In vitro* and *in vivo* Detection

In vivo detection is a technology that detects the structure and function of living bodies while *in vitro* detection is a technology that detects the blood, urine, living tissues or pathological specimens *in vitro* (Gründler, 2007). These detecting technologies are very important in clinical laboratory tests. *In vitro* analysis and detection of tissue slices and blood or gas samples, aim to quantitatively analyze the composition and quantity of those substances, and to evaluate whether they are normal or whether there are some pathological microorganisms. *In vitro* detection requires high detection accuracy, precision and quick response. Because of the variety of detection categories, multiple kinds of automatic detection are required to make the most of the samples and testing reagents. Based on the requirements above, many detection methods have been developed, and new chemical and biological sensors have been invented. In addition to the update of conventional clinical analysis detection, the following new technological fields have made great improvements (Wang and Liu, 2009a):

- Minim and tracing element detection;
- Super minim hormone detection;
- Molecular level and cellular level detecting technology;
- Biosensor micro system development and applications;
- Cancer cells self-recognition;
- Chromosome automatic classification;
- DNA automatic analysis;
- Detection of olfactory and gustatory quantities.

As with an increase in the importance of clinical biochemical analysis and in the amount of analyzing samples and contents, *in vitro* detection is becoming multi-functional, continuous and automatic. All different kinds of automatic biochemical detecting devices, using the methods of optical analysis and electrochemical analysis, will rapidly improve with the development of computer automated recognition and analysis technology.

In vitro detection mostly belongs to the biochemical quantity detection field; although part of it belongs to image detection and automatic analysis. This detection involves many fields including gene engineering, protein engineering, LB (Langmuir-Blodgett) film techniques, biosensor techniques, image analysis process and measurement.

Monitoring *in vivo* can be done by observing physiological and pathological processes in real-time, from a fixed point over a long period of time (Hauser and Fhrs, 2009). *In vivo* monitoring provides important information which cannot be obtained in other ways. Along with the progress in emerging sensor technology, there are also wide ranges of monitoring technologies: implanted sensors can send information from inside to outside the body, and the catheter sensors can continuously detect gas/ion in intravascular blood or the heart. The main problem of *in vivo* monitoring is how to improve the compatibility between the organs and the issues (Vo-Dinh and Cullum, 2000).

Brain-computer interface (BCI), sometimes called a brain-machine interface, is a direct communication pathway between a brain and an external device. BCIs are often aimed at assisting, augmenting or repairing human cognitive or sensory-motor functions. Neuroprosthetics is an area of neuroscience concerned with neural prostheses, using artificial devices to replace the function of impaired nervous systems or sensory organs. Neuroprosthetics typically connect the nervous system to a device, whereas BCIs usually connect the brain (or the nervous system) with a computer system. Practical neuroprosthetics can be linked to any part of the nervous system—for example, peripheral nerves, while the term "BCI" usually designates a narrower class of systems which interface with the central nervous system.

Moreover, the sensor in a molecular system can identify proteins. The processor can ascertain the structure of the gene and the actuator can cut or unite the gene, which is the molecular system that can control and modify the gene and affect the life course. The design and synthesis of molecular systems is a new task for medicine and pharmacology. The research into anti-cancer drugs is moving in this direction.

The research has made achievements in two respects: one is the property of all magnetic waves and infrared light passing through skin and human tissue; the other is the coupling method of internal and external information. The common method for internal and external exchange is an echo response which implants the energy needed for *in vivo* detection and the controlling device from an *in vitro* body, and sends the detected signal *in vivo* to the body outside for later processing (Ricci et al., 2010). Another form is to send a stimulus or program-controlled signal into the body and to pick up the signal outside the body using a coupling coil. Take an implantable temperature detecting device as an example. A quartz crystal should be implanted *in vivo* to measure the temperature and a magnetic coupling coil should be placed externally. Linear FM (frequency modulation) signals are supplied *in vitro*, and the temperature is measured by using the linear relation between the crystal resonance frequency and temperature. Measurement error can be controlled under 0.1 °C and the method is quite stable over a long time.

1.4.4 Intelligent Artificial Viscera

The invention of an intelligent artificial pancreas provides a reference to intelligent artificial organs. There are a lot of relationships between the viscera and other organs. Actual artificial organs just have one function, so they cut down all the connections to other organs. Intelligent artificial viscera, which are equipped with a sensor system, microsystems or a molecular system, are intended to have all the functions of normal organs. Xenotransplantation is faced with problems of insurmountable rejection, so it will be an effective way to equip an anti-rejection molecular system on the transplanted organs

(Nakamura and Terano, 2008).

Detecting is one of the most important development trends. Usually the whole process including sampling, submitting and reporting takes over half an hour, which is highly disadvantageous in saving time and doing good surgery. To solve the problem, bedside monitoring sensors have been developed. Bedside monitoring sensors should be simple, durable, lightweight and be in a continuous or semi-continuous operation for the convenience of medical professionals (Ricci et al., 2009).

At the same time, wearable devices developed quickly in recent years. Smart textiles using fabric-based sensors have been utilized in biomedical applications, such as monitoring gesture, posture or respiration. Most of fabric-based sensors were fabricated by either coating piezo-resistive materials on a fabric or directly knitting conductive fibers into fabrics. The sensors used are generally physical sensors like resistance sensors, capacitance sensors and inductance sensors in bio-monitoring, rehabilitation, and telemedicine.

1.4.5 Micro-Nano Systems

Modern sensors have changed from traditional structure design and manufacturing technique to micromation. Micro-sensors are made by micro-mechanic technology, including photoengraving and corrosion. Its sensitive component is micron-level small.

Micro-sensors can enter part of the human body such as the inside of the viscera and disease focus that is unreachable for traditional sensors to get information. In addition, since the micro-sensor is very small, it largely reduces the impedance and effect on normal physiological activity, which makes the measured value more genuine and reliable.

Using silicon technology, it is possible to integrate a CPU and a miniature sensor on the same silicon chip, which promotes intellectualized micro-sensor technology. The microsystem is a silicon chip integrated micro-sensor, microprocessor and micro-actuator. Now comes the molecular biomedical era. At the level of the system, organs, tissues, cells and macromolecules, sensor detection types are changing with developments in life science research, from mechanical sensors, physical sensors, chemical sensors, and biosensors to molecular sensors (Table 1.4.1).

Table 1.4.1 Development history of sensors

Time	Electronic technology	Sensor technology
1960s	Vacuum tube	Normal sensor (cm)
1970s	Transistor	Small sensor (mm)
1980s	Microelectronics	Micro-sensor (μm)
1990s	Cell-molecular electronics	Cell-molecular sensor (μm, nm)
2000s	Nano-molecular electronics	Nano-sensor (nm)
2010s	Nano electro-mechanical system	Nano-sensor (nm)

On a technological level, sensor miniaturization and electronic devices miniaturization are being carried out at the same pace. A nanode, whose tip-diameter is of nanometer size, already exists. Moreover, in-nucleus detection has been on the agenda for some time.

Nano-technology involves many disciplines across advanced technologies that study the structure and property of substances at the level of 1 – 100 nm. The key to this technology is the study of how to make molecules produce substances and how to control the process, which is also called the molecular production process. Therefore, nano-technology brings opportunities to functionally sensitive materials in biosensors and compatible nano-devices. It also provides hope for molecular detection and diagnoses (Kricka, 2001).

Because of the specificity of its structure, nano-material has some specific effects, mainly the micro-size effect and interface effect. Nano-biosensors will play an important role in sensing and detection technology as it is different from typical sensors and has specific biological effects.

1.4.6 Biochips and Microfluidics

Current biochemical analyzers in the laboratory departments of domestic hospitals are large in size and expensive (thousands of dollars) and nearly all are dependent on import. According to the aim of developing economical biomedical engineering, both at home and abroad, low-input and high-output detecting devices must be emphasized (Mohanty and Kougianos, 2006; Whitesides, 2006). This has several advantages such as economic, easy to operate. Therefore the performance price ratio is much higher than large precision instruments of the same kind.

Early diagnoses should not depend too much on imaging apparatus. Biochemical changes take place earlier than organic changes, and immune sensors can quickly detect a-FT.

The gene controls cell activities and the processes of men's life. Gene detection is viewed as one of the core techniques in modern life science. Gene detection uses traditional biochemical methods and gene probes at present. The disadvantages of these methods are complicated operations and low efficiency, and an effective solution is to develop a DNA/RNA sensor. These researches are actively proceeding.

The cell is the basic unit of the human body, and the main human physiological and biochemical processes take place in the cells. It is a hot topic in life sciences to monitor ion incidents and molecular events in cells. Ion-selective microelectrode technology which is used for monitoring ion incidents is becoming more and more mature. And ion-selective microelectrode technology which is used to monitor molecular events is being researched.

1.4.7 Biomimetic Sensors

There are various sensors in the human body, which have good features such as high sensitivity, good selectivity, and high density. The development of biomimetic sensors is an important direction in biomedical sensor technology. There are many kinds of receptor sensors, nerve cell sensors, and biomimetic nerve cell sensors. The main problem with directly using biomaterials is that when the sensors leave their original environment they will lose their activity. The main solution is to use biomimetic chemistry to modify or synthesize sensitive materials (Zeravik et al., 2009; Wang and Liu, 2009b).

As two of the basic senses of human beings, olfactory and taste play a very important role in daily life. These two types of chemical sensors are important for recognizing environmental conditions. Electronic nose and electronic tongue, which mimic animals' smell and taste to detect odors and chemical components, have been carried out due to their potential commercial applications for biomedicine, food industry and environmental protection. The biomimetic artificial nose and tongue will be presented. Firstly, the smell and taste sensors mimicking the mammalian olfaction and gustation will be described, and then, some mimetic applications with the signal processing methods for odorants and tastants detecting will be developed. Finally, olfactory and gustatory biosensors are presented as the developing trends of this field.

References

Gründler P., 2007. *Chemical Sensors*. Springer, Germany.

Harsányi G., 2000. *Sensors in Biomedical Applications*. Technomic Publishing Company, Inc., 65-68.

Hauser R.G. & Fhrs F., 2009. Development and industrialization of the implantable cardioverter-defibrillator: A personal and historical perspective. *Cardiac Electrophysiology Clinics*. 1(1), 117-127.

Kricka L.J., 2001. Microchips, microarrays, biochips and nanochips: personal laboratories for the 21st century. *Clinica Chimica Acta*. 307(1-2), 219-223.

Mohanty S.P. & Kougianos E., 2006. Biosensors: a tutorial review. *Potentials, IEEE*. 25, 35-40.

Nakamura T. & Terano A., 2008. Capsule endoscopy: past, present, and future. *Journal of Gastroenterol*. 43, 93-99.

Ricci R.P., Morichelli L. & Santini M., 2009. Remote control of implanted devices through Home Monitoring™ technology improves detection and clinical management of atrial fibrillation. *Europace*. 11, 54-61.

Ricci R.P., Morichelli L., Quarta L., Sassi A., Porfili A., Laudadio M.T., Gargaro A. & Santini M., 2010. Long-term patient acceptance of and satisfaction with implanted device remote monitoring. *Europace*. 12(5), 674-679.

Talary M.S., Dewarrat F., Huber D. & Caduff A., 2007. In vivo life sign application of dielectric spectroscopy and non-invasive glucose monitoring. *Journal of Non-Crystalline Solids*. 353, 4515-4517.

Tigaran S., Rasmussen V., Dam M, Pedersen S., Hegenhaven H. & Friberg B., 2009. ECG changes in epilepsy patients. *Acta Neruologica Scandinavica*. 96(2), 72-75.

Vo-Dinh T. & Cullum B., 2000. Biosensors and biochips: advances in biological and medical diagnostics. *Fresenius' Journal of Analytical Chemistry*. 366, 540-551.

Wang P. & Liu Q., 2009 a. *Cell-based Biosensors: Principles and Applications*. Artech House Publishers, New York, USA.

Wang P. & Liu Q., 2009 b. Progress of biomimetic artificial nose and tongue. Olfaction and electronic nose: *Proceedings of the 13th International Symposium on Olfaction and Electronic Nose*. Brescia, Italy.

Wang P. & Ye X., 2003. *Modern Biomedical Sensing Technology* (in chinese). Zhejiang University Press, Hangzhou, China, 181-196.

Whitesides G.M., 2006. The origins and the future of microfluidics. *Nature*. 442, 368-373.

Wu Y., Wang P., Ye X., Zhang Q., Li R., Yan W. & Zheng X., 2001. A novel microphysiometer based on MLAPS for drug screening. *Biosensors & Bioelectronics*. 16, 277-286.

Zeravik J., Hlavacek A., Lacina K. & Skládal P., 2009. State of the art in the field of electronic and bioelectronic tongues—towards the analysis of wines. *Electroanalysis*. 21(23), 2509-2520.

Chapter 2

Basics of Sensors and Measurement

This chapter introduces basic sensor microfabrication technology, characteristics of sensors and measurement technology as well as common measuring methods and systems. Moreover, the serious concern about devices that comes from direct contact with the human body, which termed as "biocompatibility", will be discussed here.

2.1 Introduction

Biomedical sensing and measurement involves sensitive technology and measurement techniques. Different sensors are adopted to detect different types of measuring objects. Due to the diversity of the measurements, there are different sensor types, working principles and sensor structures. Despite the differences of various sensors, the evaluation methods of sensor characteristics are almost the same. Sensors working under certain environment will become less sensitive, which is related to the lifetime of sensors. Sensors produce a uniform response to the same input, which is related to repeatability of sensors. Such kinds of features are categorized into static characteristics and dynamic characteristics, which provide unified criteria for evaluation of different sensors. The most important and commonly mentioned properties of sensors are "4s": selectivity, sensitivity, stability, and safety. "Safety should be entirely considered in design and operation." Likewise, the measurement technology for different sensors is similar. A sensor detection system usually consists of sensors, measuring circuits and an output system. For different measuring purposes, circuits are designed differently such as parameter converting circuits, computing circuits and so on. The measurement methods can be classified by the measuring object's features, such as direct and indirect, active and passive, invasive and non-invasive, wired and wireless. For some implantable biomedical sensors in particular biological measurement environment, the compatibility between biological tissues and biomedical sensors is one of fundamental problems. This issue has been extended to sensor detection *in vitro*.

Therefore, this chapter tries to make a concrete introduction to the common parts of the biomedical sensors: the basic microfabrication technology, basic sensor characteristics, and measuring methods and systems. We also intend to emphasize the biocompatibility design of biomedical sensors because of its unique and important status. The following chapters will utilize the basic concepts and basic knowledge of biomedical sensors mentioned in this chapter.

2.2 Sensor Characteristics and Terminology

Since a sensor influences the characteristics of the whole measurement system, it is important to describe its performance. The characteristics of a sensor may be classified either static or dynamic, and these parameters are essential in describing its behavior. Static characteristics can be measured after all transient effects have stabilized to their final or steady state, while dynamic characteristics describe the sensor's transient properties, and they can be measured in the sensor's responses to the time-varying inputs.

2.2.1 Static Characteristics

Static characteristics are measured under the standard static condition, which means no acceleration, no vibration or shock (unless shock is the measurand), and that the temperature is (25 ± 5) °C, the relative humidity is lower than 85%, and the atmospheric pressure is (101.3 ± 8) kPa.

Under the standard static condition, using instruments with higher accuracy to measure the sensor's output repeatedly, the static calibration curve can be plotted depending on the measured data. The static characteristics can be obtained from the static calibration curve.

Because of the specificities of biomedical sensors, some static characteristics (such as lifetime, sensitivity, selectivity, etc.) are particularly important for them, and these characteristics are highlighted in this section.

Lifetime is the length of time that sensors remain sensitive under normal operational conditions. As biomedical sensors may sometimes be implanted into a body, they should be able to withstand the internal environment of the human body and maintain the normal functions meanwhile.

Selectivity is the ability of sensors to measure a single component in the presence of others. For example, a calcium ion-selective microelectrode that does not show a response to other ions such as Na^+, K^+ and Mg^{2+}, may be considered selective.

Sensitivity is the ratio between output signal and measured property. For example, if a gas sensor's output voltage increases 1 V when the oxygen concentration increases 1,000 ppm, then the sensitivity is 1 mV/ppm. In many cases, the signals measured by biomedical sensors are relatively weak. Therefore, sensitivity is a very meaningful parameter for biomedical sensors, which is especially significant in the measurement of electrophysiological signals.

A linear sensor's sensitivity is a constant, while for a nonlinear one, its sensitivity changes with inputs (Fig. 2.2.1). The static sensitivity can be obtained from the static calibration line, and for a linear sensor, its static sensitivity is the slope of the static calibration line.

Detection Limit is the minimum detectable signal that can be extracted in a sensing system when noise is taken into account. If the noise is large relative to the input, it is difficult to extract a clear signal from the surrounding noise. As the human body is a very complicated system, various signals impact each other. So for biomedical sensors, how to extract target signals from the complex noise background is an important issue.

Apart from the points described above, there are many other static characteristics extremely important.

Chapter 2 Basics of Sensors and Measurement

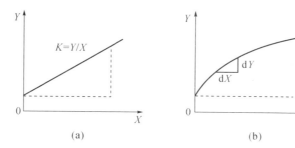

Fig. 2.2.1. Static sensitivity curves: (a) Linear sensor; (b) Nonlinear sensor

Linearity is the closeness of a sensor's calibration curve to a specified straight line, and the specified straight line is usually the sensor's theoretical behavior or its least-squares fit. Linearity is expressed as nonlinear error, which is the maximum deviation (Δ_{max}) of any calibration point from the corresponding point on the specified straight line, and it is usually denoted as ε:

$$\varepsilon = \frac{\pm \Delta_{max}}{Y_{F.S.}} \times 100\% \qquad (2.2.1)$$

Hysteresis is an error caused by when the measured property reverses direction, but there is some finite lag in time for the sensor to respond, creating a different offset error in one direction than in the other. Fig. 2.2.2 shows hysteresis loop of sensors. The maximum difference of the output ΔH_{max} over the full range is usually given to describe the hysteresis of sensors:

$$\delta_H = \pm \Delta H_{max} / Y_{F.S.} \times 100\% \qquad (2.2.2)$$

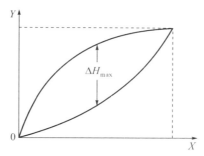

Fig. 2.2.2. Hysteresis curve

Repeatability is the sensor's ability to produce the same response for successive measurements of the same input, when all operating and environmental conditions remain constant.

Error is defined as the difference between the measured result and the true value, which is always inevitable in any measurement. But it is true that for all measurements, errors must be reduced by improvements in techniques and ideas, then the validity of results can be estimated.

Accuracy defines how correct the sensor output is, compared to the true value, and it is related to the bias of a set of measurements. Accuracy is measured by the absolute and relative errors, which are calculated by:

$$absolute\ error = result - true\ value \qquad (2.2.3)$$

$$\text{relative error} = \frac{\text{absolute error}}{\text{true value}} \qquad (2.2.4)$$

In order to assess the accuracy of a sensor, either the measurement should be calibrated against a standard measurand or the output should be compared with a measurement system with a known accuracy (Wilson, 2005).

Precision signifies the number of decimal places to which a measurand can be reliably measured. It refers to how carefully the final measurement can be read, not how accurate the measurement is. And it implies the agreement between successive readings, not the closeness to the true value.

Resolution signifies the smallest incremental change in the measurand that will result in a detectable increment in the output signal. It is the minimal change of the input necessary to produce a detectable change of the output.

Drift is the gradual change in the sensor's response while the measurand remains constant. Drift is the undesired and unexpected change that is unrelated to the input. It may be attributed to aging, temperature instability, contamination, material degradation, etc. (Kalantar-zadeh and Fry, 2008).

Dynamic range or span is the range of input signals that will result in a meaningful output for the sensor. All sensors are designed to perform over a specified range. Signals outside of this range may cause large inaccuracies, and may even result in damage to the sensor.

2.2.2 Dynamic Characteristics

Dynamic characteristics are the response characteristics to time-varying inputs. The response comprises two parts: the transient one and the steady-state one. The transient response is the process from an initial state to a final state. And the steady-state response is the output state when time is moving to infinity. Since the majority of physiological signals detected by biomedical sensors are functions of time, in order to obtain the real body information, the biomedical sensors should possess both good static and dynamic characteristics, which means the sensor's output curves should be the same or similar to the measurand curves during the same time period.

It is advantageous to use the linear differential equations with constant coefficients for sensing systems. Such representations have been widely studied and used, and they are easy to extract information from and give an overall vision about the sensing systems. The relationship between the input and output of any linear time invariant measuring system can be written as:

$$a_n \frac{d^n y(t)}{dt^n} + a_{n-1} \frac{d^{n-1} y(t)}{dt^{n-1}} + \ldots + a_1 \frac{dy(t)}{dt} + a_0 y(t) \\ = b_m \frac{d^m x(t)}{dt^m} + b_{m-1} \frac{d^{m-1} x(t)}{dt^{m-1}} + \ldots + b_1 \frac{dx(t)}{dt} + b_0 x(t) \qquad (2.2.5)$$

where $x(t)$ is the measured quantity (input signal) and $y(t)$ is the output reading and $a_0, \ldots, a_n, b_0, \ldots, b_m$ are constants.

According to the order of Eq. (2.2.5), there are three basic types of sensing systems: zero-order, first-order and second-order systems, which not only because these three basic types could describe most common biomedical sensing systems, but the more complex ones can also be approximately represented with these three types. In analysis of their dynamic characteristics, the step and sinusoidal signals are often regarded as typical experimental signals.

2.2.2.1 Zero-order sensors

The differential equation of zero-order system is:

$$a_0 y(t) = b_0 x(t) \tag{2.2.6}$$

and the transfer function is:

$$H(s) = b_0 / a_0 = K \tag{2.2.7}$$

where K is a constant and is defined as the static sensitivity. A zero-order sensor sensing system has ideal dynamic characteristics.

2.2.2.2 First-order sensors

They can be represented by such differential equations:

$$a_1 \frac{dy(t)}{dt} + a_0 y(t) = b_0 x(t) \tag{2.2.8}$$

or

$$\tau \frac{dy(t)}{dt} + y(t) = Kx(t) \tag{2.2.9}$$

and the transfer function is:

$$H(s) = K / (1 + \tau s) \tag{2.2.10}$$

where $K = b_0 / a_0$ is the static sensitivity, and $\tau = a_1 / a_0$ is the time constant.

The unit step response of a first-order sensor is:

$$y(t) = K(1 - e^{-t/\tau}) \tag{2.2.11}$$

which is shown in Fig. 2.2.3, and τ is the time constant when the output value reaches 63.7% of its steady-state value.

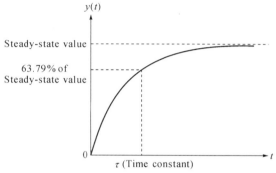

Fig. 2.2.3. Unit step response curve of a first-order sensor

2.2.2.3 Second-order sensors

On the other hand, the representation of a second-order system can be defined as either

$$a_2 \frac{d^2 y(t)}{dt^2} + a_1 \frac{dy(t)}{dt} + a_0 y(t) = b_0 x(t) \tag{2.2.12}$$

or

$$\frac{1}{\omega_0^2} \frac{d^2 y(t)}{dt^2} + \frac{2\xi}{\omega_0} \frac{dy(t)}{dt} + y(t) = Kx(t) \tag{2.2.13}$$

where $w_0 = \sqrt{a_0/a_2}$ is defined as natural frequency, $\xi = a_1/(2\sqrt{a_0 a_2})$ is damping ratio of the sensor, and $K = b_0/a_0$ is static sensitivity of the sensor. And the transfer function is:

$$H(s) = \frac{\omega_0^2 K}{s^2 + 2\xi\omega_0 s + \omega_0^2} \tag{2.2.14}$$

From the transfer function presented above, its amplitude-frequency and phase-frequency characteristics can be written as:

$$A(\omega) = \frac{K}{\sqrt{[1-(\omega/\omega_0)^2]^2 + 4\xi^2(\omega/\omega_0)^2}} \tag{2.2.15}$$

and

$$\Phi(\omega) = -\arctan\left[\frac{2\xi(\omega/\omega_0)}{1-(\omega/\omega_0)^2}\right] \tag{2.2.16}$$

Figs. 2.2.4 and 2.2.5 show the amplitude-frequency and phase-frequency characteristics of second-order sensors respectively, in which the damping ratio ξ has a great influence on the frequency characteristics. When $\xi > 1$, the amplitude-frequency curve would be decreasing without any peak. Thus the width of the amplitude-frequency curve's flat part is closely related to ξ, and when $\xi = 0.707$ around, the flat part has the largest width. On the other hand, an under-damped system ($0 < \xi < 1$) is faster to reach the steady state than a critical damped one ($\xi = 1$), while the over-damped system ($\xi > 1$) is slow to respond. Therefore, under-damped systems are generally used, and the value of ξ usually is 0.6 – 0.8.

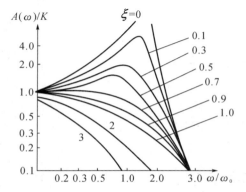

Fig. 2.2.4. Amplitude-frequency characteristics

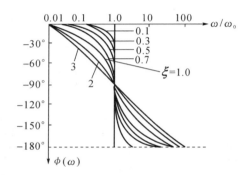

Fig. 2.2.5. Phase-frequency characteristics

Fig. 2.2.6 shows the response of a high-order system to the unit step input. Some parameters in time-domain can be defined as following:

Rise time t_r: the required time that the steady-state value rises from 10% to 90%.

Response time t_s: the required time that the output response maintains stability in the $\pm \Delta\%$ error tolerances.

Overshoot α: the value that the maximum output exceeds the steady-state value.

Delay time t_0: the required time that the output first reaches 50% of steady-state value.

Attenuation ψ: the percentage of height decline between two adjacent peaks (or troughs),
$$\psi = (\alpha - \alpha_1)/\alpha \times 100\%.$$

Among the performance parameters in time-domain above, τ, t_s and t_r reflect the response rate of a system, but ψ and α indicate the relative stability of a system. And it should be noted that these performance parameters are not all used under any case; for example, in a damping system, attenuation and overshoot are not necessarily utilized.

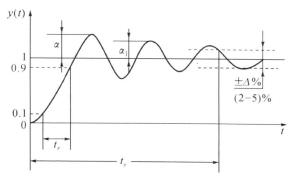

Fig. 2.2.6. Unit step response curve of a high-order sensor

2.3 Sensor Measurement Technology

The sensor measurement technology is used to detect signals. According to different kinds of classifications, the measurement methods can be classified as direct and indirect, active and passive, invasive and non-invasive, wired and wireless. To detect signals, a proper sensor system is needed. A sensor system should include the sensor interface, signal preprocessing circuits, as well as other computer-aided Digital Signal Processing hardware and software and so on. Sometimes, in some specific environments, it is difficult for a sensor system to detect signals. In these cases, some other methods are implemented to improve the performance of the sensor system.

2.3.1 Measurement Methods

In this Section we will introduce the following measurement methods: direct and indirect measurement, active and passive measurement, invasive and non-invasive measurement, and wired and wireless measurement.

2.3.1.1 Direct and indirect measurement

Direct measurement refers to measuring exactly the desired objects or things. For example, the measurement of a circuit by an electromagnetic current meter, and the measurement of pressure by a bourdon tube pressure gauge are both direct measurements.

Indirect measurement means that measuring something is found by measuring something else using proportions. In other words, you figure out the result from some measured parameters since the thing is a little harder to measure. For example, the measurement of inductance by measuring the resonant frequency of the circuit is an indirect measurement. This method has more steps and requires longer time, and generally, it is only used when direct measurement is inconvenient.

2.3.1.2 Active and passive measurement

According to the power supply of the measurement system, measurement is divided into active measurement and passive measurement.

- *Active measurement*

The structure of the active measurement system is shown in Fig. 2.3.1. It provides power to the measured object. For example, when measuring the impedance, we need to supply an electrical voltage perturbation to the measured sample.

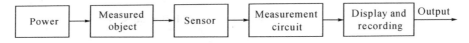

Fig. 2.3.1. Active measurement system

- *Passive measurement*

The structure of the passive measurement system is shown in Fig. 2.3.2. The passive measurement system does not need outside power supplies.

Fig. 2.3.2. Passive measurement system

2.3.1.3 Invasive and non-invasive measurement

Invasive measurement uses methods that would influence or even injure the objects while the non-invasive measurement would have tiny or no influence on the objects. For example, there are many methods to detect cancers, including gene detection (Aaroe et al., 2010), blood detection (Oono et al., 2010), PET-CT (Barbara et al., 2009), e-Nose (Wang et al., 2009) and so on. Among these methods, only the electronic nose is non-invasive as it just collects the patients' exhaled without any injuries. This is thought to be a great advantage of the electronic nose compared with other methods.

2.3.1.4 Wired and wireless measurement

In recent years, wireless measurement has developed very fast and is being used in more and more fields such as communication technology, biomedical fields and so on. For example, a RFID-Based Closed-Loop Wireless Power Transmission System has been reported in use in the biomedical field, especially for inductively powering implantable biomedical devices. In this system, the transmitter and receiver coils are in a wireless power transmission. Any changes in the distance and misalignment would cause a significant change in the received power. As it is a closed loop, this change would get detected. This system can be used on the implanted chips to detect diseases better (Mehdi and Maysam, 2010).

2.3.2 Sensor Measurement System

With the development of micro-processing technology, the composition of the sensor detecting system has also developed significantly. Smart sensor detecting systems with a microprocessor are the direction for modern sensor detecting technology. The basic components are shown in Fig. 2.3.3.

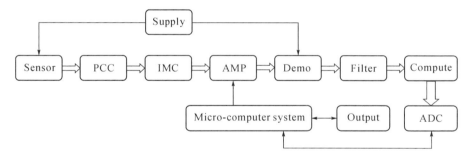

Fig. 2.3.3. Fundamental composition of sensor detecting system. PCC: parameter converting circuit; IMC: impedance converting circuit; AMP: amplifier; Demo: demodulation

The detecting system is generally composed of sensors, measuring circuits (sensor interface and signal pre-processing circuits) and the output system. The sensors detect the measured signals in the measurement environment, and convert them into electric signals. Sensors are generally called the primary instrument, while the following measuring circuits and output circuits are called the secondary instrument.

In the process of measurement, the measuring circuits can be divided into the sensor interface circuit and the pre-processing circuit. The sensor interface circuit connects the sensor and the pre-processing circuits. It is usually composed of the parameter conversion (basic conversion circuit), sensor output signal modulation and impedance matching circuit. The sensor interface circuit extracts the measured signal. The signal pre-processing circuit is usually composed of operations, demodulation, filtering, A/D and D/A. It detects the measured signal, and conducts the discrete signal processing if necessary. Intelligent sensing detection systems cannot only detect and process signals, but also self-diagnose and self-recover.

- *Parameter converting circuits*

The measured signal makes electrical parameters of the sensors change, such as resistance, inductance and capacitance. In general, these changes are converted into voltage or frequency signals that are proportional to them. Therefore, there must be parameter converting circuits (PCC).

The commonly used parametric conversion circuits include: the bridge circuit (converting resistance, inductance or capacitance into voltage); current-voltage (IV) converting circuit; resistance, inductance, capacitance (RLC) oscillation circuit (converting resistors, inductors, capacitors into voltage or frequency digital signal), etc.

- *Impedance converting circuits*

Sensors can be seen as a signal source with certain output impedance, while the following interface circuit has certain input impedance. In order to decrease the effects of the input impedance on the output signal, the impedance transformation is used in the design of the sensor interface circuit. By taking high input impedance amplifiers, the high output impedance of sensors is converted into low output impedance. For example, in the design of measuring circuits and the voltage equivalent circuit using piezoelectric sensor, voltage amplifier has the effects of impedance conversion and signal amplification.

- *Computing circuits*

In the measurement circuits, operation circuits mainly consist of ratio circuits, addition and subtraction circuits, integral circuits, differential circuits, logarithmic circuits, exponential circuits, multiplication and division circuits, etc.

Ratio computing: The outputs of the sensor are generally relatively small, usually in millivolt level which should be amplified. The integrated operational amplifier can be used as a signal amplification circuit device. According to different ways of op amp feedback connection, the integrated operational amplifier can be divided into the non-inverting proportional circuit (constituting voltage amplifier) and inverting proportional circuit (constituting current-voltage conversion amplifier).

Addition and subtraction circuits: To achieve the addition and subtraction of different signals and adjust the amplitude of the signal bias. Differential amplifier circuit is one of the basic types of such circuits.

Logarithmic and exponential circuits: To achieve the non-linear operation in the circuits. For example, logarithmic circuits could linearize the output of the exponential signals.

- *Analogue filter circuit*

In the measurement system, the analog filter in the signal pre-processing circuit is very important as it could select the needed frequency components. Filtering circuits could pass specific frequency components but greatly decrease other frequency components, so as to filter the noise. Considering the amplitude-frequency characteristics, filters are generally divided into four categories: low pass, high pass, band pass and band stop. Considering the components, filters can be divided into RC, LC, and crystal resonator filters. Considering the components of the filters, they can be divided into two types: active and passive filters. Considering the characteristics of amplitude-frequency and phase-frequency near the cut-off frequency, filters can be divided into Butterworth filters, Chebyshev filters, Bessel filters and so on.

- *DAC and ADC interface circuits*

Both the signals from the sensor to the sensor interface and those the signal pre-processing circuits involve are analog signals. In order to adopt DSP in the measurement system, analog signals must be converted to digital signals so that computers can handle them. A/D converters can convert analog signals to digital signals. The high frequency of sampling and the required accuracy of the measurements are the main technical requirements to choose the suitable A/D converter and D/A converter. A/D interface, D/A interface and the computer constitute an intelligent closed-loop control system.

- *Digital signal processing*

The use of computer and Digital Signal Processing technology greatly advance the measurement system. Computers can analyze, judge the signals, and display the results. For example, in the measurement and control system, the zero point error correction would be very simple under the microchip's control. In a system without a microchip, the only choice to reduce the zero-point error caused by temperature drift is to choose components with low drift coefficients and to design a temperature compensation circuit.

Digital Signal Processing has the advantages of stability and good repeatability, compared with analogue signal processing. To improve detection accuracy and reduce the cost of the hardware circuit, sensor linearization software is very applicable in engineering practices. With the development of advanced high-speed computer and the increase of storage capacity, technologies such as the multi-sensor detection have been widely used.

2.3.3 Signal Modulation and Demodulation

If the sensor's output values are relatively small, the noise voltage of the amplifier, the DC amplification temperature drift of the measuring circuits, zero-point drift and inter-stage coupling would result in serious measuring errors. In order to improve the anti-interference capability, the output signal of the sensor is modulated in a parametric conversion circuit so that the sensor would output an AC signal, with the change of the signal corresponding to signal amplitude, frequency or phase changes.

In the measuring circuit, narrow-band filter demodulation and the relevant demodulation can both be used to detect the measured signal. Generally the slow-varying signals which control the high-frequency oscillation are called modulation wave (measured signal); the high-frequency oscillatory signals containing the variables are called the carrier (the excitation signal in the parameters conversion circuit); and modulated high-frequency oscillatory waves are called the modulated wave (the sensor output signal). Demodulation processes the modulated wave to gain the slowly changing measured signal.

2.3.3.1 Signal modulation

Generally a high-frequency sinusoidal signal is used as the carrier signal. A sinusoidal signal has three parameters: amplitude, frequency, and phase, which can be modulated, referred to as amplitude modulation (AM), frequency modulation (FM) and phase modulation (PM) respectively. Pulse signals can also be used as the carrier signals. Different features of the pulse signal can be modulated. The most commonly used method is the pulse width modulation, called PWM.

- *Amplitude modulation*

Amplitude modulation (AM) uses the modulation signal to control the amplitude of the high-frequency carrier signal. A commonly used method is the linear amplitude modulation, that is, the amplitude of the AM signal changes with the modulation signal linearly. The process of AM is as follows: The modulation signal is multiplied by the high-frequency carrier signal, thus forming the AM signal. In this case, the amplitude of the modulation signal is contained in the AM signal.

Supposing the expression of the carrier signal is $u_c = U_0 \cos\omega_c t$, and the expression of the modulation signal is $u_x = X_0 \cos\Omega t$. The AM signal is generally expressed as follows:

$$u_s = U_0(1 + m\cos\Omega t)\cos\omega_c t$$
$$= U_0 \cos\omega_c t + \frac{mU_0}{2}\cos(\omega_c + \Omega)t + \frac{mU_0}{2}\cos(\omega_c - \Omega)t \tag{2.3.1}$$

where ω_c is the angular frequency of the carrier signal, m is the degree of modulation, $m = X_0/U_0$.

We can see from Eq. (2.3.1), u_s contains three kinds of frequencies: the angular frequency ω_c of carrier wave, $\omega_c+\Omega$ and $\omega_c-\Omega$ of side frequency signal. The bi-sideband signals can be extracted from the AM signal by removing the carrier signal, which can be expressed as follows:

$$u'_s = \frac{mU_0}{2}\cos(\omega_c + \Omega)t + \frac{mU_0}{2}\cos(\omega_c - \Omega)t \tag{2.3.2}$$

Fig. 2.3.4 shows the process of bi-sideband amplitude modulation.

- *Frequency modulation*

Frequency modulation (FM) uses the modulation signal to control the frequency of the high-frequency carrier signal. A commonly used method is linear FM, that is, the frequency of the FM signal changes linearly with the modulation signals. Thus, in the demodulation process, by measuring the change of the frequency of the FM signal, the modulation signal is detected. FM generally expressed as follows:

$$u_s = U_0 \cos(\omega_c + mu_x)t \tag{2.3.3}$$

where ω_c is the angular frequency of the carrier signal, U_0 is the amplitude of the carrier signal, m is the degree of modulation, u_x is the measured signal. Fig. 2.3.5 shows the schematic process of frequency modulation.

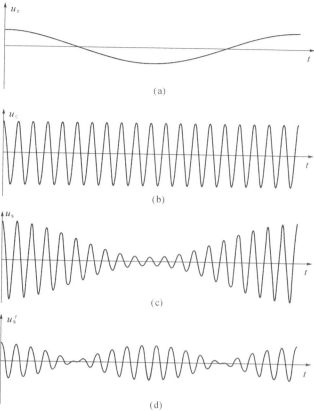

Fig. 2.3.4. The schematic process of bi-sideband modulation: (a) Modulation wave; (b) Carrier wave; (c) AM wave containing carrier wave; (d) Bi-sideband AM wave

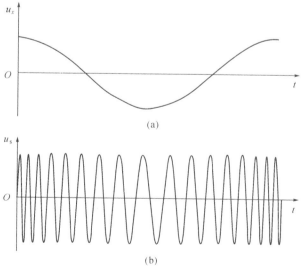

Fig. 2.3.5. The schematic process of frequency modulation: (a) Modulation signal u_x; (b) FM wave

To avoid frequency aliasing and be more convenient to demodulate the signal, ω_c should be far higher than the frequency of u_x.

Biomedical Sensors and Measurement

- *Phase modulation*

Phase modulation uses the modulation signal to control the phase of the high-frequency carrier signal. A commonly used method is the linear phase modulation, that is, the phase of the PM signal changes linearly with the modulation signal. In the demodulation process, the modulation signal would be detected by measuring the phase change of the PM signal. Phase modulation is generally expressed as follows:

$$u_s = U_0 \cos(\omega_c t + m u_x) \quad (2.3.4)$$

where ω_c is the angular frequency of the carrier signal, U_0 is the amplitude of the carrier signal, m is the degree of modulation, u_x is the measured signal. Fig. 2.3.6 shows the schematic process of phase modulation. The phase difference between PM signal and carrier signal varied with u_x. If $u_x < 0$, the phase of PM signal is behind that of the carrier signal, otherwise when $u_x > 0$, it is before the carrier signal.

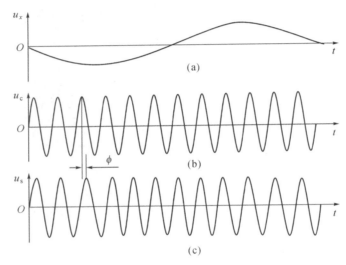

Fig. 2.3.6. The schematic process of phase modulation. (a) Modulation signal u_x; (b) The carrier wave; (c) PM wave

- *Pulse-width-modulation*

Pulse-width-modulation (PWM) is widely used in pulse modulation. PWM uses the modulation signal to control the width of the high-frequency pulse signal. The most commonly used method is the linear PWM, that is, the width of the PWM signal is linear with the amplitude of the modulation signal. The mathematic expression is as follows:

$$B = b + mx \quad (2.3.5)$$

where b is a constant, m is the degree of modulation, $\dot{x}$ is the modulation signal. Fig. 2.3.7 shows the schematic process of PM wave, (a) shows the waveform of the modulation signal x, (b) shows the waveform of the PWM signal, and T is the period of the pulses. We can see from Fig.2.3.7 that, when the amplitude of the modulation signal gets larger, the width of the PWM signal becomes larger accordingly, and vice versa. In other words, the duty cycle has been used to convey the information of the modulation signal. Through PWM modulation, the frequency band of PWM is among the narrow range with the frequency of the carrier signal as the center. For this reason, the noises with a variety of frequencies have little effect on the signal, which is also considered to be white noise. Thus, PWM signal has increased noise immunity.

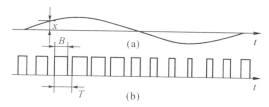

Fig. 2.3.7. The schematic diagram of PWM process. The waveform (a) of the modulation signal; (b) PWM

2.3.3.2 Signal demodulation or detection

The process in which the measured signal is extracted from the modulated signal is called demodulation. Amplitude modulation makes the amplitude of the AM signal change with that of the modulation signal, so that the envelope shape is consistent with that of the modulation signal. Demodulation is achieved as long as the envelope curve can be detected. This method is called envelop detection.

- **Envelope detection**

Envelope detection could demodulate the modulation signal by detecting the envelope. Fig. 2.3.8 shows the process of the envelope detection. Envelope detection is based on the principle of rectification. First, the bi-sideband AM signal is rectified with half-wave rectification circuit. Next, the high-frequency signal is filtered by low-pass filter, so that the low-frequency modulation one is obtained.

- **Phase-sensitive detection**

A phase-sensitive detection circuit can identify the phase of the modulation signal. There are two issues around envelope detection.

First, the main process of demodulation is the half-wave or full-wave rectification of the AM signal, in which the phase of the modulation signal cannot be detected from the output of the detector. For example, when an inductive transducer is used to detect the displacement of work pieces, when the core moves up and down for the same amount by its equilibrium position, the amplitudes of the sensor's output signal are the same, while the phase difference is 180°. From the output of the envelope detection circuit, it cannot be determined whether the core moves up or down.

Second, the envelope detector circuit itself cannot differentiate signals with different carrier frequencies. The envelope detection circuit rectifies carrier signals with different frequencies in the same way to restore the modulation signal, that is to say, it is not able to identify signals. In this case, a phase-sensitive detection circuit is adopted, so that the detection circuit can distinguish phases and frequencies of signals.

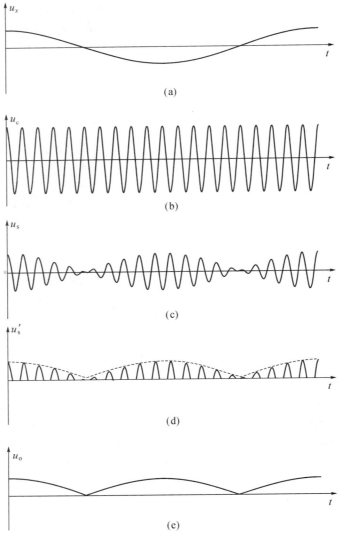

Fig. 2.3.8. Envelope detection: (a) Modulation signal; (b) Carrier wave; (c) Bi-sideband AM signal; (d) The waveform of signal with half-wave recticification; (e) The output waveform with envelop detection

Fig. 2.3.9 shows the schematic process of the modulation and phase-sensitive detection. Supposing the modulation signal $u_x = X_0 \cos\Omega t$ (Fig. 2.3.9a), the carrier signal $u_c = U_0 \cos\omega_c t$ (Fig. 2.3.9b), the expression of the bi-sideband AM signal (Fig. 2.3.9c) is as follows:

$$u_s = \frac{mU_0}{2}\cos(\omega_c + \Omega)t + \frac{mU_0}{2}\cos(\omega_c - \Omega)t \qquad (2.3.6)$$

where ω_c is the angular frequency of the carrier signal, m is the degree of modulation, $m = X_0/U_0$.

We can obtain the output signal u_{o1} (Fig. 2.3.9d) through multipling u_s by the carrier wave $\cos\omega_c t$:

$$u_{o1} = u_s \cos \omega_c t = \frac{mU_0}{2} \cos \Omega t + \frac{mU_0}{4} \cos(2\omega_c + \Omega)t + \frac{mU_0}{4} \cos(2\omega_c - \Omega)t$$
$$= \frac{X_0}{2} \cos \Omega t + \frac{X_0}{4} \cos(2\omega_c + \Omega)t + \frac{X_0}{4} \cos(2\omega_c - \Omega)t \qquad (2.3.7)$$

Finally, we can get the signal u_{o2}, by filtering u_{o1} with a low-pass filter. The result (Fig. 2.3.9e) contains the information of u_x, the expression is as follows:.

$$u_{o2} = \frac{X_0}{2} \cos \Omega t \qquad (2.3.8)$$

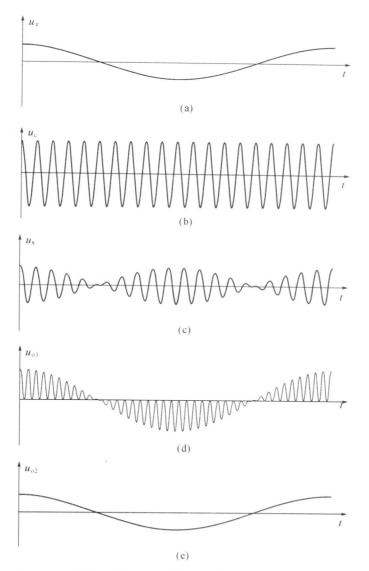

Fig. 2.3.9. The schematic process of AM modulation and Phase-sensitive detection. (a) Modulation signal; (b) Carrier wave; (c) AM signal; (d) The waveform of signal with phase-sensitive detection; (e) The demodulation wave

2.3.4 Improvement of Sensor Measurement System

The overall errors in the sensor detecting system are integrated by errors of three basic parts. Thus, to improve the detection performance of the system, the improvement of the measurement precision is one of the main tasks.

The sensor is the first input of a detection system, whose errors would be passed throughout the entire system. The relative errors caused by the sensor itself can be seen as an input to the following parts. Therefore, improving the accuracy of the sensor is of vital importance.

The design of sensors is the integration of mechanics, electricity, chemistry, biology, materials science and many other aspects. Thus, it is difficult to manufacture a sensor with good performance in all these aspects. However, in the actual application there may be some aspects that do not have influences on a particular signal. Therefore, when design and choose sensors, the characteristics of the measured signals which determine sensor performance should first be considered. It is not necessary to pursue the perfect performance. For example, in a static test, there is no need to pursue the dynamic performance. The following focuses on how to improve the sensor's performance:

- *Choosing proper sensors*

When design and choose sensors, the following requirements should be considered: the detecting environment, the characteristics of the detected signals, and the requirements for the test accuracy. With all the above aspects fulfilled, the type, structure, material, size, weight and life expectancy of the sensors should also be considered.

- *Stabilization technology*

The various materials and components that constitute the sensor would change with time and environment. Then the performance of the sensor would change, resulting in instability. In this case, it is necessary to stabilize materials, components or even the overall sensors (also known as the aging processing). Electrical materials, such as magnetic materials, conductive materials and insulation materials, are first stored and even exposed to a certain voltage for a period of time to stabilize their performance and then the good ones among them are chosen.

- *Linearization technology*

The non-linearity of the sensor may be caused by the non-linearity of the input-output model or a poor sensor manufacture process. Thus, in the practical application, the input and output of the sensor is hardly linear. To improve the linearity of the sensor, generally some methods such as the differential design and the sensor linearization correction circuits (related to the design of sensor interface circuit) are used.

- *Averaging technique*

The averaging technique uses a certain number of sensor units to measure the signal at the same time, whose outputs are the addition of these sensor units. The outputs are composed of the signal and the noise. According to the probability theory, the noise of each sensor can be considered to be unrelated and isolated, so the addition of the outputs of the certain number sensor units would eliminate the noise and keep the signal at the same time.

2.4 Biocompatibility Design of Sensors

The term "biocompatibility" is used extensively within biomaterial science. Sensors used for biological detection and medical diagnosis must seriously consider biocompatibility in order to avoid hemo- or histo- problems with direct contact. This Section introduces the concept, the classification and the evaluation of biocompatibility. The topic of biocompatibility related to the sensor design focuses on two aspects—invasive sensors for implanting and monitoring physiological environment *in vivo*, and non-invasive sensors for monitoring cell or tissue *in vitro*.

2.4.1 Concept and Principle of Biocompatibility

The following content will include the concept of biocompatibility, classification of biocompatibility and evaluation, which provides essential rules for biomedical sensor fabrication.

2.4.1.1 Biocompatibility

Biomedical sensors implanted in animal or human bodies or in direct contact with tissue *in vitro* must be non-toxic, non-allergenic, non-irritating, non-genetic toxic and non-carcinogenic. Besides, they are not supposed to make the animal or human tissues, blood and immune systems produce any adverse reaction. Sensors should be compatible with the chemical composition in organisms. Therefore, the biocompatibility of the material is a priority to be considered. David (2008) enhances the list of events that has to be avoided. He believes that the key to understanding biocompatibility is the determination of that chemical, biochemical, physiological, physical or other mechanisms become operative, under the highly specific conditions associated with contact between biomaterials and the tissues of the body, and what are the consequences of these interactions. Accordingly, not only the mechanisms by which materials and human tissues respond to each other but also the particular natural processes in different applications have to be carefully evaluated.

Biocompatibility is a serious problem, especially when it comes to biomedical sensors for practical clinic use. There is a large battery of *in vitro* tests that are used in accordance with ISO 10993 to determine if certain material is biocompatible. These tests constitute an important step towards animal testing and finally towards clinical trials by implanting medical devices. Indeed, since the immune response and repair functions in the body are so complicated, it is not adequate to describe the biocompatibility of a single material in relation to a single cell type or tissue. Generally speaking, in the design of implantable biomedical sensors, the following factors should be considered:

- Sensors are not supposed to be corroded or toxic. In this way, they must be compatible with the chemical composition in organism.
- The shape, size and structure of sensors implanted in the human body should adjust to the anatomical structures of the measuring object. What's more, sensors should never injure the tissue.
- Sensors should be solid enough. When implanted, the sensor should not be damaged.
- Sensors must be electrically insulated. Once the sensor is damaged in a human, it should be certain that the voltage imposed on the human is under the safe security value.
- Sensors should neither place physical activities a burden, nor interfere with the normal physiological function.
- The *in vivo* sensors used for long term implantation should not cause any vegetation.
- The structure of the sensor should be easily disinfected.

The biomedical materials used as mediators between sensors and measuring objects should also be investigated. The major interaction between biomedical materials and the human body includes mechanical interaction (friction, impact, bending etc.), physical-chemical interaction (dissolution, absorption, permeability, degradation etc.) and chemical interaction (decomposition, modification etc.). The interaction causes corresponding changes both in biomedical materials and organisms, which are listed in Table 2.4.1.

Table 2.4.1 The corresponding changes in biomedical materials and organisms

Biomedical materials changes		Response and changes in organisms	
Physical changes	Size, shape, flexibility, rigid, plasticity, brittleness, relative density, melting point, conductivity, and thermal conductivity	Acute systemic toxicity	Allergic reactions, toxicity, hemolytic reaction, fever, paralysis
		Chronic systemic toxicity	Toxicity, teratogenesis, immune response, dysfunction
Biochemical changes	Hydrophilic, hydrophobic, pH value, adsorption, dissolution, permeability	Acute local reactions	Inflammation, thrombosis, necrosis, rejection
		Chronic local reaction	Carcinogenic, calcification, inflammation, ulceration

Factors causing biomedical materials to change:
- The dynamic mechanics of bones, joints and muscles in physical activities;
- Cellular bioelectricity, magnetic fields, electrolysis and oxidation in cells;
- Biochemistry and enzyme-catalyzed reaction in metabolic procedure;
- Adhesion and phagocytosis of cells;
- Biodegradation by various enzymes, cytokines, proteins, amino acids, peptides, and free radicals' in body fluids.

Factors causing organism reactions:
- The residual toxic low-molecular-weight substances in the materials;
- The residual toxic, irritating monomer in polymerization process;
- The adsorbed chemical agents and the split products caused by high temperature in the sterilization process;

- The shape, size, surface smooth level of the materials and products;
- The pH value of the materials.

2.4.1.2 Classification of biocompatibility

- ***Blood compatibility***

Biomedical sensors used in the cardiovascular system are in direct contact with blood. Thus, the reactions between them should be carefully considered. Implanted devices need to coexist with blood; in that case, they can function under the intended purpose without reactions to blood and tissues. Many reactions can be triggered by catheter insertion, which would cause sensor failure or tube plug due to the accumulation of fiber and blood clots. In addition, the existence of devices may put patients at risk, which may be due to toxic products being released into the blood or body fluids, or the increasing thrombosis caused by the sensor's charge or the external leakage current of the sensor. Blood clotting time is the most obvious indicator to show biological incompatibility of materials exposed to blood, which has been widely used in biomedical sensor technology. PTFE (Poly Tetia Fluoro Ethylene), which was considered to be the most inert and most compatible biomedical polymer, has a blood clotting time of 11 – 13 min, which is regarded as 100%. While other ordinary inorganic silicon and glasses have very short blood clotting time, usually less than 10 min. This is why the procedure composition of sensors and the catheter surface, as well as surface modifying, become critical issues in invasive sensor research and development.

- ***Histocompatibility***

Biomedical sensors implanted out of cardiovascular system mainly focus on the interaction between the materials and the tissues or organs. The immune system at first makes antibodies for all sorts of antigens, including those it has never been exposed to, but stops making them to antigens present in the body. If the body is exposed to foreign antigens, it attacks the foreign materials. Therefore, the materials in sensors should be histocompatible to avoid immune system attack.

2.4.1.3 Evaluation of biocompatibility

The quality and biocompatibility of biomedical materials and all kinds of medical equipment made of them are directly related to the safety of patients. Thus, authorities should set a standard registration and approval procedure for such products like ISO10993. Biomaterials and medical equipment must pass the biological evaluation in either research or manufacture to ensure their safety. Generally speaking, biomedical materials safety contains the following four aspects: physical properties, chemical properties, biological properties and clinical researches. While in a narrow sense, biomaterials and medical device safety evaluation refers to the biological assessment.

Biological assessment generally contains the following aspects:
- Contact tissues: body surface and body tissues, bones, teeth, blood;
- Contact means: direct contact and indirect contact;
- Contact time: Temporary contact is less than 24 h, short and medium-term exposure is longer

than 24 h but shorter than 30 d, and long-term exposure is more than 30 d;
- Purpose: general function, reproductive and embryonic development, biodegradation.

For example, silk fibroin materials used for constructing artificial nerve grafts is biocompatible and evaluated with tissues and cells (Yang et al., 2007). The physical and biological characterizations of the materials were studied. Biological assessment directly contacting with cells was also investigated. The outgrowth of rat dorsal root ganglia cells was observed by light and electron microscopy coupled with immunocytochemistry. The morphology and proliferation of Schwann cells were tested by MTT (a yellow tetrazole, its main application is to assess the viability and the proliferation of cells) test and cell cycle analysis.

2.4.2 Biocompatibility for Implantable Biomedical Sensors

The state-of-the-art implantable biomedical sensors fall into the category of biophysical sensors for bioelectric detection and biochemical sensors for certain metabolite detection, for example, sensors for detecting electrical signals in the brain (Schneider and Stieglitz, 2004) and sensors for monitoring bio-analytes in the brain (Hu and Wilson, 2002).

For the sake of causing less chemical reactions in metallic systems, the corrosive plain carbon and vanadium steels were replaced by stainless steels, then by the strongly passivated cobalt-chromium alloys, titanium alloys and platinum group metals. In the case of cardiac pacemakers and implantable cardioverter defibrillators, which help to manage a wide variety of cardiac rhythm disorders, the active components are encapsulated in sealed titanium can. Leads transmit pulses for both sensing and delivery purposes from the can to the electrode placed at the relevant site on the heart. The electrode, which requires the delivery of the electrical impulse, should be concentrated on high electrical conductivity, and fatigue resistant and corrosion resistant alloys such as the platinum group metal alloys or the cobalt-chromium group (Crossley, 2000).

With polymer, the readily available and versatile nylons and polyesters have been replaced by the more degradation resistant PTFE, PMMA, polyethylene and silicones. However, with the development of degradable implantable materials and systems, the concept of biocompatibility enriches its content. The degrading material performs a function before or during a process with acceptable host response. For example, the drug delivery system untilizes degradable polymers such as polylactide and polyglycolides (Shive and Anderon, 1997) or poly (methylidene malonate) (Founier et al., 2006) to form microspheres.

The biocompatibility design for biomedical sensors *in vivo* involves several problems: surface materials, device morphology, infection, toxicology and host response affections toward the responses of the device (Reach et al., 1994). The chemically modified working electrode must have a compatible surface to permeate the desired solute as well as reject interferent species. The structure and properties of the sensor's outer membrane, which is in direct contact with the host tissue or blood, is essential to consider carefully to provide high transport selectivity and rate for analytes to be measured and to avoid leakage of potentially toxic components. The morphological characteristics also influence the sensitivity of the sensor. If the membrane is subject to mechanical damage or with stripping from the underlying enzyme layer, the sensor will operate abnormally. The adherence of bacteria to the surface of the sensor (foreign body) and toxicity to the host are the problems to overcome. Recently, progress has been made in the development of clinically important chemical sensors, such as catheter-type sensors placed within the radial artery of hospitalized patients for blood-gas measurements such as pH, P_{O_2}, and P_{CO_2}. This type of sensor is substantially replaced by the non-invasive pulse-oximeters, which make the same measurement through the skin (Collison and Meyerhoff, 1990). The needle-type glucose sensor is constructed by multiple membrane coating and electrodeposition of the glucose oxidase layer (Robert et al., 1989). Moreover, ion-selective electrodes and ion-selective fields effect transistors for monitoring electrolytes such as Na^+, K^+ and Ca^{2+} is invented (Kimura, 2001).

How to evaluate the biocompatibility of the sensor? Yang et al. (2003) developed an optic micropressure sensor and evaluated its biocompatibility using the International Organization for Standardization (ISO) test standard 10993-6, "Biological Evaluation of Medical Devices Part 6: Tests for local effects after implantation." The sensor consists of an optical fiber and a diaphragm. The optical fiber is made of inert silica glass and the diaphragm is made from polyimide. The sensor is a flat-shaped cylinder, 10 mm in length and 360 µm in diameter (see Fig. 2.4.1). The *in vivo* biocompatibility of the material can be evaluated by analyzing the cell population present, measuring the mediator and metabolite cells excreted, and analyzing the morphologic characteristics of the tissue and the capsule thickness around the implant. Here, the capsule thickness was analyzed to evaluate the host response. Twelve healthy white rabbits were divided into three groups for different implantation periods (1, 4 and 12 weeks), with four animals tested during each time period. After the surgical procedure, the sensor was implanted in the paravertebral muscles on one side of the spine about 2 – 5 cm from the midline and parallel to the spinal column. The Section orientation in relation to the implant dimensions and implant orientation was recorded. The biological response parameters assessed and recorded included: (a) the extent of fibrous capsule and inflammation; (b) the degeneration determined by changes in tissue morphology; (c) the number and distribution of inflammatory cells types, namely macrophages, polymorphonuclear leukocytes, lymphocytes, plasma cells, giant cells, as a function of distance from the material/tissue interface; (d) the presence of necrosis determined by nuclear debris and/or capillary wall breakdown; (e) the other parameters such as material debris, fatty infiltration and presence of granulomata. The one-week group results show that the inflammatory reaction was mild (see Fig. 2.4.2). The capsule wall was composed of three layers. The inner layer was mostly composed of macrophages and monocytes. Eosinophils were located in the middle layer. The outer layer was mainly composed of fibroblasts. The four-week group results show that the capsule thickness decreased sharply. From the four- week to twelve-week groups, the capsule thickness decreased slightly. In conclusion, the biocompatibility of implanted sensors requires histologic and morphologic *in vivo* examinations, even when inert materials are selected.

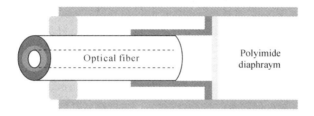

Fig. 2.4.1.   An optic micropressure sensor structure

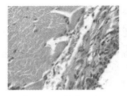 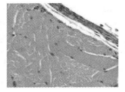 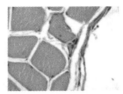

One-week (test)　　　　Four-week (test)　　　　Twelve-week (test)

Fig. 2.4.2. View of implanted site at 1, 4 and 12 weeks after implantation (reprinted from (Yang et al., 2003), Copyright 1995, with permission from Elsevier Science B.V.)

2.4.3 Biocompatibility for *in vitro* Biomedical Sensors

Biocompatibility has been traditionally concerned with implantable devices that have been intended to remain within an individual for a long period of time. Biochemistry reaction in the organism is mysterious and complex. It is a tough way to approach the principle and mechanism of these reactions through direct invasive measurements. And it is a big challenge to analyze the signals collected from the complex environment inside the organism. As a result, biomedical samples such as proteins, cells and tissues are often extracted from the primary source and studied *in vitro*. The biomedical sensor is a good choice for biological samples in research of *in vitro*. At first, biocompatibility of the biomedical materials on sensor surfaces must be enhanced to maintain the activity and viability of these samples.

The biocompatibility for biomedical sensors *in vitro* is an essential part for the whole biological sample hybrid system, which should be evaluated completely in biomedical sensor design. The biocompatibility of sensor surface materials affects the biological sample's viability and activity. Furthermore, it is one of the crucial factors of the final output of the system to control the quality of sample immobilization. It is an important concept. But there still exist many uncertain methods and mechanisms within biomedical sensor design. The surface material is selected on the basis of non-toxic, non-immunogenic, non-thrombogenic, non-carcinogenic, non-irritant and so on, which must be tested *in vitro* according to biocompatibility requirements in biological evaluation of medical devices in ISO 10993.

Surface materials used in biomedical sensors are usually of two types: passivated materials (e.g., silicon, glass, polymers and photoresist) or metal and metal oxide materials. These materials are quite different from the environment inside biological organism in physical and chemical properties. Therefore, some methods are employed to improve the surface characteristics in hydrophilicity improving, roughness modifying and surface chemical coating. Nowadays, more and more innovative applications are involved in this area.

Functional proteins such as enzyme, antigen and antibody play an important role as receptors in biomedical sensor systems. Receptor parts contain the biologically active components that are capable for specific chemical reactions with the analytes. Various physical and chemical methods are utilized to immobilize these proteins onto sensor substrate surfaces, which enhance the biocompatibility of the surface and maintain the samples good activity and stability. Various approaches are used to modify the characteristics in electrostatics, physical adsorption, and surface chemical group immobilization of the biomedical sensor surfaces. Besides, these samples are very sensitive to temperature and pH value changes.

Isolated cells and tissue could still provide physiological responses, which can be detected by biomedical sensors. Hydroxyl ion implantation into the silicon surface can significantly affect the structural properties of the surface. It is well known that a surface with better hydrophilicity facilitates cell and tissue adherence and proliferation (Fan et al., 2000). Studies have shown that neurons can adhere and spread on the silicon surface with appropriate roughness. Silicon chips are dipped into hydrofluoric acid, which efficiently enhances the roughness to 25 nm from the original 3 nm (Fan et al., 2002). To coat a biocompatible material on the whole surface is the most convenient way to enhance the functional groups on the biomedical sensor surface. As shown in Fig. 2.4.3, cells and tissues are respectively immobilized on glass, silicon, and metal substrates (Bergen et al., 2003; Jungblut et al., 2009; Krause et al., 2006; Zeck and Fromherz, 2001).

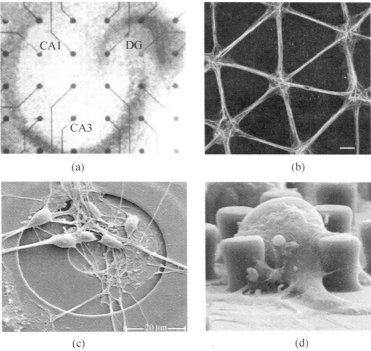

Fig. 2.4.3. Cells and tissues immobilized on different substrates: (a) Slice of hippocampal tissue immobilized on sensor, including CA1, CA3 and DG regions (reprinted from (Bergen et al., 2003), Copyright 2003, with permission from Elsevier Science B.V.); (b) Artificial triangular patterned neuron network formation on glass. Scale bar: 10 μm (reprinted from (Jungblut et al., 2009), Copyright 2009, with permission from Springer); (c) SEM image of passive palladium electrodes with cultured neuronal cells after 3 d *in vitro* with a developing glial carpet (reprinted from (Krause et al., 2006), Copyright 2005, with permission from Elsevier B.V.); (d) A single neuron trapped on silicon chip surface (reprinted from (Zeck and Fromherz, 2001), Copyright 2001, with permission from the National Academy of Sciences)

Cell- and tissue-based biosensors, which treat living cells and tissue as sensing elements, could reflect the functional information of bioactive analytes. Cells and tissue can be extracted from primary sources and successfully cultured *in vitro*. The special activities relating to cellular functions can be detected by biomedical sensors (Gross et al., 1995; Maher et al., 1999; Kovacs, 2003; Stett et al., 2003).

The biosensor composed of light-addressable potentiometric sensors (LAPS), taste or olfactory neurons, which can detect the extracellular potentials and is sensitive to tastant or odorous change. In the manner of invasive and light addressable, cell response to living environment changing was monitored. Taste and olfactory cells are difficult to immobilize onto sensors. For the uniform surface of a LAPS chip, using chemical coating was one of the most efficient methods. Before cell culture, a thin layer of poly-L-lysine and laminin was deposited on the LAPS chip, which makes better biocompatibility for cell immobilization (Zhang et al., 2008; Wu et al., 2009). The cell immobilization result is shown in Fig. 2.4.4.

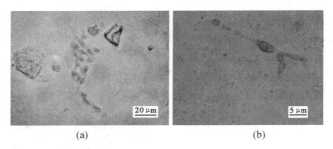

Fig. 2.4.4. Taste and olfactory cells immobilized on LAPS: (a) Taste receptor cells were cultured on LAPS chip surface. (reprinted from (Zhang et al., 2008), Copyright 2007, with permission from Elsevier B.V.); (b) Olfactory cells were immobilized on LAPS chip surface (reprinted from (Wu et al., 2009), Copyright 2008, with permission from Elsevier B.V.)

Microelectrode array (MEA) can record the multisite potentials simultaneously and allow for the ability of long-term recording of the firing of neural networks *in vitro*. In the present study, we managed to combine the intact olfactory tissue slice with MEA (Liu et al., 2006). Compared to the cultured olfactory cells, the olfactory tissue can be obtained conveniently with the primary cell structure well preserved. Mimicking the *in-vivo* process of gas sensing, it is a good candidate for the biological elements of bioelectronic nose. MEA can record the extracellular potentials of the olfactory receptor neurons in the tissue slice. The multi-channel signal analysis may reveal some spatial and temporal information of olfactory information processes in the olfactory bulb. The olfactory bulb slice morphology structure is well reserved, along with tissue immobilization on the MEA chip shown in Fig. 2.4.5.

Fig. 2.4.5. Olfactory tissue immobilized on MEA chip: (a) Olfactory bulb tissue morphology structure; (b) Olfactory bulb tissue slice immobilized on MEA chip

And various methods in biofunctionalized polymer surfaces analysis are well considered in surface characterization evaluation, including spectral methods (X-ray photoelectron spectroscopy, Fourier transform infrared spectroscopy, Atomic Force Microscopy (AFM), and others) as well as non-spectral methods (water contact angle, dye assays, biological assays, and zeta potential).

Besides the factors of the materials mentioned above, environment control is another important aspect for *in vitro* cell and tissue culture on biomedical sensors. It is essential to keep the activity of the living cell and tissue during real-time analysis. Relative parameters including temperature, medium components, pH, osmotic pressure, gas pressure, etc., are required to be well controlled.

2.5 Microfabrication of Biomedical Sensors

The biomedical sensors are strictly required to be compatible to the objected bio-components and cause lower invasive injury in *in-vivo* measurements. Additionally, the development of microfabrication technology excites the potential of biomedical sensors with characteristics of low cost, low consumption of samples, high-throughput performance and the ability can be easily integrated into systems. Thus in this section, the fundamental microfabrication including lithography, film forming and etching is introduced and an example of fabrication procedure of a typical sensor is illustrated.

2.5.1 Lithography

Lithography in MEMS is typically the transfer of a pattern to photosensitive material. The transfer of the designed image onto a resist-coated wafer is very critical in sensor fabrication. Two main processes of lithography used in sensor fabrication are the photolithography and electron beam lithography.

Photolithography is the most widely used form of lithography. In photolithography, a photo mask is required, which is usually a nearly optically flat glass or quartz plate with a metal absorber pattern. The photo mask is placed into direct contact with photoresist surface coating. When the wafer is exposed to the ultraviolet radiation, the absorber pattern on the photo mask is opaque to ultraviolet light and the pattern is transferred to the photoresist based on its polarity. Photolithography has matured rapidly and has become better and better at resolving smaller and smaller features. For sensor fabrication, this continued improvement in resolution makes it very suitable to construct micro sensors in the dimension of several microns.

Electron beam lithography can overcome the limit of the resolution by light diffraction in photolithography, because the quantum mechanical wavelengths of high energy electrons are exceedingly small. In electron beam lithography, a photo mask is not required and direct writing is processed on the resist-coated surface. The major advantages of electron beam lithography are the ability to register accurately over small areas of a wafer, lower defect densities, and a large depth of focus because of continuous focusing over topography. However, electron beam lithography has some disadvantages. Electrons scatter quickly in solids, limiting practical resolution to dimensions greater than 10 nm. The resolution of electron beam lithography tools is not simply the spot size of the focused beam. Serious variations of exposure over the patterns occur when pattern geometries fall into the micrometer and sub-micrometer ranges. Electrons also need to be held in a vacuum, making the apparatus more complex and expensive than that for photolithography. Besides, the difficulties with scanning electron-beam lithography include the relatively slow exposure speed and high system cost, which keep it away from wider use.

2.5.2 Film Formation

Film formation process in micro fabrication is the growth and deposition of layers with different materials. To form films of various materials, different film formation processes are involved, such as silicon oxidation, physical vapor deposition (PVD), and chemical vapor deposition (CVD).

Silicon oxidation is a primary process in sensor fabrication, since silicon is a normally used base material in sensor fabrication. Silicon dioxide growth involves the heating of a Si wafer in a stream of steam at 1 atm wet or dry oxygen/nitrogen mixtures at elevated temperatures (600 – 1,250 °C). The high temperature aids diffusion of oxidant through the surface oxide layer to the silicon interface to form thick oxides in a short amount of time. In fabrication of most sensors based on silicon structure such as light addressable potentiometric sensor (LAPS) (Yu et al., 2009), silicon oxidation is usually a necessary procedure.

Gas phase deposition is a key technology in film formation. Two major categories of gas phase deposition can be distinguished as physical vapor deposition (PVD) and chemical vapor deposition (CVD).

Physical vapor deposition (PVD) is a direct line-of-site impingement type deposition. PVD reactors may use a solid, liquid, or vapor raw material in a variety of source configurations. Evaporation and sputtering are two major methods of PVD. Thermal evaporation represents one of the oldest thin film deposition techniques. Material is heated and boiled off onto a substrate in a vacuum. Evaporators emit material from a point source, resulting in shadowing and sometimes causing problems with metal deposition, especially on very small structures. Meanwhile, during the sputtering process, the target (a disc of the material to be deposited, i.e., Au), at a high negative potential, is bombarded with positive argon ions (other inert gases such as Xe can be used as well) created in a plasma (also glow discharge). Sputtering is preferred over evaporation in many applications due to a wider choice of materials to work with, better step coverage, and better adhesion to the substrate. Actually, sputtering is employed in laboratories and production settings, whereas evaporation mainly remains a laboratory technique (Madou, 2002).

Chemical vapor deposition (CVD) is a diffusive-convective mass transfer technique. There are several different CVD processes with different operational pressures and temperatures. In plasma-enhanced chemical vapor deposition (PECVD), an RF-induced plasma transfers energy into the reactant gases allowing the substrate to remain at lower temperature. Besides, all of the dry etching equipment can be used for PECVD as well. The most significant application of PECVD is probably the deposition of SiO_2 or Si_3N_4 over metal lines. In low pressure CVD (LPCVD) at below 10 Pa, large numbers of wafers can be coated simultaneously without detrimental effects to the film uniformity. However, LPCVD suffers from a low deposition rate and the relatively high operating temperatures.

Screen printing is essential in low cost disposable sensor fabrication. Low cost disposable sensors are an important developing trend of biosensor applications. In screen printing, a paste of ink is pressed onto a substrate through openings in the emulsion on a stainless steel screen. The paste consists of a mixture of the material of interest, an organic binder and a solvent. The lithographic pattern in the screen emulsion is transferred onto a substrate by forcing the paste through the mask opening with a squeegee. In the first step, paste is put down on the screen, and then the squeegee lowers and pushes the screen onto the substrate forcing the paste through opening in the screen during its horizontal motion. During the last step, the screen snaps back, the thick-film paste which adheres between the screening frame and the substrate shears, and the printed pattern is formed on the substrate. The resolution of the process depends on the openings in the screen and the nature of the pastes.

2.5.3 Etching

Etching is a process to remove material from the unprotected area to form functional structure on a wafer. It usually follows after the process of lithography to present an accurate pattern matching with the photomask. In the MEMS fabrication, two major types of etching are wet etching and dry etching.

Wet etching is performed as a chemical reaction to selectively dissolve the film by etchant. The process usually includes three steps: reactive species diffuse from liquid bulk to the surface of the film; reaction happens at the surface to form solvable species; reaction products are diffused away from the surface of the film to liquid bulk. The etch rate is related with temperature, specific reaction and liquid concentration, which results in isotropic etching and anisotropic etching by the relative etch rates in different directions. In isotropic etching (Fig. 2.5.1b), etching progresses at the same speed in all directions and the surface of etched grooves can be atomically smooth. For the material with crystal structure such as silicon, it exhibits anisotropic etching because of the different etch rates in different directions (Fig. 2.5.1a). In wet etching, a good selectivity can often be obtained because the etch rate of target material is considerably higher than that of the mask material.

Dry etching is commonly used due to its ability to better control the etching process and eliminate handling of dangerous acids and solvents better than wet drying. It effectively etches targeted film through: (1) Chemical reactions with reactive gases or plasma; (2) Physical bombardment of atoms; (3) The combination of both physical removal and chemical reactions.

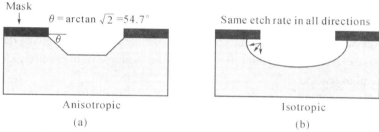

Anisotropic (a) Isotropic (b)

Fig. 2.5.1. Anisotropic and isotropic etch: (a) An anisotropic etch on a silicon wafer creates a cavity in the cross-section. The sides of cavity are <111> planes with an angle of 54.7°; (b) In isotropic etch, it has the same etch rate in different directions

Plasma etching is a purely chemical dry etching technique. In a plasma system, it ionizes and generates reactive species by using RF (radio frequency) excitations and they diffuse to the surface of the wafer. After adsorption, the chemical reactions occur between the reactive species and the material being etched. The generated byproducts in reaction are desorbed and diffused into the bulk of the gas. Sputtering is a physical dry etching technique that removes the targeted material by bombarding it with highly energetic but chemically inert species or ions. Compared with chemical methods, such a physical removal process is non-selective and it will hit the mask layer covering the targeted layer through the bombarding species. Thus it has never become popular as a dry etching technique for MEMS fabrication.

Reactive ion etching (RIE) has been a fundamental process in semiconductor fabrication. It combines chemical and physical techniques to accelerate the chemically reactive ions to bombard and etch away the targeted material. In comparison to plasma etching, RIE has a higher probability to have movement in the direction given by the electric field and owns an advantage in terms of anisotropy. In addition, the etch rate of RIE is highly controllable, the selectivity to the material is reasonable and the endpoint of the etching process is optical. RIE can etch deeply and has been developed as a subclass, the deep RIE. It is based on the "Bosch process" and there are two different gas compositions alternating in the reactor. The etch depth of hundreds of micrometers can be achieved with almost vertical sidewalls. Thus a lot of patterns in biochips are formed by RIE or deep RIE in MEMS development. Fig.

2.5.2 shows an example of a biochip fabricated by deep RIE techniques from SINTEF.

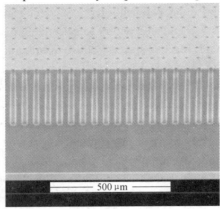

Fig. 2.5.2. DRIE for through wafer holes: Average etch rate was 12 μm/min and each hole was 400 μm deep with diameter of 50 μm holes

2.5.4 Design of the Biomedical Sensors

Generally, the procedure in the development of a biosensor includes several steps: design, fabrication, package, performance test and application. Here we take an example of microelectrode array (MEA) to describe the fabrication process.

MEA, a typical type of cell-based biosensors, is an array of metallic film disks used to detect the extracellular field potential of electrogenic components such as cardiac cells and neuronal cells (Mayer et al., 2004; Gross et al., 1995). It is generally fabricated on wafers of glass or silicon grown with silicon oxide. Before fabrication, two masks are prepared for lithography to form the patterns of metallic film layer and the passivation layer, respectively. In Fig. 2.5.3, it shows the procedure of fabrication based on the MEMS techniques such as lithography, film formation and etching. In Fig. 2.5.3, it applies the glass as the substrate of MEA because of its low dielectric constant.

- A wafer of glass is cleaned and then the Cr (30 nm) and Au (300 nm) are sequentially sputtered or evaporated onto it. The Cr layer is used to enhance the adhesion of the Au layer onto the substrate (Figs. 2.5.3.a and 2.5.3.b).
- The photoresist is then spin coated on the metallic layer. The electrodes and tracks are patterned by a mask (Fig. 2.5.3.c). When exposed to the UV light, the positive photoresist with no shield decomposed (Fig. 2.5.3d).
- The exposed part of composited metallic layer (Cr/Au) is chemically etched, while the other part covered with photoresist is retained. By washing off the photoresist, the electrodes are present (Fig. 2.5.3.e).
- The alternative $SiO_2/Si_3N_4/SiO_2$ as a passivation layer is deposited onto the surface of the chip by PECVD (Fig. 2.5.3.f).
- The photoresist is spin coated on the chip again (Fig. 2.5.3.g). Take another mask, the sensory sites are exposed to the UV light (Fig. 2.5.3 h). Next, the exposed passivation layer is etched (Fig. 2.5.3.i).
- Usually, the electrodes are in a diameter of 10-50 μm and their high impedance causes a high

noise. It needs to increase the roughness of the electrode surface to lower down the noise and stabilize the baseline. In this step, we take the TiN as the surface treatment option. It is reactive sputtered onto the chip in a nitrogen/argon atmosphere (Fig. 2.5.3.j). Then remove the TiN on photoresist by lift-off process. In this way, only the sensory sites are treated with TiN (Fig. 2.5.3.k).

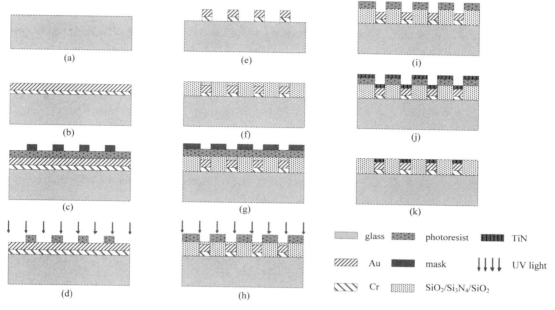

Fig. 2.5.3. The fabrication procedure of microelectrode array

If the surface is treated by platinum black, the plating process can be performed after removing the photoresist on the sensory sites and applying a constant current of 5 nA/mm^2 onto the electrodes in electrolyte of chloroplatinic acid for about 30 s.

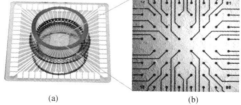

Fig. 2.5.4 Microelectrode array: (a) The picture of packaged MEA; (b) Rat hippocampal cells are culture on the MEA for 12 d

After fabrication, a group of chips on the same wafer, usually with a diameter of 4 inches, are scribed into individual ones. For the chips with small size of several millimeters, they are usually fixed onto a PCB board and then pads of chip are bonded with pads of PCB board by gold wire. The signals detected by electrodes on chips are conducted into external setup through the circuits on the PCB. For the chips with the size of about two centimeters, the larger pads on the chips are pressed with spring tips to conduct the signal to the external setup. Finally, a chamber is fixed onto a chip to form a room for the cell culture. The chip in Fig. 2.5.4a is a packaged commercial MEA (Multi Channel Systems MCS, GmbH) whose pads contact with the spring tips to create output signals. The electrode arrays are located in the center of the chamber to help to effectively concentrate the cells onto the surface of the electrodes. Fig. 2.5.4b shows that the rat hippocampal cells are cultured on the MEA and have shown good biocompatibility and formed a communicated network on the chip. By detecting the extracellular field potential in different sites in parallel, MEA can provide valuable information about the spike patterns and communication in the neuronal network.

In summary, a series of sensors could be fabricated by combining different fundamental techniques. The development of fabrication helps to further explore chips with refined structures and small sizes in nanometers. Thus various sensors with multi-channels and good compatibility can be designed to study molecules, receptors, cells and organs.

References

Aaroe J., Lindahi T., Dumeaux V., Sabo S., Tobin D., Hagen N., Skaane P., Lonneborg A., Sharma P. & Borresen-Dale A., 2010. Gene expression profiling of peripheral blood cells for early detection of breast cancer. *Breast Cancer Research*. 12, R701-R711.

Barbara F., Ulrik L., Jann M., Søren L., Annika L., Anne B., Jesper R., Paul C., Asbjørn H., Klaus L., Torben R., Susanne K., Asger D., Oke G., Birgit S., Ida S., Hanne H., Peter V., Grete J., Vibeke B., Niels M., Jesper P., Henrik M., Henrik N. & Liselotte H., 2009. Preoperative staging of lung cancer with combined PET-CT. *New England Journal of Medicine*. 361, 32-39.

Bergen A., Papanikolaou T., Schuker A., Moller A. & Schlosshauer B., 2003. Long-term stimulation of mouse hippocampal slice culture on microelectrode array. Brain Research Protocols. 11, 123-133.

Collison M.E. & Meyerhoff M.E., 1990. Chemical sensors for bedside monitoring of critically ill patients. *Analytical Chemistry*. 6, 425A-437A.

Crossley G.H., 2000. Cardiac pacing leads. *Cardiology Clinics*. 18, 95-112.

David F.W., 2008. On the mechanisms of biocompability. *Biomaterials*. 29, 2941-2953.

Fan Y., Cui F., Chen L., Zhai Y. & Xu Q., 2000. Improvement of neural cell adhesion to silicon surface by hydroxyl ion implantation. *Surface & Coatings Technology*. 131, 355-359.

Fan Y., Cui F., Hou S., Xu Q., Chen L. & Lee I., 2002. Culture of neural cells on silicon wafers with nano-scale surface topograph. *Journal of Neuroscience Methods*. 120, 17-23.

Founier E., Passirani C., Colin N., Sagodira S., Menei P., Benoit J.P. & Montero-Menei C.N., 2006. The brain tissue response to biodegradable poly(methylidene malonate 2.1.2)-based microspheres in the rat. *Biomaterials*. 27, 4963-4974.

Gross G.W., Rhoades B.K., Azzazy H.M. & Wu M.C., 1995. The use of neuronal networks on multielectrode arrays as biosensors. *Biosensors & Bioelectronics*. 10, 553-567.

Hu Y. & Wilson S.G., 2002. Rapid changes in local extracellular rat brain glucose observed with an in vivo glucose sensor. *Journal of Neurochemistry*. 68, 1745-1752.

Jungblut M., Knoll W., Thielemann C. & Pottek M., 2009. Triangular neuronal networks on microelectrode arrays: an approach to improve the properties of low-density networks for extracellular recording. *Biomedical Microdevices*. 11, 1269-1278.

Kalantar-zadeh K. & Fry B., 2008. Nanotechnology-enabled sensors. Springer Science+Business Media, *LLC*. 13-15.

Kimura K., 2001. Design of biocompatible ion sensors. In: A.G. Volkov (ed.), *Liquid Interfaces in Chemical, Biological & Pharmaceutical Applications 95*, Marcel Dekker, Inc., New York, 585-607.

Kovacs G.T.A., 2003. Electronic sensors with living cellular components. *Proceedings IEEE*. 91, 915-929.

Krause G., Lehmann S., Lehmann M., Freund I., Schreiber E. & Baumann W., 2006. Measurement of electrical activity of long-term mammalian neuronal networks on semiconductor neurosensor chips and comparison with conventional microelectrode arrays. *Biosensors & Bioelectronics*. 21, 1272-1282.

Liu Q., Cai H., Xu Y., Li Y., Li R. & Wang P., 2006. Olfactory cell-based biosensor: a first step towards aneurochip of bioelectronic nose. *Biosensors & Bioelectronics*. 22, 318-322.

Madou M.J., 2002. *Fundamentals of Microfabrication*, 2nd ed. CRC Press, Boca Raton, FL.

Maher M.P., Pine J., Wright J. & Tai Y., 1999. The neurochip: a new electrode device for stimulating and recording from cultured neurons. *Journal of Neuroscience Methods*. 87, 45-56.

Mehdi K. & Maysam G., 2010. An RFID-based closed-loop wireless power transmission system for biomedical applications. *IEEE Transactions on Circuits and Systems*. 57, 260-264.

Meyer T., Boven K. H., Gunther E. & Fejtl M., 2004. Micro-electrode arrays in cardiac safety pharmacology, a novel tool to study QT interval prolongation. *Drug Safety*. 27, 763-772.

Oono Y., Iriguchi Y., Doi Y., Tomino Y., Kishi D., Oda J., Takayanagi S., Mizutani M., Fujisaki T., Yamamura A., Hosoi T., Taguchi H., Kosaka M. & Delgado P., 2010. A retrospective study of immunochemical fecal occult blood testing for colorectal cancer detection. *Clinica Chimica Acta*. 411, 802-805.

Reach G., Feijen J. & Alcock S., 1994. BIOMED concerted action chemical sensors for in vivo monitoring. The biocompatibility issue. *Biosensors & Bioelectronics*. 9, 21-28.

Robert S., Marie-Bernadette B., Laurent G., Daniel R.T., Dilbir S.B., George S.W., Gilberto V., Philippe F. & Gerard R., 1989. Study and development of multilayer needle-type enzyme-based glucose microsensors. *Biosensors*. 4, 27-40.

Schneider A. & Stieglitz T., 2004. Implantable flexible electrodes for functional electrical stimulation. *Medical Device Technology*. 1, 16-18.

Shive M.S. & Anderon J.M., 1997. Biodegradation and biocompatibility of PLA and PLGA microspheres. *Advanced Drug Delivery Reviews*. 28, 5-24.

Stett A., Egert U., Guenther E., Hofmann F., Meyer T., Nisch W. & Haemmerle H., 2003. Biological application of microelectrode arrays in drug discovery and basic research. *Analytical and Bioanalytical Chemistry*. 377, 486-495.

Wang D., Wang L., Yu J., Ping W., Yanjie H. & Kejing Y., 2009. A study on electronic nose for clinical breath diagnosis of lung cancer. *Olfaction and Electronic Nose, Proceedings*. 1137, 314-317.

Wilson J.S., 2005. *Sensor Technology Handbook*. Elsevier Inc. 21-23.

Wu C., Chen P., Yu H., Liu Q., Zong X., Cai H. & Wang P., 2009. A novel biomimetic olfactory-based biosensor for single olfactory sensory neuron monitoring. *Biosensors & Bioelectronics*. 24, 1498-1502.

Yang C., Zhao C., Word L. & Kaufman R.K., 2003. Biocompatibility of a physiological pressure sensor. *Biosensors & Bioelectronics*. 19, 51-58.

Yang Y., Chen X., Ding F., Zhang P., Liu J. & Gu X., 2007. Biocompatibility evaluation of silk fibroin with peripheral nerve tissues and cells *in vitro*. *Biomaterials*. 28, 1643-1652.

Yu H., Cai H., Zhang W., Xiao L., Liu Q. & Wang P., 2009. A novel design of multifunctional integrated cell-based biosensors for simultaneously detecting cell acidification and extracellular potential. *Biosensors & Bioelectronics*. 24, 1462-1468.

Zeck G. & Fromherz P., 2001. Non-invasive neuroelectronic interfacing with synaptically connected snail neurons immobilized on a semiconductor chip. *Proceedings of the National Academy of Sciences of the United States of America*. 98, 10457-10462.

Zhang W., Li Y., Liu Q. Xu Y., Cai H. & Wang, P., 2008. A Novel experimental research based on taste cell chips for taste transduction mechanism. *Sensors & Actuators B-Chemical*. 131, 24-28.

Chapter 3

Physical Sensors and Measurement

Physical sensors have been widely used in the biomedical field. The commonly used sensors include resistance sensors, inductive sensors, capacitive sensors, piezoelectric sensors, electromagnetic sensors, photoelectric sensors, and thermoelectric sensors. Physical sensors will have more significant applications in biomedicine, especially with the development of MEMS technology for developing more precise and compact sensors, as well as the development of the novel measuring technology.

3.1 Introduction

The nature of physical phenomena includes mechanical, thermal, electrical, magnetic, atomic and nuclear properties. Each of them holds the properties of bodies or physical systems. Based on these natural properties, some physical effects become part of the fundamentals of physical sensors for measuring physical quantities and converting them into signals which can be read out by an observer or an instrument. For example, a thermocouple converts temperature to an output voltage which can be read out by a voltmeter.

Taking into account that the output signal is definitely determined by the input signal, and basing our criteria on the differences of measuring objects in the biomedicine field, physical sensors can be classified as pressure sensors, displacement sensors, speed sensors, acceleration sensors, flow sensors, and temperature sensors. Whereas another criterion based on different physical effects is also significant. In this case, there are resistance sensors, inductive sensors, capacitive sensors, piezoelectric sensors, electromagnetic sensors, photoelectric sensors, and thermoelectric sensors. It is a well-known fact that customers of sensors will choose a sensor in connection with the physical nature of information to be obtained about a phenomenon or a physical system.

As the most widely used measuring devices, physical sensors have already made a great contribution to the development of industry, agriculture, military and aerospace. In addition, they are also playing an important role in our daily life. Along with the rapid development of biomedicine in the latter half of the 20th century, physical sensors have become a kind of measurement instruments of paramount importance in medical diagnosis and therapy. Combining the good features of materials like optical fiber, superconductor or nanophase materials and the semiconductor micro fabrication

Biomedical Sensors and Measurement

technology, the possibility of multifunction, high precision, and integration for physical sensors is guaranteed.

3.2 Resistance Sensors and Measurement

As a primary kind of resistance sensor, the resistance strain sensor is capable of converting strain into a resistance variation. Another type of resistance sensor, known as the piezoresistive sensor, is based on the piezoresistive effect and has the advantages of high sensitivity, good resolution and smaller size. Both of them are widely used in the measurement of blood pressure, pulse and intraocular pressure, intracranial pressure, and eyelid pressure (Shaw et al., 2009) in the biomedical field.

3.2.1 Resistance Strain Sensors

It is a basic phenomenon that deformations of elastic elements bring about resistance change of strain sensitive materials under functions of tested physical parameters. The most commonly used sensing element is resistance strain gage.

3.2.1.1 Strain effect and characteristics

- **Strain effect**

With regard to the working principle of resistance strain gage, strain effect means resistance value changes with mechanical deformation of elastic elements. As is shown in Fig. 3.2.1, metal resistance wire will elongate along the axial direction and shorten along the radial direction when subjected to force in its elastic range.

Fig. 3.2.1. The schematic diagram of strain effect: L is the initial length of resistance wire, dL is the increment of length, r is the radius of cross section, dr is the increment of radius, and F is force

As is showing in Fig. 3.2.1, when a metal resistance wire is not subjected to force, its original resistance can be calculated as follows:

$$R = \frac{\rho L}{S} \tag{3.2.1}$$

where ρ is the resistivity, L is the length, S is the cross sectional area of the resistance wire.

When resistance wire is subjected to tension F, its length will increase ΔL, the cross sectional area will decrease ΔS, and the variation of resistivity caused by distortion of crystal lattice will be $\Delta \rho$, so the relative variation of resistance value can be calculated as follows:

Chapter 3 Physical Sensors and Measurement

$$\frac{\Delta R}{R} = \frac{\Delta L}{L} + \frac{\Delta \rho}{\rho} - \frac{\Delta S}{S} \qquad (3.2.2)$$

where $\Delta L/L$ is relative variation of length, which can be expressed as follows:

$$\varepsilon = \frac{\Delta L}{L} \qquad (3.2.3)$$

$\Delta S/S$ is relative variation of cross sectional area of rounded resistance wire, which can be expressed as follows:

$$\frac{\Delta S}{S} = \frac{2\Delta r}{r} \qquad (3.2.4)$$

We can know from material mechanics that metal resistance wire will elongate along the axial direction and shorten along the radial direction when subjected to force in elastic range, so the relationship between axial strain and radial strain can be expressed as follows:

$$\frac{\Delta r}{r} = -\mu \frac{\Delta L}{L} = -\mu\varepsilon \qquad (3.2.5)$$

where μ is Poisson ratio of resistance wire, and the minus implies the direction of strain is reversed.

Then we can conclude that:

$$\frac{\Delta R}{R} = (1+2\mu)\varepsilon + \frac{\Delta \rho}{\rho} \qquad (3.2.6)$$

or

$$\frac{\Delta R/R}{\varepsilon} = (1+2\mu) + \frac{\Delta \rho/\rho}{\varepsilon} \qquad (3.2.7)$$

Generally the variation of resistance value caused by unit strain is called sensitivity coefficient of resistance wire, which actually is the relative variation of resistance value caused by unit strain in physical meaning. It can be expressed as follows:

$$K = (1+2\mu) + \frac{\Delta \rho/\rho}{\varepsilon} \qquad (3.2.8)$$

Sensitivity coefficient is affected by two factors: one is the change of material geometry size when bearing force, that is $(1+2\mu)$; the other is the change of material resistivity when bearing force, that is $(\Delta\rho/\rho)/\varepsilon$. As for metal resistance wire, the value $(1+2\mu)$ in sensitivity coefficient is much higher than $(\Delta\rho/\rho)/\varepsilon$, but much lower than $(\Delta\rho/\rho)/\varepsilon$ for semiconductor materials. Many experiments show that relative variation of resistance is proportional to strain in the stretch limit of resistance wire, that is to say, K is constant.

When using strain gage to measure strain or stress, small mechanical distortion will be brought to the measurand, and strain gage will change with it, then its resistance value will change. If variation ΔR is tested in strain gage, the stain of measurand can be obtained. According to the relationship between stress and strain, stress value σ can be calculated as follows:

$$\sigma = E \cdot \varepsilon \qquad (3.2.9)$$

where σ is the stress of specimen, ε is the strain, E is the modulus of elasticity of specimen material.

Therefore, stress σ is proportional to strain ε, and the specimen strain ε is proportional to the change of resistance. So, stress σ is proportional to the change of resistance. That is the basic principle of strain measurement using strain gage.

- **Categories of resistance strain gage**

Among the great varieties of forms, strain gages can be generally classified as bonded foil/non-foil types, thin film types, and semiconductor types.

Bonded foil types are made from alloy foil like Constantan or Karma with a thin polyamide or cast epoxy backing. Bonded non-foil types are very similar to bonded foil types except metal wire is used as the gage element.

Thin film types, made by sputtering insulation and gage elements directly on a polished sensing area, have eliminated instability caused by adhesive bonds. With merits of greater sensitivity coefficient, greater allowable current density, and a large working range, they are widely used.

Semiconductor strain gage can be formed as bonded, unbonded, or integrated strain gage units. They have a high gage factor and are suitable for dynamic measurements. However, they are more temperature-sensitive and inherently more nonlinear than metal strain gages.

Its working principle is based on piezoresistive effect of semiconductor materials. Piezoresistive effect refers to the phenomenon that resistivity ρ changes when semiconductor materials are subjected to axial force.

When semiconductor strain gage is subjected to axial force, the relative change of resistance can be expressed as follows:

$$\frac{\Delta R}{R} = (1+2\mu)\varepsilon + \frac{\Delta \rho}{\rho} \tag{3.2.10}$$

where $\Delta\rho/\rho$ is the relative change of resistivity of semiconductor strain gage. The relationship between $\Delta\rho/\rho$ and axial stress of semiconductor sensitive element can be expressed as follows:

$$\frac{\Delta \rho}{\rho} = \pi_1 \sigma = \pi_1 E \varepsilon \tag{3.2.11}$$

where π_1 is the piezoresistive coefficient of semiconductor materials, E is the modulus of elasticity.

From Eq. (3.2.10) and Eq. (3.2.11), we can conclude that:

$$\frac{\Delta R}{R} = (1+2\mu+\pi_1 E) \cdot \varepsilon \tag{3.2.12}$$

Experiments show that $\pi_1 E$ is several hundred folds than $(1+2\mu)$, which can be ignored, so sensitivity coefficient of semiconductor strain gage can be expressed as follows:

$$K = \frac{\Delta R / R}{\varepsilon} = \pi_1 E \tag{3.2.13}$$

The outstanding advantages of semiconductor strain gage is high sensitivity, which is 50~80 times higher than that of metal wire, and small dimension, little transverse effect, and good dynamic response. Its disadvantages are large temperature coefficient and severe nonlinear properties when strain appears.

New materials are designed to improve security either in rehabilitation or health monitoring. Elastic resistance strain gages with flexible conductive elastomers (such as electrically conductive liquid silicon rubber) have been used. They are easy to deform, and have good mechanical, electrical, ageing, fast vulcanization properties (Martineza et al., 2009) and biocompatibility, and have great potentials in biomedical measurement.

Chapter 3   Physical Sensors and Measurement

- *Temperature error*

Temperature error of strain gage refers to the additional error brought by the temperature change in the measuring circumstance. The cause for temperature error includes:

a) Affection of the resistance temperature coefficient. Resistance of the sensing element changes with temperature. Their relationship can be expressed as follows:

$$R_t = R_0(1+\alpha_0 \Delta t) \qquad (3.2.14)$$

where R_t is resistance at t °C. R_0 is resistance at t_0 °C, α_0 is resistance temperature coefficient of resistance wire, Δt is change of temperature, and $\Delta t = t - t_0$.

If the change of temperature is Δt, the change of resistance can be expressed as follows:

$$\Delta R_t = R_t - R_0 = R_0 \alpha_0 \Delta t \qquad (3.2.15)$$

b) Affection of different linear expansion coefficients of tested pieces and resistance sensing materials. Once temperature changes, distortion will happen, thus additional resistance is produced.

If linear expansion coefficients of tested piece and resistance wire materials are the same, distortion of resistance wire does not bring additional deformation, just like in free condition, and it does not change with temperature.

If linear expansion coefficients of tested piece and resistance wire materials are different, distortion of resistance wire appears as a result of temperature change, thus additional resistance is produced.

Suppose the length of resistance wire and tested piece is L_0 at 0 °C their linear expansion coefficients are β_s and β_g respectively, and they are not pasted together, then their length can be expressed as follows:

$$L_s = L_0(1+\beta_s \Delta t) \qquad (3.2.16)$$

$$L_g = L_0(1+\beta_g \Delta t) \qquad (3.2.17)$$

If the two are pasted together, additional deformation of resistance wire is ΔL, additional strain is ε_β and change of additional resistance is ΔR_β, which can be expressed as follows:

$$\Delta L = L_g - L_s = (\beta_g - \beta_s)L_0 \Delta t \qquad (3.2.18)$$

$$\varepsilon_\beta = \frac{\Delta L}{L} = (\beta_g - \beta_s)\Delta t \qquad (3.2.19)$$

$$\Delta R_\beta = K_0 R_0 \varepsilon_\beta = K_0 R_0 (\beta_g - \beta_s)\Delta t \qquad (3.2.20)$$

Then we can calculate relative change of strain gage resistance caused by temperature as follows:

$$\frac{\Delta R_t}{\Delta R_0} = \frac{\Delta R_\alpha + \Delta R_\beta}{\Delta R_0} = \alpha_0 \Delta t + K_0(\beta_g - \beta_s)\Delta t = [\alpha_0 + K_0(\beta_g - \beta_s)]\Delta t = \alpha \Delta t \qquad (3.2.21)$$

It can be converted into additional strain or fictitious strain ε_t, which can be expressed as follows:

$$\varepsilon_t = \frac{\Delta R_t / R_0}{K_0} = \left[\frac{\alpha_0}{K_0} + (\beta_g - \beta_s)\right]\Delta t = \frac{\alpha}{K_0}\Delta t \qquad (3.2.22)$$

Form Eq. (3.2.21) and Eq. (3.2.22) we know that relative resistance change caused by temperature change is related to performance parameter (K_0, α_0, β_s) of strain gage and linear expansion coefficient β_g of tested piece, as well as environment temperature.

3.2.1.2 Measurement

As mechanical strain is generally small, we need to measure out small resistance changes caused by small strain, at the same time convert relative resistance change $\Delta R/R$ into variation of voltage or current. So, a special measuring circuit used to measure resistance change caused by strain change is needed. Usually we adopt direct current bridge and alternating current bridge.

- **Direct current bridge**

a) Equilibrium condition of direct current bridge

Electric bridge is shown in Fig. 3.2.2, where E is power supply, R_1, R_2, R_3 and R_4 are resistances of bridge arms, R_L is load resistance. When $R_L \to \infty$, output voltage of bridge can be expressed as follows:

$$U_0 = E\left(\frac{R_1}{R_1+R_2} - \frac{R_3}{R_3+R_4}\right) \tag{3.2.23}$$

when electric bridge is balanced, that is to say, $U_0 = 0$, then

$$R_1 R_4 = R_2 R_3$$

or

$$\frac{R_1}{R_2} = \frac{R_3}{R_4} \tag{3.2.24}$$

Eq. (3.2.24) is called the balance condition of bridge. This shows that if we want the bridge to be balanced, the ratio of two contiguous resistors must be equal, in other words, the product of two opposite resistance must be equal.

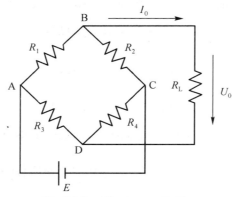

Fig. 3.2.2. Direct current bridge

b) Voltage sensitivity

In strain gage's working process, change of resistance is small, and output voltage of bridge is relatively small, so amplifier is needed to amplify signals. As the input resistance of amplifier is much greater than the output resistance of bridge, we consider the bridge open-circuit. When strain appears, and change of strain gage resistance is ΔR, other bridge-arms are fixed, and output voltage of bridge $U_0 \neq 0$, then output voltage of unbalanced bridge is calculated as follows:

$$U_0 = E\left(\frac{R_1 + \Delta R_1}{R_1 + \Delta R_1 + R_2} - \frac{R_3}{R_3 + R_4}\right)$$

$$= E\frac{\Delta R_1 R_4}{(R_1 + \Delta R_1 + R_2)(R_3 + R_4)} \quad (3.2.25)$$

$$= E\frac{\frac{R_4}{R_3}\frac{\Delta R_1}{R_1}}{\left(1 + \frac{\Delta R_1}{R_1} + \frac{R_2}{R_1}\right)\left(1 + \frac{R_4}{R_3}\right)}$$

Suppose bridge arm ratio $n=R_2/R_1$, for $\Delta R_1 \ll R_1$, $\Delta R_1/R_1$ in denominator can be ignored. Considering the balance condition $R_2/R_1=R_4/R_3$, Eq. (3.2.25) can be rewritten as follows:

$$U_0 = E\left(\frac{n}{(1+n)^2}\right)\frac{\Delta R_1}{R_1} \quad (3.2.26)$$

Bridge voltage sensitivity is defined as follows:

$$K_U = \frac{U_0}{\Delta R_1 / R_1} = E\frac{n}{(1+n)^2} \quad (3.2.27)$$

From which we can see that:

Bridge voltage sensitivity is proportional to power supply voltage of bridge. The higher the power supply voltage, the higher voltage sensitivity of bridge. But increase of power supply voltage is restricted to allowable power of strain gage, so it must be chosen properly.

Bridge voltage sensitivity changes with bridge arm ratio n. If bridge arm ratio n is chosen properly, higher voltage sensitivity would be guaranteed.

When E is confirmed, how to take value of n to make K_U the highest?

From $\frac{dK_U}{dn} = 0$, we can calculate the maximum of K_U as follows:

$$\frac{dK_U}{dn} = \frac{1-n}{(1+n)^3} \quad (3.2.28)$$

When $n = 1$, K_U takes maximum value. That is to say, if bridge voltage is confirmed, and $R_1=R_2=R_3=R_4$, bridge voltage sensitivity is the highest, so:

$$U_0 = \frac{E}{4}\frac{\Delta R_1}{R_1} \quad (3.2.29)$$

$$K_U = \frac{E}{4} \quad (3.2.30)$$

From the above we know that when power supply voltage E and relative variation of resistance $\Delta R_1/R_1$ are confirmed, output voltage of bridge and its sensitivity is a fixed value, which is unrelated to the resistance of bridge arms.

c) Nonlinear error and compensation method

The output voltage calculated from Eq. (3.2.25) is ideal value because of ignoring $\Delta R_1/R_1$ in denominator. The actual value is calculated as follows:

$$U'_0 = E \frac{n \frac{\Delta R_1}{R_1}}{\left(1+n+\frac{\Delta R_1}{R_1}\right)(1+n)} \tag{3.2.31}$$

Nonlinear error is:

$$\gamma_L = \frac{U_0 - U'_0}{U_0} = \frac{\frac{\Delta R_1}{R_1}}{1+n+\frac{\Delta R_1}{R_1}} \tag{3.2.32}$$

For balanced bridge, $R_1 = R_2 = R_3 = R_4$, we know that:

$$\gamma_L = \frac{\Delta R_1 / 2R_1}{1 + \Delta R_1 / 2R_1} \tag{3.2.33}$$

For common strain gage, strain ε is usually below 5×10^{-3}. If K=2, we can see $\Delta R_1/R_1 = K\varepsilon = 0.01$. From Eq. (3.2.33) we know that nonlinear error is 0.5%. If $K_U = 130$ and $\varepsilon=1\times 10^{-3}$, we know $\Delta R_1/R_1=0.130$, so nonlinear error is 6%. In a word, if nonlinear error can't satisfy the requirements of the measurement, it needs to be eliminated.

To reduce and overcome nonlinear error, we generally use differential bridge which is shown in Fig. 3.2.3. Two working strain gages are installed into tested piece: one is subjected to tensile strain, the other bears compressive strain. If this connects to adjacent bridge arm, so called half bridge differential circuit is formed. Its output voltage is expressed as follows:

$$U_0 = E\left(\frac{\Delta R_1 + R_1}{\Delta R_1 + R_1 + R_2 - \Delta R_2} - \frac{R_3}{R_3 + R_4}\right) \tag{3.2.34}$$

If $\Delta R_1 = \Delta R_2, R_1 = R_2, R_3 = R_4$, we know that:

$$U_0 = \frac{E}{2} \cdot \frac{\Delta R_1}{R_1} \tag{3.2.35}$$

From Eq. (3.2.35) we know that U_0 has a linear relation with $(\Delta R_1 / R_1)$. Differential bridge has no nonlinear error, its voltage sensitivity $K_U=E/2$ is two times of that of single arm, and has effect of temperature compensation.

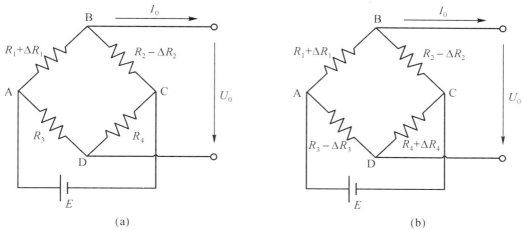

Fig. 3.2.3. Differential bridge: (a) Half bridge; (b) Full bridge

Four strain gages are connected to four arms of electric bridge, as shown in Fig. 3.2.3b. That is to say, two of them are subjected to tensile strain, the others are subjected to compressive strain. Put two strain gages with the same strain direction into opposite bridge arms to form full bridge differential circuit, if $\Delta R_1 = \Delta R_2 = \Delta R_3 = \Delta R_4$ and $R_1 = R_2 = R_3 = R_4$, we can conclude that:

$$U_0 = E \frac{\Delta R_1}{R_1} \tag{3.2.36}$$

$$K_U = E \tag{3.2.37}$$

Here full bridge differential circuit has no nonlinear error, and voltage sensitivity is four times of that of single gage. At the same time it also has effect of temperature compensation.

- *Alternating current bridge*

From the above analysis of direct current bridge we know that strain bridge needs amplifier because the output voltage is small. But direct current amplifier can easily bring zero drift, so alternating current bridge is commonly used in strain bridge.

Fig. 3.2.4 shows alternating current bridge, where $\dot{U}$ is an alternating voltage, $\dot{U}_0$ is output voltage of open circuit. For the power supply is AC power source, and distributed capacitance of lead wire makes bridge with two arms perform as complex impedance. It's like the two strain gages are parallel connected to a capacitor respectively. Thus complex impedance of every bridge arm is expressed as follows:

$$\begin{cases} Z_1 = \dfrac{R_1}{1+j\omega R_1 C_1} \\ Z_2 = \dfrac{R_2}{1+j\omega R_2 C_2} \\ Z_3 = R_3 \\ Z_4 = R_4 \end{cases} \tag{3.2.38}$$

where C_1, C_2 is the distributed capacitance of strain gage. From analysis of alternating current circuit we know that:

$$\dot{U}_0 = \frac{\dot{U}(Z_2 Z_3 - Z_1 Z_4)}{(Z_1 + Z_2)(Z_3 + Z_4)} \tag{3.2.39}$$

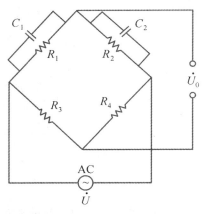

Fig. 3.2.4. Alternating current bridge

To satisfy conditions for electric balance, which means $\dot{U}_0 = 0$, we can conclude the following expression:

$$Z_1 Z_4 = Z_2 Z_3 \tag{3.2.40}$$

Let be $Z_1 = Z_2 = Z_3 = Z_4$, we have the following equation:

$$\frac{R_1}{1 + j\omega R_1 C_1} R_4 = \frac{R_2}{1 + j\omega R_2 C_2} R_3 \tag{3.2.41}$$

or

$$\frac{R_3}{R_1} + j\omega R_3 C_1 = \frac{R_4}{R_2} + j\omega R_4 C_2 \tag{3.2.42}$$

where real parts and imaginary parts are equal respectively. From all above we can express the balance condition of alternating current bridge as follows:

$$\frac{R_2}{R_1} = \frac{R_4}{R_3}$$

or

$$\frac{R_2}{R_1} = \frac{C_1}{C_2} \tag{3.2.43}$$

For alternating current capacitance bridge, capacitance balance conditions as well as resistance balance conditions must be satisfied. So balance regulation function of resistance and capacitance is added to bridge.

When change of tested stress lead to change of Z_1 and Z_2, which is expressed as $Z_1 = Z_0 - \Delta Z$, $Z_2 = Z_0 + \Delta Z$, the bridge output can be expressed as follows:

$$\dot{U}_0 = \dot{U}\left(\frac{Z_0 + \Delta Z}{2 Z_0} - \frac{1}{2}\right) = \frac{\dot{U}}{2} \frac{\Delta Z}{Z_0} \tag{3.2.44}$$

3.2.1.3 *Biomedical applications*

• *Blood pressure measurement*

As a primary indicator of physiological distress, blood pressure is very important in determination of the functional integrity of the cardiovascular system.

Non-invasive blood pressure measurement (Zhang and Wu, 2003), like cuff-based blood pressure measurement, which has the advantages of facility, safety, painlessness and more acceptance, is generally used in home health monitoring (Jobbágy et al., 2007) and conventional physical examination. As "normal" blood pressure varies during the day, with age, state of health and clinical situation, and also has beat-to-beat variations, sometimes the non-invasive blood pressure monitoring is not possible or likely to be inaccurate.

Continuous, invasive blood pressure monitoring (Hu and Wang, 2008) is the gold standard of blood pressure measurement giving accurate beat-to-beat information. As it allows repetitive and convenient sampling for blood gases analysis, it is also used when long-term measurement in critically ill patients is required, avoiding the problem of repeated cuff inflation, which would cause localized tissue damage.

A typical method for invasive blood pressure measurement uses for the extravascular system (Peng, 2000), as shown in Fig. 3.2.5. A catheter is placed in the artery or vein and is connected to a 3-way stopcock and the pressure sensor. In this system a catheter couples a flush solution (heparinized saline) through a disposable pressure sensor with an integral flush device to the sensor port. The 3-way stopcock is used to take blood samples and initialize the pressure sensor. The catheter must be flushed frequently (every few minutes) to prevent blood clotting at the catheter tip. The catheter is inserted by a surgical cut-down or by a percutaneous insertion (surgical needle and a guide wire). Blood pressure is transmitted via the catheter to the sensor's diaphragm. The strain gage sensor has several types, as shown in Fig. 3.2.6.

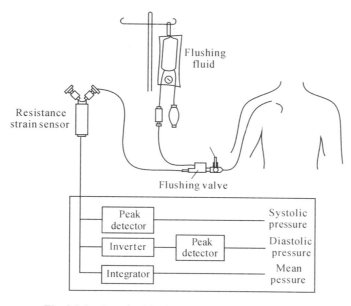

Fig. 3.2.5. Invasive blood pressure measurement system

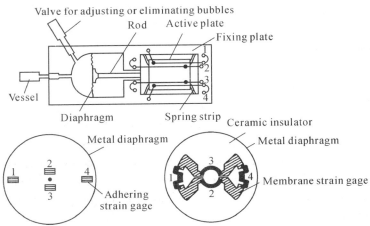

Fig. 3.2.6. Resistance strain sensor applied in invasive blood pressure measurement system: (a) Four wire type strain gages with the same resistance are connected into a balance bridge; (b) Schematic diagram of adhering strain gage to metal diaphragm, the adhering position is determined via mechanical analysis for the site of generating the maximum strain, and the four strain gages are connected to a balance bridge; (c) Another way (vacuum deposition) of depositing membrane with strain effect directly on the surface of metal diaphragm

- **Bladder volume measurement in patients with urinary dysfunction**

Millions of people have been persecuted by urinary bladder dysfunction, which leads to loss of voluntary control over the bladder muscles and cuts off sensorial feedback to the central nervous system (Gaunt and Prochazka, 2006). The prevalent therapeutic method in clinical practice is stimulating the sacral root at the base of the spine to produce microstimulation. In the past few years, direct sacral nerve stimulation, using a dual implantable stimulator (Ba and Sawan, 2003), has proved to be clinically feasible. The stimulation can be permanent, selective, or involve conversion between the two types. But the best choice is the process which creates the ability to trigger emptying of the bladder in response to maximal bladder volume, which is similar to the automatic sensorial feedback. So bladder volume detection becomes the key concern.

There are some traditional ways to measure bladder volume, such as using a pressure sensor, ultrasound measurements, and bioelectric impedance measurements. However, they are not entirely satisfactory because of some potential defects or unwanted interference. A new method reported by Rajagopalan et al. (2008) is employing an implantable polypyrrole-based strain sensor, using a conductive polymer as the sensing device. The conducting polymer-polypyrrole (PPY) is coated on a flexible fabric and inserted over the upper portion of the bladder (Fig. 3.2.7a).

Like most soft tissues in the body, the urinary bladder wall is non-linear, viscoelastic, and anisotropic. The collagen fibers are kinked and coiled when the bladder is relaxed and begins to stretch during filling. Correspondingly, the collagen fibers allow for high strain, which means that the urinary bladder can cater for a volume of up to 11 times its resting volume.

Fig. 3.2.7b shows the implantable measuring circuit, which can read out the changes in resistance. The sensing current proportional to the sensing resistor is extracted through the clocking system (Sw1, Sw2) which is amplified in the current mirror block and then integrated using a capacitor (C_{in}). The output of the Schmitt Trigger block drives the digital counter which outputs a value proportional to the input resistance value. The circuit can provide continuous resistance outputs for a given input voltage.

This resistance reading can then be transmitted wirelessly to a wearable display positioned just outside the body.

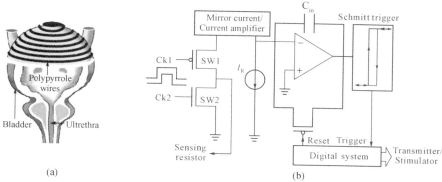

Fig. 3.2.7. Bladder volume measurement: (a) Illustration of bladder covered by stock with strip lines of PPY; (b) Interface circuit (reprinted from (Rajagopalan et al., 2008), Copyright 2008, with permission from MDPI Publishing)

3.2.2 Piezoresistive Sensors

Piezoresistive sensors are based on piezoresistive effect of monocrystalline silicon and are made up by using an integrated circuit technique. They are generally used in measuring pressure and some other physical parameters that can be converted into pressure. Microfabrication and MEMS technology have made the miniaturization of piezoresistive sensors possible and applicable in practice.

3.2.2.1 Piezoresistive effect

The piezoresistive effect describes the changing electrical resistance of a material due to applied mechanical stress. It only causes a change in resistance and does not produce an electric potential. If a mechanical stress σ is applied on a resistor, the resistivity change $\Delta\rho/\rho$ can be calculated as:

$$\Delta\rho/\rho = \Pi\sigma \quad (3.2.45)$$

where Π is the piezoresistivity coefficient. The sensitivity of piezoresistive devices is characterized by the gage factor:

$$K = (\Delta R/R)/\varepsilon \quad (3.2.46)$$

where ΔR is the change in resistance due to deformation, R is the original resistance and ε is the strain.

As shown in Eq. (3.2.46), the resistance of silicon changes is affected not only by the stress dependent change of geometry, but also by the stress dependent resistivity of the material. The latter is the dominant factor, which results in sensitivity to orders of magnitudes larger than those observed in metals. This effect is present in materials like germanium, polycrystalline silicon, amorphous silicon, silicon carbide, and single crystal silicon, which are among the several types of semiconductor materials. Since silicon is today the material of choice for integrated digital and analog circuits the use of piezoresistive silicon devices has been of great interest. Many commercial devices such as pressure sensors and acceleration sensors employ the piezoresistive effect in silicon.

3.2.2.2 Measurement

The measuring circuits for piezoresistive sensors are similar to strain sensor measurement. A Wheatstone bridge is primarily used in piezoresistive sensor devices, but with different signal conditioners. As mentioned in Subsection 3.2.1, semiconductor materials are more temperature-sensitive and nonlinear than metal materials. Piezoresistive pressure sensors are usable only after corrections have been made for offset and compensation for temperature.

For medium-accuracy sensors, a resistor network can compensate for offset, offset drift, and FSO (full-scale output) drift (Fig. 3.2.8). Zero trim resistors adjust for initial offset. But the bridge resistors have a positive temperature coefficient that causes the bridge voltage to rise with temperature, so resistor R_{TS} is used to stabilize the sensitivity by shunting an increasing amount of excitation current as temperature rises. Besides, resistor R_{TZ} works against the change of offset with temperature.

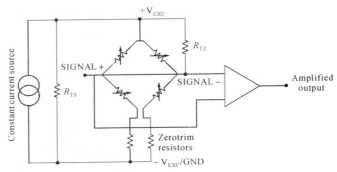

Fig. 3.2.8. Interaction of the three compensation mechanisms in a conventional resistive compensation circuit (R_{TS} for sensitivity drift, R_{TZ} for offset drift, and zero trim resistors)

But biomedical detection needs better precision. Fig. 3.2.9 shows an intergrated compensation circuit which includes two main functional blocks: a controlled current source for driving the sensor, and a programmable-gain amplifier (PGA, implemented in switched- capacitor technology and virtually free of offset). The numerous external resistors and voltage dividers are commonly realized with hybrid technology and adjusted with laser trimming. The temperature drift is adjusted by feeding back the sensor's drive voltage (from the BDRIVE pin) to the ISRC pin. The circuit's initial sensitivity is adjusted at the FSOTRIM pin of MAX1450. Compensation of offset and offset drift is accomplished at the PGA and decoupled from the sensitivity compensation. The key function, however, is the controlled current source, which implements a unique algorithm for compensating the sensitivity drift.

3.2.2.3 Biomedical applications

Piezoresistive sensors are widely used to measure pressure in biomedicine. To reduce the unavoidable nonlinearity increase with sensitivity, Marco et al. (1996) used thin structured membranes in piezoresistive pressure sensors to obtain high-performance. A generally used Si/Porous-Si membrane is shown in Fig. 3.2.10. Pramanik and Saha (2006) also used various diaphragms to realize low pressure measurement in biomedicine, such as respirators, ventilators and spirometers. Among the diaphragms, use of nanocrystalline silicon, which is a three-phase mixture of silicon, silicon oxide, and voids and formed by electrochemical etching of silicon, increases the sensitivity almost three times to that of conventional piezoresistive pressure sensors of similar dimensions. Besides, silicon nanowire can be used to enhance sensitivity of piezoresistive sensors (Kim et al., 2009).

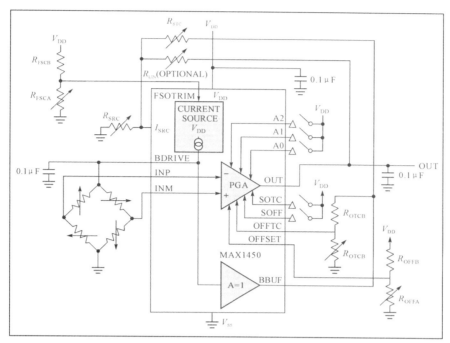

Fig. 3.2.9. Laser-trimmed resistor dividers in the MAX1450 signal conditioner provide better than 1% compensation full scale over temperature

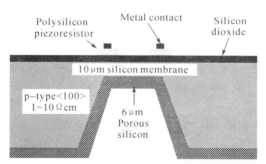

Fig. 3.2.10. Pressure sensor with Si/Porous-Si composite membrane

With the development of MEMS technology, ultra-miniaturized piezoresistive pressure sensors (Gowrishetty et al., 2008) are developed to monitor intra-cranial pressure during neurosurgery, air pressure for respiratory disease, blood pressure during surgery/intensive care, intra-uterine for obstetrics, and abdominal/urinary pressure for the diagnosis of respective disorders. According to Gowrishetty, the dimensions of the fabricated sensor are 650 μm×230 μm×150 μm (length, width, thickness) with 2.5 μm thick diaphragms, and sensitivity of the sensors with half Wheatstone bridge configuration is determined to be 27 – 31 μV/V/mmHg.

Apart from pressure sensing, a piezoresitive sensor is also applied in a position-sensing system for a MEMS-based cochlear implant (Wang et al., 2005). Fig. 3.2.11a shows the prototype of a fully-implantable thin-film cochlear prosthesis (Kensall et al., 2008). A polysilicon piezoresistive position-sensing system is monolithically integrated into the electrode arrays in order to provide real-time visualization of array position for the surgeon. The sensing array of the cochlear microsystem consists of the electrode array, flip-chip bonded to a signal-processing chip. As shown in Fig. 3.2.11b, the position sensors with each formed using two strain-sensing resistors in a half-bridge configuration are implemented underneath electrode sites using piezoresistive polysilicon strain gages and extra passivating dielectrics (Kensall et al., 2008). Polysilicon wall-contact and strain gages are buried under IrO sites distributed along the shank.

A microprocessor gives instructions to the circuit chip to select the addressed sensor, and the addressed sensor bridge is connected to the positive input of an instrumentation amplifier (Wang et al., 2005), which is referenced to a voltage generated by the DAC (Fig. 3.2.11c). Then the bridge output signal is amplified and filtered to determine local bending.

Another development of piezoresistive sensors is as an application in wearable devices with smart textiles to monitor gesture, posture, or respiration. The piezoresistive sensors can be yarn-based (Huang et al., 2008), and can measure respiratory rate even in the rapid running motion (Jeong et al., 2009).

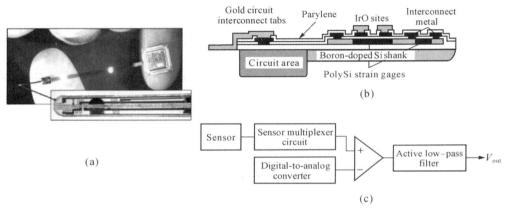

Fig. 3.2.11. MEMS-based cochlear implant: (a) The prototype of a fully-implantable thin-film cochlear prosthesis; (b) Cross-sectional diagram of an electrode array with position sensors; (c) Selected sensor output is read out using an instrumentation amplifier and a LP filter (reprinted from (Kensall et al., 2008), Copyright 2008, with permission from Elsevier)

3.3 Inductive Sensors and Measurement

Based on the electromagnetic induction principle, non-electric quantities, such as displacement, stress, flux, vibration, can be converted into variations of self-inductance L or mutual inductance M of the coil, which would be finally output as voltage or current through a measuring circuit. This kind of devices is called inductance sensors, which has reliable performance and high measurement accuracy. The main shortcoming is that its sensitivity, linearity and measurement range restrict each other, and the frequency response does not apply to rapid dynamic measurement.

3.3.1 Basics

Based on the conversion mode from non-electric parameters to voltage, inductive sensors can be classified as self-inductance sensors and mutual inductance ones.

3.3.1.1 *Variable reluctance sensors*

A variable reluctance sensor is a typical self-inductance sensor. It consists of a coil, a core and an armature. The coil and armatures are made of permeable magnetic materials like silicon steel sheet and permalloy. There exists an air-gap with a thickness δ between them. Once the core moves, the thickness δ would change, which results in the change of reluctance in the magnetic circuit, and the inductance value of the inductance coil. If the coil is placed in an AC circuit, the change of inductance can be used to change the voltage drop across the inductor or it can be used in an oscillator circuit to change the frequency of the circuit.

- *Working principle*

The structure of variable reluctance transducer is shown in Fig. 3.3.1. It consists of a coil, a core and an armature. Coil and armature is made of permeable magnetic materials like silicon steel sheet and permalloy. There exists air-gap with thickness δ between core and armature. The moving part of sensor is connected to armature, so when the core moves, thickness δ will change, which will cause the change of reluctance in magnetic circuit, thus the inductance value of inductance coil will change. Once inductance value is measured, the size and direction of displacement of armature is determined.

According to definition of inductance, the inductance of coil can be expressed as follows:

$$L = \frac{\psi}{I} = \frac{\omega\phi}{i} \qquad (3.3.1)$$

where ψ is the total magnetic linkage of coil, I is the current of coil, ω is the coil turn, ϕ is the coil flux.

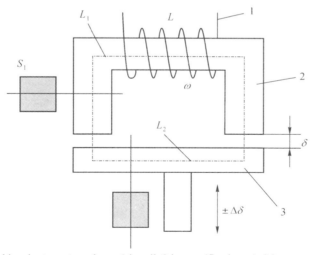

Fig. 3.3.1. Variable reluctance transducer. 1 is coil; 2 is core (fixed core); 3 is armature (movable core)

According to magnetic Ohm's Law:

$$\phi = \frac{I\omega}{R_m} \qquad (3.3.2)$$

where R_m is total reluctance of magnetic circuit. For variable gap sensor has small air-gap, the magnetic in air-gap can be considered uniform. Ignoring core loss of magnetic circuit, total reluctance of magnetic circuit can be expressed as follows:

$$R_m = \frac{L_1}{\mu_1 S_1} + \frac{L_2}{\mu_2 S_2} + \frac{2\delta}{\mu_0 S_0} \tag{3.3.3}$$

where μ_1 is the magnetic conductivity of core materials, μ_2 is the magnetic conductivity of armature materials, L_1 is length of flux through core, L_2 is length of flux through armature, S_1 is sectional area of core, S_2 is sectional area of armature, μ_0 is magnetic conductivity of air, S_0 is sectional area of air-gap, δ is thickness of air-gap.

Usually reluctance of air-gap is much greater than that of core and armature,

$$\frac{2\delta}{\mu_0 S_0} \gg \frac{L_1}{\mu_1 S_1}, \quad \frac{2\delta}{\mu_0 S_0} \gg \frac{L_2}{\mu_2 S_2} \tag{3.3.4}$$

So Eq. (3.3.3) can be approximated as follows:

$$R_m = \frac{2\delta}{\mu_0 S_0} \tag{3.3.5}$$

From Eq. (3.3.1), Eq. (3.3.2) and Eq. (3.3.5), we can conclude that:

$$L = \frac{\omega^2}{R_m} = \frac{\omega^2 \mu_0 S_0}{2\delta} \tag{3.3.6}$$

The upper equation shows that reluctance L only changes with reluctance R_m when coil turn is constant. Change of δ or δ_0 will cause change of reluctance, so variable reluctance transducer is divided into two classes: variable air-gap thickness δ and variable air-gap area S_0, the former is the most widely used.

- **Output characteristic**

Suppose the initial air gap of inductance sensor is δ_0, the initial inductance value is L_0 and variation of air-gap caused by armature displacement is $\Delta\delta$, from Eq. (3.3.6) we know that L has a nonlinear relation with δ. The characteristic curve is shown in Fig. 3.3.2. The initial inductance value is expressed as follows:

$$L_0 = \frac{\omega^2 \mu_0 S_0}{2\delta} \tag{3.3.7}$$

When armature moves upwards by distance of $\Delta\delta$, air gap of sensor will reduce $\Delta\delta$, that is $\delta = \delta_0 - \Delta\delta$, then the output inductance value L is expressed as $L=L_0+\Delta L$. From Eq. (3.3.6) we know that:

$$L = L_0 + \Delta L = \frac{\omega^2 \mu_0 S_0}{2(\delta_0 - \Delta\delta)} = \frac{L_0}{1 - \Delta\delta/\delta_0} \tag{3.3.8}$$

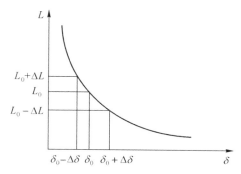

Fig. 3.3.2. $L-\delta$ characteristics of variable gap inductance sensor

If $\Delta\delta/\delta_0 \ll 1$, the series form of upper equation is as follows:

$$L = L_0 + \Delta L = L_0\left[1 + \left(\frac{\Delta\delta}{\delta_0}\right) + \left(\frac{\Delta\delta}{\delta_0}\right)^2 + \left(\frac{\Delta\delta}{\delta_0}\right)^3 + \cdots\right] \quad (3.3.9)$$

From above we can calculate inductance increment ΔL and relative increment $\Delta L/L_0$ as follows:

$$\Delta L = L_0\frac{\Delta\delta}{\delta_0}\left[1 + \left(\frac{\Delta\delta}{\delta_0}\right) + \left(\frac{\Delta\delta}{\delta_0}\right)^2 + \left(\frac{\Delta\delta}{\delta_0}\right)^3 + \cdots\right] \quad (3.3.10)$$

$$\frac{\Delta L}{L} = \frac{\Delta\delta}{\delta_0}\left[1 + \left(\frac{\Delta\delta}{\delta_0}\right) + \left(\frac{\Delta\delta}{\delta_0}\right)^2 + \left(\frac{\Delta\delta}{\delta_0}\right)^3 + \cdots\right] \quad (3.3.11)$$

Similarly, when armature moves downwards by $\Delta\delta$, the following relation can be concluded:

$$\Delta L = L_0\frac{\Delta\delta}{\delta_0}\left[1 - \left(\frac{\Delta\delta}{\delta_0}\right) + \left(\frac{\Delta\delta}{\delta_0}\right)^2 - \left(\frac{\Delta\delta}{\delta_0}\right)^3 + \cdots\right] \quad (3.3.12)$$

$$\frac{\Delta L}{L} = \frac{\Delta\delta}{\delta_0}\left[1 - \left(\frac{\Delta\delta}{\delta_0}\right) + \left(\frac{\Delta\delta}{\delta_0}\right)^2 - \left(\frac{\Delta\delta}{\delta_0}\right)^3 + \cdots\right] \quad (3.3.13)$$

Linear processing of Eq. (3.3.11) and Eq. (3.3.13) by ignoring high order terms come to the following relation:

$$\frac{\Delta L}{L} = \frac{\Delta\delta}{\delta_0} \quad (3.3.14)$$

The sensitivity is:

$$K_0 = \frac{\Delta L/L}{\Delta\delta} = \frac{1}{\delta_0} \quad (3.3.15)$$

Therefore, the measuring range of variable gap inductance sensor is in contradiction with sensitivity and linearity. So, it can measure micro-displacement precisely. To reduce nonlinear error, differential variable gap inductance sensor is widely used in actual measurement.

Fig. 3.3.3 shows the structure of differential variable inductance sensor. From the figure we know that this kind of sensor is made up of two identical inductance coil and magnetic circuit. In measurement, armature is connected to tested displacement through guide rod. When tested piece moves ups and downs, the guide rod drive armature to move by the same displacement, thus the reluctance change of the two magnetism loop appears, with equal size but opposite direction, which leads to inductance increase in one coil, but decrease in the other. That's how differential type forms.

When armature moves upwards by $\Delta\delta$, inductance variation of two coils ΔL_1 and ΔL_2 are expressed in Eq. (3.3.10) and Eq. (3.3.12). When differential type is taken into action, the two inductance coil is connected to form adjacent arms of alternating current bridge, while the other arms are formed by resistors. The relationship between output voltage of bridge and ΔL is expressed as follows:

$$\Delta L = \Delta L_1 + \Delta L_2 = 2L_0 \frac{\Delta\delta}{\delta_0}\left[1+\left(\frac{\Delta\delta}{\delta_0}\right)^2+\left(\frac{\Delta\delta}{\delta_0}\right)^4+\cdots\right] \quad (3.3.16)$$

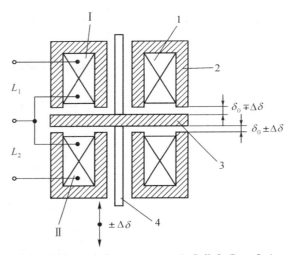

Fig. 3.3.3. Differential variable gap inductance sensor. 1: Coil; 2: Core; 3: Armature; 4: Guide bar

The following equation is obtained through linear processing of upper expression by ignoring high order terms:

$$\frac{\Delta L}{L} = \frac{2\Delta\delta}{\delta_0} \quad (3.3.17)$$

Sensitivity K_0 is expressed as follows:

$$K_0 = \frac{\Delta L/L}{\Delta\delta} = \frac{2}{\delta_0} \quad (3.3.18)$$

By comparing characteristics of single coil and differential variable gap inductance sensor, the following conclusion can be made:

(1) Sensitivity of differential type is 1 time higher than that of single coil type.

(2) Nonlinear term of differential type equals to that of single type multiplied by factor $\Delta\delta/\delta_0$. As $\Delta\delta/\delta_0 \ll 1$, linearity of differential type is improved obviously.

To improve output characteristics effectively, the two variable gap inductance sensors, which are used to form differential type, should be totally uniform in structure size, materials and electrical parameters.

- *Measuring circuit*

Classification of measuring circuit of inductance sensor includes AC bridge type, AC transformer type and resonant type.

a) AC bridge type measuring circuit

Fig. 3.3.4 shows an AC bridge measuring circuit, which uses two coils as two arms of bridge Z_1 and Z_2, while the other two arms are replaced by pure resistors. We can know from Eq. (3.2.44), for differential inductance sensor with high Q value ($Q = \omega L / R$), the output voltage can be expressed as follows:

$$\dot{U}_0 = \frac{\dot{U}_{AC}}{2} \frac{\Delta Z_1}{Z_1} = \frac{\dot{U}_{AC}}{2} \frac{j\omega \Delta L}{R_0 + j\omega L_0} \approx \frac{\dot{U}_{AC}}{2} \frac{\Delta L}{L_0} \qquad (3.3.19)$$

where L_0 is inductance of single coil when armature is at neutral position, ΔL is inductance value difference of two coils.

For $\Delta L = 2L_0 (\Delta \delta / \delta_0)$, we can conclude from Eq. (3.3.19) that $\dot{U}_0 = \dot{U}_{AC}(\Delta \delta / \delta_0)$. Output voltage of bridge is related to $\Delta \delta$. $Z_1 = Z_2 = Z$, $Z_3 = Z_4 = R$, $L_1 = L_2 = L$, $Z_1 = R_0 + j\omega L_1$.

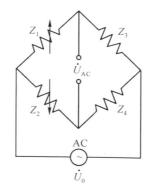

Fig. 3.3.4. AC bridge measuring circuit

b) Transformer AC bridge

Transformer AC bridge measuring circuit is shown in Fig. 3.3.5. Two arms of bridge Z_1 and Z_2 are coil impedance of sensor, while the other two arms is half of secondary coil impedance. When load impedance is infinite, the output voltage of bridge can be expressed as follows:

$$\dot{U}_0 = \frac{Z_1 \dot{U}}{Z_1 + Z_2} - \frac{\dot{U}}{2} = \frac{Z_1 - Z_2}{Z_1 + Z_2} \frac{\dot{U}}{2} \qquad (3.3.20)$$

When armature is at neutral position, that is to say, $Z_1 = Z_2 = Z$, then $\dot{U}_0 = 0$, the bridge reaches balance.

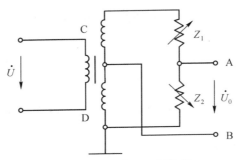

Fig. 3.3.5. Transformer AC bridge

When armature moves upwards, that is to say, $Z_1 = Z + \Delta Z, Z_2 = Z - \Delta Z$, then:

$$\dot{U}_0 = \frac{\dot{U}}{2}\frac{\Delta Z}{Z} = \frac{\dot{U}}{2}\frac{\Delta L}{L} \qquad (3.3.21)$$

When armature moves downwards, that is to say, $Z_1 = Z - \Delta Z, Z_2 = Z + \Delta Z$, then:

$$\dot{U}_0 = -\frac{\dot{U}}{2}\frac{\Delta Z}{Z} = -\frac{\dot{U}}{2}\frac{\Delta L}{L} \qquad (3.3.22)$$

From Eq. (3.3.21) and Eq. (3.3.22) we know that when upwards and downwards movement distance of armature is the same, size of output voltage is also the same, but in opposite direction. As $\dot{U}_0$ is alternating voltage, displacement direction cannot be told from the output. Using phase-sensitive detection circuit is a way to solve this problem.

c) **Resonant measuring circuit**

Resonant measuring circuit is classified into resonant amplitude modulation circuit, shown in Fig. 3.3.6, and resonant frequency modulation circuit, shown in Fig. 3.3.7. In amplitude modulation circuit, inductor L, capacitor C and transformer original side are in series connection. Accessing AC power supply, transformer sub-side will output voltage $\dot{U}_0$, whose frequency is identical to power frequency, but amplifier will change with inductance L. Fig. 3.3.6b shows the relationship between output voltage $\dot{U}_0$ and inductance L, where L_0 is inductance value at resonance point. The sensitivity of this circuit is very high, but linearity is poor, so it is applicable to situation that doesn't require high linearity.

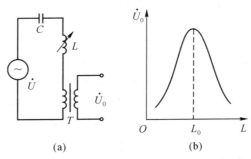

Fig. 3.3.6. Resonant amplitude modulation circuit

The basic principle of frequency modulation circuit is that change of inductance of sensor L causes change of frequency of output voltage. Usually inductor L and capacitor of sensor are connected to an oscillating circuit with oscillation frequency $f = 1/[2\pi(LC)^{1/2}]$. Oscillation frequency changes with L, so we can measure out the value being tested based on value of f. Fig. 3.3.7b shows characteristics of f and L, from which we can see that the two have a nonlinear relationship.

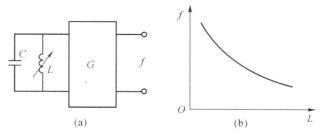

Fig. 3.3.7. Resonant frequency modulation circuit

3.3.1.2 Eddy current sensors

According to Faraday's law of electromagnetic induction, when massive metal conductor is placed in mutative magnetic field or move in magnetic field cutting magnetic lines, induced eddy-electric current will be produced in conductor, which is called eddy current. The phenomenon is called eddy current effect.

Sensors made on the basis of eddy current effect are called Eddy-Current Transducer. According to the way eddy current flows through conductor, the sensor can be classified as reflective high-frequency type and low-frequency penetrable type. They are still similar in the way of basic work principle.

The most obvious features of this kind of sensors is that it can continuously and non-contacting measure displacement, thickness, surface temperature, velocity, stress and material damage. It also has characteristics of small size, high sensitivity, and wide frequency response. So it has extremely wide applications.

- **Working principle**

Fig. 3.3.8 shows the principle diagram of eddy current sensor, in which sensor coils and tested conductor make up a coil-conductor system.

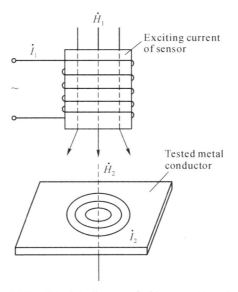

Fig. 3.3.8. Principle diagram of eddy current transducer

According to Faraday's law, when sensor coil is provided sinusoidal alteration current $\dot{I}_1$, sinusoida alteration magnetic field $\dot{H}_1$ will be produced in space surrounding coil, and this will make the metal conductor in the magnetic field induct eddy current $\dot{I}_2$, then $\dot{I}_2$ will produce new alternating magnetic field $\dot{H}_2$. According to Lenz's law, the action of $\dot{H}_2$ will resist the action of original magnetic field $\dot{H}_1$, which leads to the change of equivalent impedance of coil. So we can know from the above that the change of coil impedance totally depends on eddy current effect of tested metal conductor. Eddy current effect has relations with not only resistivity ρ, magnetic permeability μ, and geometric shape, but also geometric parameters, exciting current frequency in coil and distance between coil and conductor. Therefore, when affected by eddy current the equivalent impedance Z of coil is as follows:

$$Z = F(\rho, \mu, r, f, x) \qquad (3.3.23)$$

where r is the size factor of coil and tested conductor.

If only one parameter in the above expression changes, coil impedance Z is just single-valued function of the parameter. Using measuring circuit combined with sensor to detect variation of impedance Z, we can measure out this parameter.

- **Basic characteristics**

A simplified model of eddy current transducer is shown in Fig. 3.3.9, in which the eddy current produced in tested metal conductor is equivalent to a short circuit ring. Suppose eddy current is only distributed in the loop body, h can be calculated as follows:

$$h = \left(\frac{\rho}{\pi \mu_0 \mu_r f} \right)^{1/2} \qquad (3.3.24)$$

where f is the excited magnetic electric current frequency.

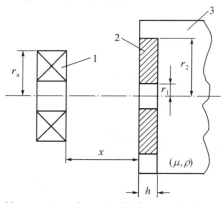

Fig. 3.3.9. Simplified model of eddy current transducer. 1: coil of sensor; 2: short circuit ring; 3: tested metal conductor

Based on the simplified model, equivalent circuit diagram can be drawn as Fig. 3.3.10 shows. R_2 is equivalent resistance of eddy-current short circuit ring, it can be expressed as follows:

$$R_2 = \frac{2\pi \rho}{h \ln \frac{r_a}{r_1}} \qquad (3.3.25)$$

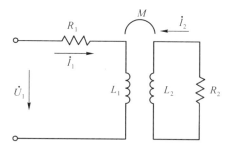

Fig. 3.3.10 Equivalent circuit of eddy current transducer. 1: sensor coil; 2: eddy current short circuit ring

According to Kirchoff's second law, the following equation can be deduced:

$$R_1 \dot{I}_1 + j\omega L_1 \dot{I}_1 - j\omega M \dot{I}_2 = \dot{U}_1 \tag{3.3.26}$$

$$-j\omega M \dot{I}_1 + R_2 \dot{I}_2 + j\omega L_2 \dot{I}_2 = 0 \tag{3.3.27}$$

where ω is the angular frequency of excited magnetic electric current, R_1, L_1 are resistance and inductance of coil, L_2 is equivalent inductance of short circuit ring, R_2 is equivalent resistance of short circuit ring.

From Eq. (3.3.26) and Eq. (3.3.27) we can calculate the equivalent impedance as follows:

$$Z = \frac{\dot{U}_1}{\dot{I}_1} = \frac{R_1 + \omega^2 M^2 R_2}{R_2^2 + (\omega L_2)^2} + j\omega \frac{L_1 - \omega^2 M^2 L_2}{R_2^2 + (\omega L_2)^2} = R_{eq} + j\omega L_{eq} \tag{3.3.28}$$

where $R_{eq} = (R_1 + \omega^2 M^2 R_2)/[R_2^2 + (\omega L_2)^2]$, $L_{eq} = (L_1 - \omega^2 M^2 L_2)/[R_2^2 + (\omega L_2)^2]$, R_{eq} is the equivalent resistance of coil effected by eddy current, and L_{eq} is the equivalent inductance of coil effected by eddy current.

The equivalent quality factor Q of coil can be expressed as follows:

$$Q = \frac{\omega L_{eq}}{R_{eq}} \tag{3.3.29}$$

In a word, according to simplified model and equivalent circuit of eddy current circuit transducer, the basic characteristics of eddy current is shown in Eq. (3.3.28) and Eq. (3.3.29) deduced from basic methods of electric circuit analysis.

- *Forming range of eddy current*

a) **Radial formation range of eddy current**

Density of eddy current produced by coil-conductor system is not only a function of distance between coil and conductor, but also a function of coil radius r. When x is identified, the relationship curve of eddy current density J and radius r is shown in Fig. 3.3.11. It can be known that:

(1) Eddy current radial formation is within the scope of about 1.8–2.5 times of coil's outside radius r_m, and is uneven distributed.

(2) Eddy current density is zero when the radius of short-circuiting ring is zero, i.e. $r=0$.

(3) The maximum of eddy current exists within a narrow region near $r=r_m$.

(4) We can use a short circuit ring with average radius of r_m ($r_m=(r_1+r_a)/2$) to show dispersed eddy current.

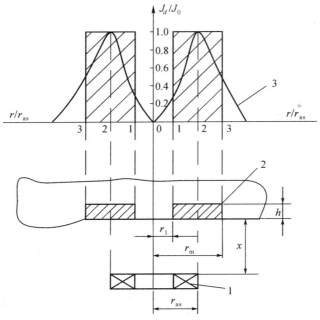

Fig. 3.3.11. Relationship curve of eddy current density J and radius r

b) **Relationship between eddy current intensity and distance**

Theoretical analysis and experiments have proved that eddy current density changes when x changes, that is to say, eddy current intensity changes with distance x. According to electromagnetic action of coil-conductor system, we can calculate the eddy current intensity on the conductor surface as follows:

$$I_2 = I_1 \frac{1-x}{\left(x^2 + r_m^2\right)^{1/2}} \quad (3.3.30)$$

where I_1 is the excitation current, I_2 is equivalent current of conductor, x is the distance between coil and conductor surface, r_m is the outside radius of coil.

Normalization curve is shown in Fig. 3.3.12.

The above analysis indicates that:

(1) Eddy current intensity has nonlinear relation with distance x, and decreases when x/r_m increases.

(2) In measuring displacement using eddy current transducer, better linearity and high sensitivity could be reached only when $x/r_m \ll 1$ (usually 0.05–0.15).

- *Eddy current's axial depth of penetration*

Because of skin effect, the vertical distribution H_1 of eddy current is uneven, and has exponential decay, which can be expressed as follows:

$$J_d = J_0 e^{-d/h} \quad (3.3.31)$$

where d is the distance between some point and conductor surface, J_d is the eddy current density along the axial direction of H_1 towards d, J_0 is the eddy current density on surface, which is also the maximum of eddy current density, h is the axial depth of penetration (skin depth) of eddy current.

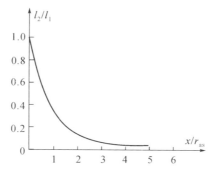

Fig. 3.3.12. Normalization curve of eddy current density and distance

Fig. 3.3.13 shows the axial distribution curve of eddy current density, from which we can see that eddy current density is mainly distributed near the surface.

From the analysis above we know that the bigger resistivity the tested conductor has, the bigger the relative magnetic permeability will be, and the frequency of excitation current of coil is lower, thus the axial depth h of penetration of eddy current is bigger.

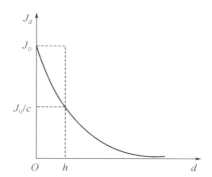

Fig. 3.3.13. The axial distribution curve of eddy current density

3.3.1.3 Differential transformer sensors

Differential transformer transducer is a kind of sensor converts non-electrical variation into change of mutual inductance, which is made up based on the basic principle of transformer. The secondary coil is connected in differential forms, so it is also called differential transformer transducer.

Differential transformer transducer have a variety of structures, including variable gap type、 variable area type and solenoid type. Their work principle is similar. In testing non-electrical quantity, solenoid type is the most widely used. It can measure the mechanical displacement within the scope of 1 – 100 mm, and has many advantages like high accuracy, high sensitivity, simple structure, and reliable performance.

- **Working principle**

The structure of solenoid type differential transformer transducer is shown in Fig. 3.3.14. It consists of a primary coil, two secondary coils, and an inserted columniform core in the middle of coil.

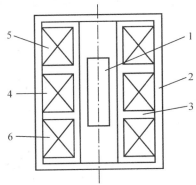

Fig. 3. 3.14. Structure of solenoid type differential transformer transducer. 1: movable armature; 2: magnetic shell; 3: skeleton; 4: primary winding with turns of w_1; 5: secondary winding with turns of w_{2a}; 6: secondary winding with turns of w_{2b}

According to arrange style of coil winding, solenoid type differential transformer transducer is classified into one segment, two segments, three segments, four segments and five segments type, which is shown in Fig. 3.3.15. One segment type has high sensitivity, while three segments type has small residual voltage at zero. Usually we use two and three segments types.

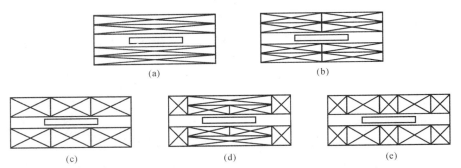

Fig. 3.3.15. Arrangement of coil winding: (a) One segment; (b) Two segments; (c) Three segments; (d) Four segments; (e) Five segments

Differential transformer transducer's two secondary coils are in series opposing connection. Under ideal condition of ignoring iron loss, magnetizer's reluctance, and distributed capacitance of coil, its equivalent circuit is shown in Fig. 3.3.16.

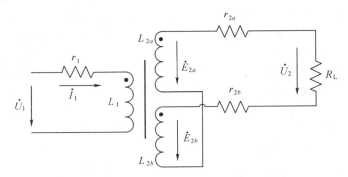

Fig. 3.3.16. Equivalent circuit of differential transformer

According to work principle of transformer, when excitation voltage $\dot{U}_1$ is applied to primary winding w_1, induced potential $\dot{E}_{2a}$ and $\dot{E}_{2b}$ will be produced in secondary winding w_{2a} and w_{2b}. If completely symmetry of transformer structure is guaranteed in process, the relationship between the mutual inductance $M_1 = M_2$ can be obtained when movable armature is at initial balance position. According to electromagnetic induction principle, $\dot{E}_{2a} = \dot{E}_{2b}$. From the above we know that $\dot{U}_2 = \dot{E}_{2a} - \dot{E}_{2b} = 0$, that is to say, output voltage of differential transformer is zero.

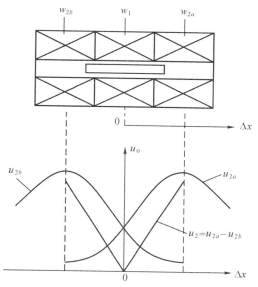

Fig. 3.3.17. Characteristic curve of output voltage of differential transformer

When movable armature moves upwards, flux in w_{2a} will be bigger than that of w_{2b} because of effect of reluctance, then $M_1 > M_2$. So $\dot{E}_{2b}$ decreases while $\dot{E}_{2a}$ increases, and vice versa. As $\dot{U}_2 = \dot{E}_{2a} - \dot{E}_{2b}$, $\dot{U}_2$ will change with x when $\dot{E}_{2a}$ and $\dot{E}_{2b}$ change with displacement x of armature. Fig. 3.3.17 shows relationship curve between output voltage $\dot{U}_2$ of transformer and displacement x of movable armature. Actually, when armature is at medial position, output voltage of differential transformer is not zero, thus we call it zero point remainder voltage $\dot{U}_x$. The output characteristic curve won't cross zero point because of existence of $\dot{U}_x$, which makes actual characteristics inconsistent with theoretical characteristics. Cause of residual voltage at zero mainly lies in electrical parameters and size asymmetry of two secondary winding, and nonlinear problem of magnetic materials. Waveform of residual voltage at zero is very complex, mainly consisting of fundamental wave and high harmonics. Cause of fundamental wave mainly lies in electrical parameters and size asymmetry of two secondary winding, which leads to different amplitude and phase of induced voltages. However we adjusted the position of armature, induced voltages of two coils wouldn't counteract completely. High harmonics mainly refers to third harmonic, whose cause is nonlinearity (magnetic saturation and hysteresis) of magnetic materials magnetization curve. Zero point remainder voltage is generally below tens of millivolt. In real application we need to find a way to reduce $\dot{U}_x$, otherwise measuring results of transducer will be affected.

Biomedical Sensors and Measurement

- **Basic characteristics**

Equivalent circuit of differential transformer is shown in Fig. 3.3.16. When secondary is opening circuit, we have the following relation:

$$\dot{I}_1 = \frac{\dot{U}_1}{r_1 + j\omega L_1} \tag{3.3.32}$$

where ω is angular frequency of excitation voltage $\dot{U}_1$, $\dot{U}_1$ is excitation voltage of primary coil, $\dot{I}_1$ is exciting current of primary coil, r_1, L_1 are DC resistance and inductance of primary coil.

According to electromagnetic induction principle, induced voltages in secondary winding can be expressed as follows:

$$\dot{E}_{2a} = -j\omega M_1 \dot{I}_1 \tag{3.3.33}$$

$$\dot{E}_{2b} = -j\omega M_2 \dot{I}_1 \tag{3.3.34}$$

where M_1, M_2 are mutual inductance of primary winding and two secondary winding respectively.

As the two secondary winding are in series opposing connection, considering opening secondary circuit, we can have the following relation:

$$\dot{U}_2 = \dot{E}_{2a} - \dot{E}_{2b} = -\frac{j\omega(M_1 - M_2)\dot{U}_1}{r_1 + j\omega L_1} \tag{3.3.35}$$

Effective value of output voltage can be expressed as follows:

$$\dot{U}_2 = \frac{\omega(M_1 - M_2)\dot{U}_1}{\left[r_1^2 + (\omega L_1)^2\right]^{1/2}} \tag{3.3.36}$$

We will analyze the following three situations.

(1) When movable armature is at medial position.

$$M_1 = M_2 = M$$

So $\dot{U}_2 = 0$.

(2) When movable armature moves upwards.

$$M_1 = M + \Delta M \qquad M_2 = M - \Delta M$$

So $\dot{U}_2 = 2\omega\Delta M \dot{U}_1 \big/ \left[r_1^2 + (\omega L_1)^2\right]^{1/2}$, which has the same polarity with $\dot{E}_{2a}$.

(3) When movable armature moves downwards.

$$M_2 = M + \Delta M \qquad M_1 = M - \Delta M$$

So $\dot{U}_2 = -2\omega\Delta M \dot{U}_1 \big/ \left[r_1^2 + (\omega L_1)^2\right]^{1/2}$, which has the same polarity with $\dot{E}_{2b}$.

Chapter 3 Physical Sensors and Measurement

- *Measuring circuit of differential transformer transducer*

The output of differential transformer transducer is AC voltage. If tested by AC voltmeter, it can only reflect quantity of displacement vector, but not direction. In addition, the tested value includes zero point remainder voltage, to distinguish direction and eliminate zero point remainder voltage, we often use differential rectifier and phase-sensitive detection circuit in actual measurement.

a) Differential rectification circuit

This kind of circuit rectifies the output voltage of two secondary sides, then output the difference value of rectified voltage or current. Fig. 3.3.18 shows several typical circuit forms, where (a) and (c) are applicable to AC load impedance, while (b) and (d) are applicable to low load impedance, resistor R_0 is applicable to adjust residual voltage at zero.

The following is analysis of work principle of differential rectification, shown in Fig. 3.2.18c.

From circuit structure shown in Fig. 3.3.18c it can be known that current direction of capacitor C_1 is from 2 to 4, that of capacitor C_2 is from 6 to 8, despite of the output instantaneous voltage polarity of two secondary coils. So the output voltage of rectification circuit can be expressed as follows:

$$U_2 = U_{24} - U_{68} \qquad (3.3.37)$$

When armature is at zero position, for $U_{24} = U_{68}$, so $U_2 = 0$. When armature is above zero position, for $U_{24} > U_{68}$, so $U_2 > 0$. When it is below zero position, for $U_{24} < U_{68}$, so $U_2 < 0$.

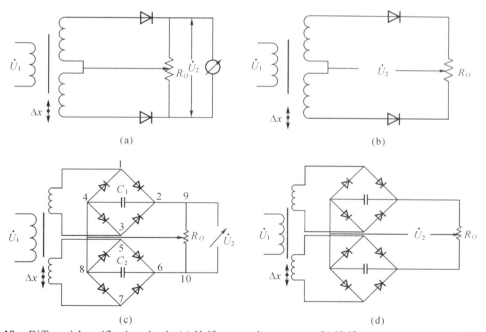

Fig. 3.3.18. Differential rectification circuit. (a) Half-wave voltage output; (b) Half-wave current output; (c) Full-wave voltage output; (d) Full-wave current output

Differential rectification circuit has advantages of simple structure, no need of considering the impact of phase adjustment and residual voltage at zero, little impact of distributed capacitance, and convenience for long-distance transmission, so it is widely used.

b) Phase-sensitive detection circuit

The circuit is shown in Fig. 3.3.19, where V_{D1}, V_{D2}, V_{D3}, V_{D4} are four diodes with the same performance, and are in series connection to form hybrid-ring in the same direction. The input signal u_2 (output AM wave voltage of differential transformer transducer) is loaded to a diagonal of the hybrid-ring through transformer T_1. Reference signal u_0 is loaded to another diagonal of the hybrid-ring through transformer T_2. Output signal u_L is tap output from the center of transformer T_1 and T_2. Balance resistor R plays the role of current limiter, avoiding excessive secondary current in transformer T_2 when diodes are driven into conduction. R_L is load resistor. The amplitude of u_0 must be much greater than that of input signal u_2, so that the on-state of the four diodes can be controlled effectively. The excitation voltage u_1 of differential transformer transducer and u_0 are powered by the same oscillator to make sure they have the same frequency and phase (or reversed phase).

From Figs. 3.3.20a, 3.3.20b and 3.3.20c we know that when displacement $\Delta x > 0$, u_2 and u_0 have the same frequency and phase, when $\Delta x < 0$, u_2 and u_0 have the same frequency but reversed phase.

When $\Delta x > 0$, u_2 and u_0 have the same frequency and phase. When u_0 and u_2 are in positive half cycle, which is shown in Fig. 3.3.19a, diodes V_{D1}, V_{D4} in hybrid-ring are off, V_{D2}, V_{D3} are on. So we can get the equivalent circuit shown in Fig. 3.3.19b.

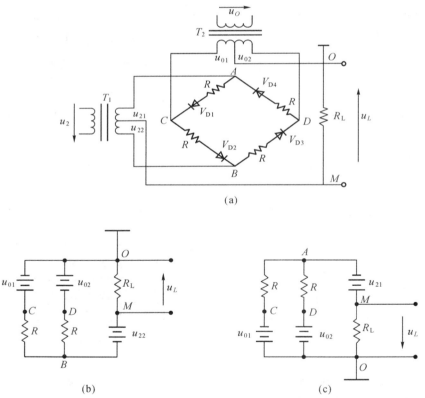

Fig. 3.3.19. Phase-sensitive detection circuit. (a) Principle diagram of phase-sensitive detection circuit; (b) Equivalent circuit when u_0 and u_2 are in positive half cycle; (c) Equivalent circuit when u_0 and u_2 are in negative half period

According to the work principle of transformer, and considering O, M are the center tap of transformer T_1, T_2 respectively, we can get the following equation:

$$u_{01} = u_{02} = \frac{u_0}{2n_2} \tag{3.3.38}$$

$$u_{21} = u_{22} = \frac{u_2}{2n_1} \tag{3.3.39}$$

where n_1, n_2 are the transformation ratio of transformer T_1, T_2. Using basic analysis method of circuit, we can calculate the output voltage u_L of circuit shown in Fig. 3.3.19b as follows:

$$u_L = \frac{R_L u_2}{n_1(R_1 + 2R_L)} \tag{3.3.40}$$

Similarly, when u_0 and u_2 are in negative half period, diodes V_{D2}, V_{D3} are off, V_{D1}, V_{D4} are on. The equivalent circuit is shown in Fig. 3.3.19c. The expression of output voltage u_L is the same as Eq. (3.3.41), which indicates that the voltage u_L of load R_L is positive as long as $\Delta x > 0$, no matter u_0 and u_2 are in positive or negative half cycle.

When $\Delta x < 0$, u_0 and u_2 have the same frequency but reversed phase. Using the same analysis method, we can easily conclude that the expression of voltage u_L of load R_L is always as follows, as long as $\Delta x < 0$, no matter u_0 and u_2 are in positive or negative half cycle.

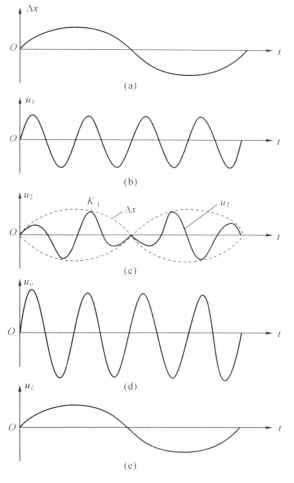

Fig. 3.3.20. Waveform graph

$$u_L = -\frac{R_L u_2}{n_1(R_1 + 2R_L)} \tag{3.3.41}$$

So the changing law of output voltage u_L of phase-sensitive detection circuit reflects the changing law of tested displacement, that is to say, u_L reflects the quantity of displacement Δx, and its polarity reflects the direction of displacement Δx.

3.3.2 Applications in Biomedicine

Because of the high sensitivity of an inductive sensor, which has a maximum resolution of 0.01 μm, an inductive sensor is mainly used to measure slight displacement in biomedical engineering. A typical application is respiratory inductance plethysmography (RIP) (Mazeika and Swanson, 2007), which is probably the most commonly accepted method for quantitative and qualitative non-invasive respiratory measurements in infants and adults. Respiratory measurements, for example, respiratory rate and tidal volume, are important indicators showing a person's health condition, so they are of great significance in first aid for the family. With the help of some techniques proposed for calibration (Poole et al., 2000), RIP may be used quantitatively, which makes respiration measurement more effective.

Since both the thoracic and abdominal area change reflects the value of minute volume, the long time measurement of a respiratory movement can be realized through measuring the variation of the cross-sectional area. In RIP, two elastic belts, into which a zigzagging (coiled) wire (for expansion and contraction) is sewn, are essential, with one worn around the chest, and the other worn around the abdomen, resembling two inductance loops (Fig. 3.3.21). Based on the principle of Faraday's Law, an alternating current applied through a loop of wire with high frequency and low amplitude generates a magnetic field normal to the orientation of the loop. According to Lenz's Law, a change in the area enclosed by the loop, which causes a variation in the self-inductance coefficient, creates an opposing current within the loop directly proportional to the change in the area. The frequency of the alternating current is set to be more than twice the typical respiratory rate in order to achieve adequate sampling of the respiratory effort waveform. In measurement, the breathing activity changes the cross-sectional area of the patient's body, and thus changes the shape of the magnetic field generated by the belt, "inducing" an opposing current that can be measured. The variation of minute volume ΔV can be calculated as follows:

$$\Delta V = K_1 \Delta L_R + K_2 \Delta L_A \tag{3.3.42}$$

where ΔL_R is the output inductance change of thoracic belt, ΔL_A is that of abdominal belt, K_1 and K_2 are volume coefficient of chest and abdomen, respectively.

With RIP, no electrical current passes through the body. Even though a weak magnetic field is present, it does not affect the patient or any surrounding equipment. Otherwise, the worry about measurement being interfered with by the surrounding environment is unnecessary. The signal produced is linear and is a fairly accurate representation of the change in cross-sectional area. Actually RIP is reliable in measuring respiratory movement, with advantages of convenience, non-invasive and ambulatory monitoring. In addition, RIP has superiority in displaying respiratory frequency, evaluating coordination of chest and abdomen respiratory movement, with no heart interference.

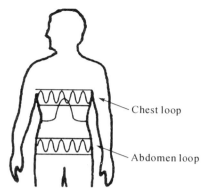

Fig. 3.3.21. Diagrammatic sketch of RIP

Over all, RIP is more widely used in sleep respiratory monitor apparatus to record chest and abdomen respiratory movement, as well as in diagnosing sleep apnea syndrome. Besides, Moreau-Gaudry et al. (2006) have demonstrated that RIP has potential in a swallowing monitor to analyze swallowing disorders and putting in place medical supervision of swallowing for individuals who might aspirate, especially in the elderly.

3.4 Capacitive Sensors and Measurement

Capacitive sensors can directly detect a variety of things- displacement, chemical and biological compositions, electric filed, and, indirectly, measure many other variables which can be converted into displacement or permittivity such as vibration, acceleration, pressure, pressure difference, and liquid level. With the rapidly development of the sensors and measurement technology, capacitive sensors will have been more widely used in the non-electrical measurements and automatic detections including biomedical fields.

3.4.1 The Basic Theory and Configuration of Capacitive Sensors

The simplest electrode configuration is two close-spaced parallel plates. The capacitance, neglecting a small fringe effect, could be calculated by

$$C = \frac{\varepsilon A}{d} \qquad (3.4.1)$$

where ε is the permittivity, $\varepsilon = \varepsilon_0 \cdot \varepsilon_r$, ε_0 is the vacuum permittivity and ε_r is the relative permittivity of media; A is the area of the parallel plates; d is the distance between the parallel plates.

Fluctuations in any of the parameters caused by the variables being detected will be revealed in the variation of the capacitance, which could be converted into electric output by the detection circuits. Thus, capacitive sensors can be divided into three types: space-variant capacitive sensors, area-variant capacitive sensors, and permittivity-variant capacitive sensors.

3.4.1.1 Space-variant capacitive sensors

The schematic representation of a space-variant capacitive sensor is given in Fig. 3.4.1a. As ε and A are constants, the initial space between the plates is d_0, and thus the initial capacitance C_0 is

$$C_0 = \frac{\varepsilon_0 \varepsilon_r A}{d_0} \tag{3.4.2}$$

If the space decreases by Δd, the capacitance will increase by ΔC, where

$$C = C_0 + \Delta C = \frac{\varepsilon_0 \varepsilon_r A}{d_0 - \Delta d} = \frac{C_0}{1 - \Delta d / d_0} = \frac{C_0(1 + \Delta d / d_0)}{1 - (\Delta d / d_0)^2} \tag{3.4.3}$$

where the output of the sensor $C = f(d)$ is nonlinear, but is hyperbolic shown in Fig. 3.4.1b.

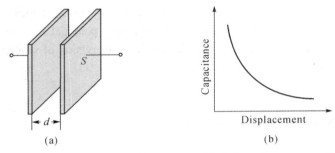

Fig. 3.4.1. The principle of space-variant capacitive sensor: (a) Schematic structure; (b) Relationship between capacitance and plate space

If $1 - (\Delta d / d)^2 \approx 1$, the formula above could be approximated by

$$C_1 = C_0 + \frac{C_0 \Delta d}{d_0} \tag{3.4.4}$$

which is a linear function. So only when the value of $\Delta d / d_0$ is small enough, do the formula has the linearity.

In addition, as shown in Eq. (3.4.4), when the value of d_0 is smaller, the sensors will be more sensitive. Therefore, media with high permittivity are placed between the parallel plates for breakdown or short-circuit protection when the plates get too close to each other. The capacitance in this case which is shown in Fig. 3.4.2 could be

$$C = \frac{A}{d_g / \varepsilon_0 \varepsilon_g + d_0 / \varepsilon_0} \tag{3.4.5}$$

where ε_g is the permittivity of mica, $\varepsilon_g = 7$; ε_0 is the permittivity of air, $\varepsilon_0 = 1$; d_0 is the thickness of air layer; d_g is the thickness of mica layer.

Chapter 3 Physical Sensors and Measurement

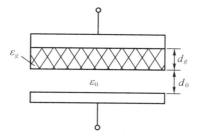

Fig. 3.4.2. Capacitor using mica as dielectric medium

- *Sensitivity and nonlinearity of capacitive sensors*

Then the relative change of capacitance is

$$\frac{\Delta C}{C_0} = \frac{\Delta d}{d_0} \frac{1}{1 - \frac{\Delta d}{d_0}} \qquad (3.4.6)$$

the formula above could be decomposed into series,

$$\frac{\Delta C}{C_0} = \frac{\Delta d}{d_0} \left[1 + \frac{\Delta d}{d_0} + \left(\frac{\Delta d}{d_0}\right)^2 + \left(\frac{\Delta d}{d_0}\right)^3 + \cdots \right] \qquad (3.4.7)$$

when $\Delta d / d_0 \ll 1$, which could be denoted by

$$\frac{\Delta C}{C_0} = \frac{\Delta d}{d_0} \qquad (3.4.8)$$

Thus the sensitivity of the sensors K is as follows:

$$K = \frac{\frac{\Delta C}{C_0}}{\Delta d} = \frac{1}{d_0} \qquad (3.4.9)$$

Note that: the relative change of capacitance (output) caused by unit change of displacement (input) is in inverse proportion to d_0.

Considering the linear and quadratic terms,

$$\frac{\Delta C}{C_0} = \frac{\Delta d}{d_0} \left(1 + \frac{\Delta d}{d_0}\right) \qquad (3.4.10)$$

The linear relative error δ could be given by

$$\delta = \frac{\left|(\Delta d / d_0)^2\right|}{\Delta d / d_0} \times 100\% = \left|\frac{\Delta d}{d_0}\right| \times 100\% \qquad (3.4.11)$$

Seen from the above, to increase the sensitivity, the initial distance d_0 should be decreased, while the nonlinear error will be increased as a result.

In actual practice, in order to increase the sensitivity and decrease the nonlinear error, differential configuration is employed, just like Fig. 3.4.3.

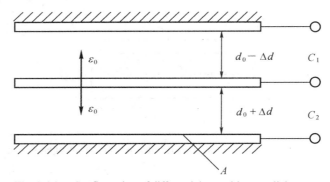

Fig. 3.4.3. Configuration of differential capacitive parallel sensor

$$C_1 = \frac{C_0}{1 - \frac{\Delta d}{d_0}} \tag{3.4.12}$$

$$C_2 = \frac{C_0}{1 + \frac{\Delta d}{d_0}} \tag{3.4.13}$$

After series expansion, we get the total change in capacitance:

$$C_1 = \frac{\Delta d}{d_0}\left[1 + \frac{\Delta d}{d_0} + \left(\frac{\Delta d}{d_0}\right)^2 + \left(\frac{\Delta d}{d_0}\right)^3 + \cdots\right] \tag{3.4.14}$$

$$C_2 = \frac{\Delta d}{d_0}\left[1 - \frac{\Delta d}{d_0} + \left(\frac{\Delta d}{d_0}\right)^2 - \left(\frac{\Delta d}{d_0}\right)^3 + \cdots\right] \tag{3.4.15}$$

$$\Delta C = C_1 - C_2 = 2C_0\left[\frac{\Delta d}{d_0} + \left(\frac{\Delta d}{d_0}\right)^3 + \left(\frac{\Delta d}{d_0}\right)^5 + \cdots\right] \tag{3.4.16}$$

The relative change of capacitance:

$$\frac{\Delta C}{C_0} = 2\frac{\Delta d}{d_0}\left[1 + \left(\frac{\Delta d}{d_0}\right)^2 + \left(\frac{\Delta d}{d_0}\right)^4 + \cdots\right] \tag{3.4.17}$$

Omitting the high-order terms, an approximated linear relationship between $\Delta C/C_0$ and $\Delta d/d_0$ could be expressed as

$$\frac{\Delta C}{C_0} \approx 2\frac{\Delta d}{d_0} \tag{3.4.18}$$

If we only consider the linear and cubic terms in Eq. (3.4.17), the nonlinear relative error δ could be approximated by

$$\delta = \frac{2|(\Delta d/d_0)^3|}{2|\Delta d/d_0|} \times 100\% = \left|\frac{\Delta d}{d_0}\right|^2 \times 100\% \tag{3.4.19}$$

Consequently, in differential capacitive sensors, the sensitivity is doubled, while the nonlinear error is decreased significantly.

3.4.1.2 Area-variant capacitive sensor

The schematic representation of an area-variant capacitive sensor is given in Fig. 3.4.4a. As these plates slide transversely, the effective coverage area A changes with motion, leading to the variation of the capacitance. The change of capacitance could be give by

$$C = C_0 - \Delta C = \frac{\varepsilon_0 \varepsilon_r (a - \Delta x) b}{d} \quad (3.4.20)$$

where $C_0 = \varepsilon_0 \varepsilon_r b a / d_0$ is the initial capacitance. The sensor in this form has a capacitance which is linear to the horizontal displacement Δx, and thus

$$\frac{\Delta C}{C_0} = \frac{\Delta x}{a} \quad (3.4.21)$$

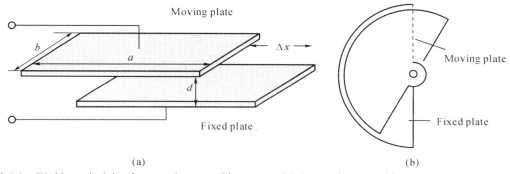

Fig 3.4.4. Working principle of area-variant capacitive sensor. (a) Area-variant capacitive sensor; (b) Angular-variant capacitive sensor

Configurations of capacitive sensors for angular displacement measurements are shown in Fig. 3.4.3b. If $\theta = 0$, then

$$C_0 = \frac{\varepsilon_0 \varepsilon_r A}{d_0} \quad (3.4.22)$$

where ε_y is the relative permittivity of media, d_0 is the distance between the parallel plates, A_0 is the initial area of the parallel plates.

Else if $\theta \neq 0$

$$C = \frac{A_0 \varepsilon_0 \varepsilon_r (1 - \theta / \pi)}{d_0} = C_0 \left(1 - \frac{\theta}{\pi}\right) \quad (3.4.23)$$

Seen from the above, the capacitance of the sensor C is linear to the angular displacement θ.

3.1.4.3 Medium-variant capacitive sensor

In Fig. 3.4.5, the capacitance could be calculated by

$$C = \frac{2\pi\varepsilon_1 h}{\ln(D/d)} + \frac{2\pi\varepsilon(H-h)}{\ln(D/d)}$$
$$= \frac{2\pi\varepsilon H}{\ln(D/d)} + \frac{2\pi h(\varepsilon_1 - \varepsilon)}{\ln(D/d)} \quad (3.4.24)$$
$$= C_0 + \frac{2\pi h(\varepsilon_1 - \varepsilon)}{\ln(D/d)}$$

where ε_0 is the permittivity of air and C_0 is the initial capacitance decided by the basic size of the transducer.

Consequently, the increment of the capacitance is in direct proportion to the height level of the liquid, h.

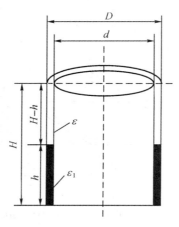

Fig. 3.4.5. Configurations of medium-variant capacitive sensor

In Fig. 3.4.6, the capacitance could be calculated by

$$C = C_1 + C_2 = \varepsilon_0 b_0 \frac{\varepsilon_{r1}(L_0 - L) + \varepsilon_{r2} L}{d_0} \quad (3.4.25)$$

where L_0 and b_0 are the length and width of the plates respectively, L is the length of the overlapping part with the inserting medium.

If $\varepsilon_{r1} = 1$, when $L=0$, the initial capacitance of the sensor is $C_0 = \varepsilon_0 \varepsilon_{r1} L_0 b_0 / d_0$. When medium ε_{r2}, which is being detected, runs in by L, the increment in the overall capacitance could be calculated by

$$\frac{\Delta C}{C_0} = \frac{C - C_0}{C_0} = \frac{(\varepsilon_{r2} - 1)L}{L_0} \quad (3.4.26)$$

The increment of capacitance is linear to the displacement of the medium.

Chapter 3 Physical Sensors and Measurement

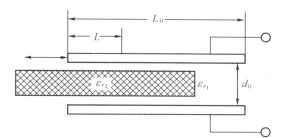

Fig. 3.4.6. Medium-variant capacitive sensor

3.4.2 Measurement Circuits

The capacitance and its changes of capacitive sensors are very small, and measuring circuits are necessary for detection. Measurement circuits mainly include: Frequency Modulation (FM) circuits, op amp circuit, diode paired T-shaped AC bridge, and pulse width modulation circuit, etc.

3.4.2.1 Frequency modulation (FM) circuit

Oscillation frequency of the FM oscillator in Fig. 3.4.7

$$f = \frac{1}{2\pi\sqrt{LC}} \tag{3.4.27}$$

where L is the inductance of oscillation circuit; C is the total capacitance of oscillation circuit, $C = C_1 + C_2 + C_0 \pm \Delta C = C_1 + C_2 + C_x$, C_1 is the capacitance of sensor, C_2 is the distributed capacitance of sensor, thus

$$f = \frac{1}{2\pi\sqrt{L(C_1 + C_2 + C_0 \pm \Delta C)}} = f_0 \pm \Delta f \tag{3.4.28}$$

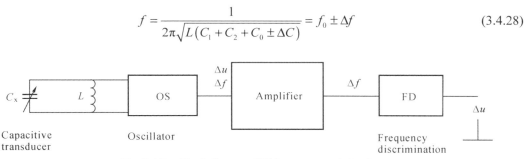

Fig. 3.4.7. Block diagram of FM measurement circuit

3.4.2.2 Op amp circuit

The schematic representation of op amp circuit is given in Fig. 3.4.8. From the principle of operational amplifiers, we get

$$\dot{U}_o = -\frac{C}{C_x}\dot{U}_i \tag{3.4.29}$$

If the sensor is a plate capacitor, rewrite Eq. (3.4.29) using $C_x = \varepsilon A / d$, we get

$$\dot{U}_o = -\dot{U}_i \frac{C}{\varepsilon A} d \qquad (3.4.30)$$

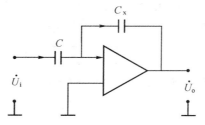

Fig.3.4.8. Schematics of Op Amp Circuit

3.4.2.3 Diode paired T-shaped AC bridge

The schematic representation of diode paired T-shaped AC bridge is given in Fig. 3.4.9.

The average output voltage in a cycle is given by:

$$U_0 = I_L R_L = \frac{1}{T}\left\{\int_0^T [I_1(t) - I_2(t)] dt\right\} R_L \approx \frac{R(R+2R_L)}{(R+R_L)^2} R_L Uf(C_1 - C_2) \qquad (3.4.31)$$

where $M = \dfrac{R(R+2R)}{(R+R_L)^2} R_L$ is a constant, thus

$$U_0 = UfM(C_1 - C_2) \qquad (3.4.32)$$

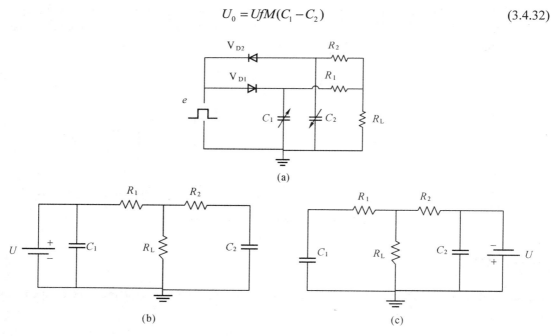

Fig. 3.4.9. Diode paired T-shaped AC bridge

3.4.2.4 Pulse width modulation circuit

Pulse width modulation circuit is to charge and discharge the capacitance of the sensor, so that output pulse width of the circuit varies along with capacitance. DC components of the signals detected could be measured through a low pass filter. Fig. 3.4.10 in the pulse width modulation circuit. Fig. 3.4.11 is the voltage waveform of pulse width modulation and demodulation circuit.

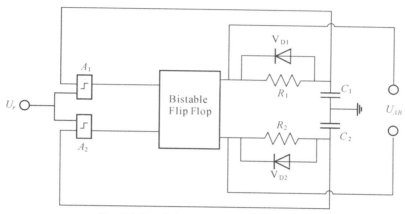

Fig. 3.4.10. Pulse width Modulation Circuit

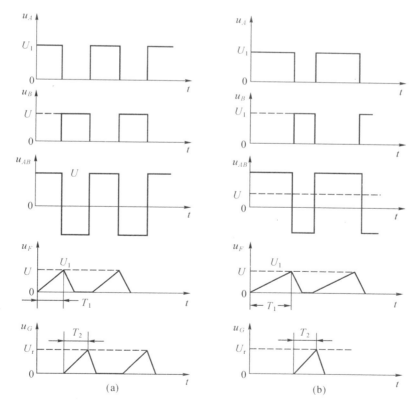

Fig. 3.4.11. Voltage waveform of pulse width modulation and demodulation circuit: (a) Initial state, (b) Working status

The average voltage of a cycle after the low-pass filter is

$$U_{AB} = U_A - U_B = U_1 \frac{T_1 - T_2}{T_1 + T_2} \qquad (3.4.33)$$

Since

$$T_1 = R_1 C_1 \ln \frac{U_1}{U_1 - U_r} \qquad (3.4.34)$$

$$T_2 = R_2 C_2 \ln \frac{U_1}{U_1 - U_r} \qquad (3.4.35)$$

Thus

$$U_{AB} = U_1 \frac{C_1 - C_2}{C_1 + C_2} \qquad (3.4.36)$$

In a space-variant case, formula could be rewritten as

$$U_{AB} = U_1 \frac{d_1 - d_2}{d_1 + d_2} \qquad (3.4.37)$$

When $C_1 = C_2 = C_0$, that is $d_1 = d_2 = d_0$, $U_0 = 0$; If $C_1 \neq C_2$, let $C_1 > C_2$, $d_1 = d_0 - \Delta d$, $d_2 = d_0 + \Delta d$, we have

$$U_{AB} = \frac{\Delta d}{d_0} U_1 \qquad (3.4.38)$$

For area-variant sensors,

$$U_{AB} = \frac{\Delta A}{A_0} U_1 \qquad (3.4.39)$$

3.4.3 Biomedical Applications

Capacitive sensors are mainly used to measure pressure in biomedical engineering.

3.4.3.1 *Capacitive pressure sensors*

Pressure sensors are required in many applications, including biomedical systems, industrial process controls, and environmental monitoring (He et al., 2007).

Capacitive pressure sensors are particularly noteworthy and can provide very high-pressure sensitivity, low power, low noise, large dynamic range, and low temperature sensitivity. Nowadays, capacitive pressure sensors have become one of the most popular more micro-electro-mechanical-systems (MEMS) sensors. In 1980, the micro capacitive pressure sensor was first fabricated by using micro machining technology (Sander et al., 1980). With a length of 3 mm and a height of 425 μm, the main structure of this sensor was a chamber and when pressure deformed the thin upper layer of the chamber, the capacitance was changed. From then on, more and MEMS capacitive pressure sensors were designed for biomedical applications.

A new capacitive pressure sensor with extremely high sensitivity (2.24 μF/kPa) (Bakhoum and cheng., 2010) is applicable to detect external pressure of human beings, such as non-invasive blood pressure measurement (Fig. 3.4.12).

Fig. 3.4.12. Non-invasive blood pressure measurement

The basic concept of the new device is to mechanically deform a drop of mercury that is separated from a flat aluminum electrode by a thin layer of a dielectric material, so as to form a parallel-plate capacitor where the electrode area is variable to a high degree. This principle is illustrated in Fig. 3.4.13. Under zero pressure, the mercury drop remains at its nearly-spherical shape. With the pressure increasing, the mercury drop is flattened against the aluminum electrode. A parallel-plate capacitor with one liquid electrode is formed.

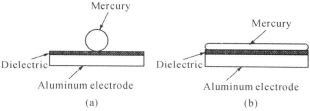

Fig. 3.4.13. The change in capacitance between the two configurations, which is proportional to the change in the contact area of the liquid electrode, can be several hundred fold: (a) A drop of mercury is flattened against an aluminum electrode that is covered with a layer of a dielectric material. A parallel-plate capacitor with one liquid electrode is formed; (b) Under zero pressure, the mercury drop returns to its nearly-spherical shape. The change in capacitance between the two configurations, which is proportional to the change in the contact area of the liquid electrode, can be several hundred folds

As shown in Fig. 3.4.14, a drop of mercury with a diameter of 3 mm is placed on top of a flat aluminum electrode that is covered with a 1-μm-thick layer of a ceramic material with a very high permittivity (specifically, $BaSrTiO_3$, with a permittivity of 12,000 – 15,000). This ceramic material was deposited on the surface of aluminum electrode by using the electrophoretic deposition technique. The drop is held in place by means of an aluminum disk that serves as the compression mechanism. The compression disk, in turn, is acted upon by means of a corrugated stainless-steel diaphragm. The compression disk is slightly curved, such that the spacing between the disk and the ceramic layer is exactly 3 mm at the center but less than 3 mm everywhere else. In this manner, the mercury drop will be forced to the center each time the stainless steel diaphragm retracts. The diaphragm is held in place by a thin aluminum ring. The entire assembly is mounted inside an open-cavity, 24-pin DIP IC package.

Since the air that surrounds the mercury droplet must be allowed to exit from the sensor and reenter as the sensor is pressurized/depressurized, an atmospheric pressure relief conduit is drilled in the IC package. In most applications, that conduit is connected to an atmospheric pressure environment via, for example, an external tube to be connected to the sensor.

As shown in Fig. 3.4.15a, the vertical pressure P changes the geometry of the drop of mercury, which leads to the variation of surface area of the electrode. Capacitance-pressure relationship in Fig. 3.4.15b can be obtained from theoretical arithmetic and experiments.

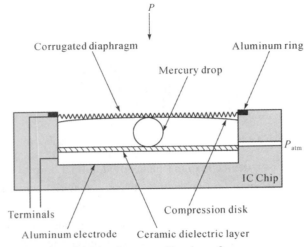

Fig. 3.4.14. A sensor with a drop of mercury

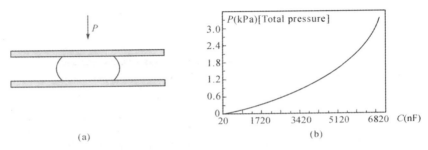

Fig. 3.4.15. A drop of mercury: (a) Pressures and geometry in the deformation; (b) Total pressure acting on the sensor as a function of the measured capacitance

Capacitive pressure sensors are also used to detect internal pressure. The micro capacitive pressure sensor shown in Fig. 3.4.16 was developed to be embedded into the cuff electrode for in situ monitoring of the interface pressure between implanted cuff and nerve tissue (Chiang et al., 2007).

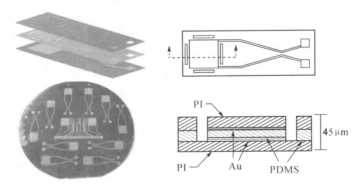

Fig. 3.4.16. Structure of the flexible capacitive pressure sensor (reprinted from (Chiang et al., 2007), Copyright 2007, with permission from Elsevier)

Cuff electrode (Fig. 3.4.17a) is an indispensable component of a neural prosthesis system. It is often employed to apply electrical stimuli on motor nerve fibers that innervate muscles or alternatively to record neural signals from the peripheral nerves. It is reported that a pressure over 20 mmHg is harmful for the nerve trunk. Therefore, measuring the interface pressure between the cuff and a nerve trunk provides a means to monitor the health of the nerve tissue.

The structure of the capacitive pressure sensor consists of two parallel electrical sensing plates, one dielectric layer sandwiched between the two sensing plates, and two outer insulating layers (Fig. 3.4.16). Polyimide (PI, Durimide 7320) is chosen as the material of the insulating layers because of its biocompatibility and insulating capability. The polydimethylsiloxane (PDMS, Sylgard 184) serves as the material of the dielectric layer. The dielectric constant of PDMS is 2.65. It is greater than the dielectric constant of air so that a larger initial capacitance and higher capacitance change can be obtained. Left of Fig. 3.4.17 also shows the fabricated array of capacitive pressure sensors before the lifting.

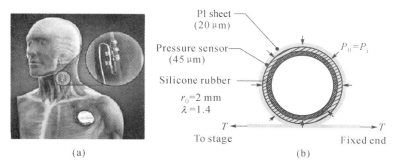

Fig. 3.4.17. The structure of the capacitive pressure sensor: (a) Cuff electrode; (b) *in vitro* circular compression test of the flexible pressure sensor

For biomedical applications, animal experiments will be conducted to test performance of the capacitive pressure sensor. In the *in-vitro* test, a calibration system developed in our previous work was employed to measure the pressure between the outer surface of a silicone rubber tube and the inner surface of a cuff made by a PI sheet. The closed silicone rubber tube was filled with water and the flexible sensor was wrapped tightly on the silicone rubber tube and the capacitive pressure sensor was further encircled by a circular loop made of PI sheet to simulate the spiral cuff electrode (Fig. 3.4.17b). The PI sheet has a thickness of 20 μm and width of 8 mm except 12 mm at center and the circular loop was formed by pulling one end of the sheet through a pre-cut small slit at the center. The other end of the circular loop is fixed on a platform and the movable end is fastened to a translation stage which can move forward to pull the PI sheet. A load cell fixed on the stage was utilized to measure the tension, T, on the PI sheet. The pressure, P_0, between the PI sheet and the silicon rubber can be formulated by

$$P_0 = \frac{T}{(1+2\pi\lambda)r_0} \tag{3.4.40}$$

where λ is the coefficient of static friction between the PI sheet and the flexible sensor, and r_0 is the outer radius of the silicon rubber. Pressure calculated from Eq. (3.4.40) is compared with the capacitance changed detected by the sensor. A straight line between the pressure applied by the PI sheet and change of the capacitance of the flexible can be fitted.

3.4.3.2 Electret microphone

An electret microphone is a type of condenser microphone, which eliminates the need for a polarizing power supply by using a permanently-charged material. An electret is a stable dielectric material with a permanently-embedded static electric charge. Electrets are commonly made by first melting a suitable dielectric material such as a plastic or wax that contains polar molecules, and then allowing it to re-solidify in a powerful electrostatic field. The polar molecules of the dielectric align themselves to the direction of the electrostatic field, producing a permanent electrostatic "bias". Electret microphones are useful in acoustic and audio applications as for example in hearing aid appliances.

Silicon micromachining enables the integration of mechanical parts with preamplifiers. A silicon-based microphone structure is shown in Fig. 3.4.18 (Schenk et al., 1996).

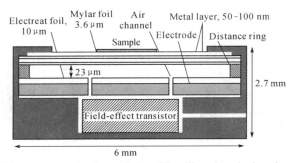

Fig. 3.4.18. Schematic cross Section and size of the silicon-based microphone (not to scale)

The membrane of the microphone consists of a 3.6 μm thick mylar foil that is coated on both sides with a layer of about 100 nm metal. The distance between the membrane and the electrets amounts to 23 μm and the thickness of the electrets foil is about 10 μm. Air channels in the stationary electrode enlarge the air volume between the membrane and the electrode. Hence, the part of the counteracting force that is generated by air compression is reduced, when the membrane is pressed down. To diminish acoustic interference for experiments with air, a small hole is drilled in the lower side of the closed microphone capsule. Interference is reduced to about 50% because disturbing sound waves act now simultaneously on both sides of the membrane. This modification additionally enables measurements with the microphone in a vacuum. The effect of this hole for the sensitivity and frequency response of this type of electret microphone has proven to be not significant. The overall size of the electret microphones used is in the range of a few millimetres. Their diameter amounts to 6.0 mm and their height to 2.7 mm. Because of the small capacitance of the system (a few pF only) a field-effect transistor is integrated into the microphone capsule acting as an impedance converter. It amplifies the very low signals induced by the movement of the membrane and, thus, the influence of the stray capacitance is reduced.

Because the membrane and electrode are electrically shortened across the finite input impedance of the FET (about 10^{12} Ω), the microphone cannot transduce static deflections of the membrane and, hence, no static measurement of the force is possible (Fig. 3.4.19a). The capacitance C_m between the membrane and the electret-electrode and the input resistance R_i of the FET forms a high pass filter with the cut-off frequency $f_g = (2\pi R_i C_m)^{-1}$ (Fig. 3.4.19b). It amounts to about 0.5 Hz for the microphone used. When the membrane is deflected, the microphone signal will be zero after $t_0 > 1/f_g$.

Vibration of the membrane will generate the voltage U_{Ri} at the input resistance of the FET, and U_{Ri}

controls, via the gate, the drain-source resistance of the FET and, hence, the microphone output voltage U_m. The drain-source resistance of the FET and the resistance R represent a voltage divider. A capacitor C decouples the microphone signal from the supply voltage U_s.

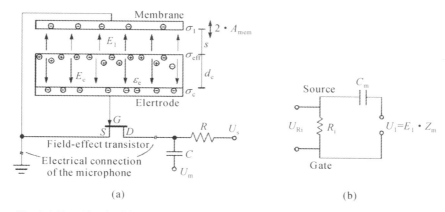

Fig. 3.4.19. Circuit of the electret microphone: (a) Electrical circuit; (b) Equivalent circuit

3.4.3.3 2D capacitive sensors

In recent years, sensor array has been one of the technology hotspots and has developed rapidly. The 2D capacitive sensor shown in Fig. 3.4.20 is a typical sensor array, which consists of an *X-Y* array of pressure capacitive sensitive cells. By detecting the capacitance change of every cell, many aspects of the force applied on the sensor can be obtained, such as magnitude, location and direction. This kind of 2D capacitive sensor is widely used in human-computer interfaces, mobile devices and robotic applications.

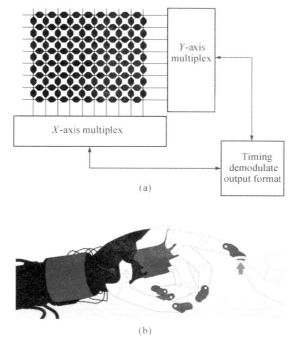

Fig. 3.4.20. A 2D capacitive sensor: (a) Schematic diagram; (b) Robotic applications

The use of a high density capacitive array, such as the sensitive part of the fingerprint sensor (Rey et al., 1997), is very suitable for the growing market of portable equipment for the low power consumption of capacitance detection. For this application, the density of sensitive cells is much higher than the density required for the tactile sensors used in robotic applications.

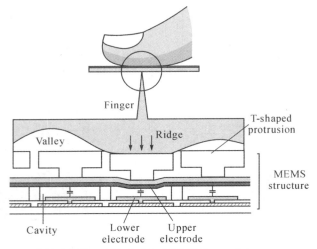

Fig. 3.4.21. A kind of the MEMS fingerprint sensor

One type of these fingerprint sensors (Sato et al., 2005) has an array of about 57,000 pixels in an area of (11.2×12.8) mm^2. Each pixel is a capacitive pressure sensor that has a MEMS cavity structure stacked on a CMOS LSI. Integrating sensing circuits, just below the MEMS cavity structures, enables a sufficiently large number of pixels. The MEMS cavity structure consists of parallel electrodes and a protrusion with a shape of a block as shown in Fig. 3.4.21. When a finger touches the sensor surface filled with a lot of protrusions, ridges of the finger deform the upper electrode via the protrusions and increase the capacitance between the upper and lower parallel electrodes. The capacitance change is detected by the sensing circuits and converted into digitized signal levels in the CMOS LSI, and the detected signals from all the pixels generate one fingerprint image in a gray-scale. Thus, the MEMS fingerprint sensor obtained shows clear fingerprint images.

3.5 Piezoelectric Sensors and Measurement

Piezoelectric sensors are based on the piezoelectric effect, which was discovered by Curie brothers in 1880s. This chapter illustrates what the piezoelectric effect is, how the piezoelectric sensors work, and where they can apply to.

3.5.1 Piezoelectric Effect

The piezoelectric effect is the phenomenon that when a pressure is applied to a piezoelectric material, it causes a mechanical deformation and a displacement of charge. The number of those charges is highly proportional to the applied pressure. In order to distinguish with converse piezoelectric effect, we called the phenomenon above the direct piezoelectric effect. By the way, the converse piezoelectric effect refers to the piezoelectric would produce some mechanical deformation when a voltage is applied to. Essentially, the piezoelectric effect is the energy conversion between the mechanical form and the electrical form, which shows in Fig. 3.5.1.

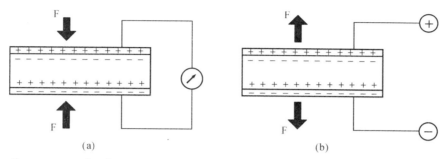

Fig. 3.5.1. Energy conversion between mechanical and electrical forms: (a) piezoelectric effect; (b) converse piezoelectric effect

Piezoelectric effect is anisotropic. As shows in Fig. 3.5.2, we usually use the digital subscript to represent directions. In detail, 1, 2, 3 represent the axis direction of X, Y, Z respectively; and 4, 5, 6 represent the tangential direction of X, Y, Z respectively.

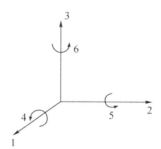

Fig. 3.5.2. The coordinate system of piezoelectric effect

The expression of digital subscripts has its own order. For example, d_{ij} means the piezoelectric coefficient between the force in j-direction and the induced charges in i-direction. Piezoelectric effect in all directions can be described in Eq. (3.5.1).

$$\Delta = D\Sigma \tag{3.5.1}$$

where

$$\Delta = [\delta_1 \ \delta_2 \ \delta_3]^{\mathrm{T}} \tag{3.5.2}$$

$$D = \begin{bmatrix} d_{11} & d_{12} & d_{13} & d_{14} & d_{15} & d_{16} \\ d_{21} & d_{22} & d_{23} & d_{24} & d_{25} & d_{26} \\ d_{31} & d_{32} & d_{33} & d_{34} & d_{35} & d_{36} \end{bmatrix} \tag{3.5.3}$$

$$\Sigma = [\sigma_1 \ \sigma_2 \ \sigma_3 \ \sigma_4 \ \sigma_5 \ \sigma_6]^{\mathrm{T}} \tag{3.5.4}$$

Δ refers to the number of induced charges in three different directions; D is called the piezoelectric coefficient matrix; Σ refers to the forces applied to piezoelectric materials in different directions.

Different piezoelectric materials have their unique piezoelectric coefficient matrix. When the matrix is known, we can calculate the number of induced charges with mechanical stress or strains applied to the piezoelectric materials. But how does the piezoelectric coefficient matrix determine? What is the physical significance? Here take quartz crystals as an example to explain it.

As shown in Fig. 3.5.3a, the positive and negative charges focus in the center of gravity without forces, and the vector sum of electric dipole moment equals to zero, that is $P_1 = P_2 = P_3$. Therefore, there is no charge on the crystal surface, and it is neutral.

As shown in Fig. 3.5.3b, with force along x-axis, there is a separation in the centre of positive and negative charges, so that the electric dipole moment along x-axis direction is non-zero due to the decreased P_1 and increased P_2 and P_3. Negative charges appear in the positive direction along x-axis, while no charges appear along y-axis direction, and the electric dipole moment along y-axis direction is zero.

As shown in Fig. 3.5.3c, with force along y-axis: similar to Fig. 3.5.3b, the increased P_1 and decreased P_2 and P_3 make the electric dipole moment non-zero. Positive charges appear in the positive direction along x-axis, while no charges appear along y-axis direction, and the electric dipole moment along this direction is zero.

With force along z-axis: since the deformation of crystal along x-axis is the same with that along y-axis, no separation in the centre of positive and negative charges is generated, and thus the vector sum of electric dipole moment equals to zero, that is, forces applied along z-axis will not generate piezoelectric effects.

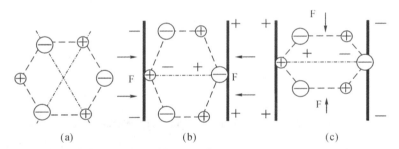

Fig. 3.5.3. Piezoelectric model of quartz crystals: (a) Without force; (b) With force along x-axis; (c) With force along y-axis

3.5.2 Piezoelectric Materials

Quartz: Although found naturally, most quartz in practical use is synthetic, *AT*-cut single crystal, right handed, and α-phase. Below the Curie temperature of 573 °C, quartz has a trigonal structure and above that temperature, it becomes β-quartz with a hexagonal structure (Bottom, 1982). Although the *AT*-cut is used due to its near zero temperature coefficient of frequency, other cuts such as the *Y*-cut or the dual mode *SC*-cut with a high sensitivity of the resonance frequency with respect to temperature can be used accurately for temperature measurement (Goyal et al., 2005). Quartz is only useful in single crystal form and to achieve high resonance frequency, the thickness of this crystal quartz must be minimized. Using micromachining techniques, resonators with thicknesses less than 10 μm and diameters of less than 100 μm have been realized for quartz crystal microbalance (QCM) applications and chemical sensing techniques.

Langasite: Another material, which has similar temperature coefficients as quartz, but has a quality factor five times higher and a piezoelectric coupling coefficient three times higher, is langasite ($La_3Ga_5SiO_{14}$). A relatively new non-ferroelectric piezoelectric material, langasite single crystals have been grown using the Czochralski method and single crystal thin films have been grown using the liquid phase epitaxy technique. Langasite does not experience phase transitions up to the melting point and has low acoustic wave propagation losses.

Lithium niobate and tantalate: Lithium niobate and lithium tantalate are well-known ferroelectric crystals discovered in 1949 and have been successfully grown into single crystals from melting by the Czochralski technique since 1965 (Xu, 1991). Both are important in surface acoustic wave (SAW) devices and high-frequency filter applications. Like quartz, these materials must be grown in bulk and have different properties based on their cuts. The cuts usually used are *Y*-cut for $LiNbO_3$ and *X*-cut for $LiTaO_3$.

Lead zirconate titanate (also called PZT): PZT are the most widely used ferroelectric materials. The higher electromechanical coefficient of PZT makes it a very attractive material for actuator and sensor applications. PZT films at the morphotropic phase boundary with a Zr/Ti ratio of 52/48 have been shown to exhibit a maximum in the piezoelectric response and are typically used in MEMS device applications.

Relaxor ferroelectrics: Lead-titanate-relaxor-based ferroelectric systems, PZN–PT, PMN–PT and PYN–PT, show extremely large electromechanical coupling coefficients and piezoelectric coefficients compared to PZT, but have integration issues that have limited their use such as a lower dielectric constant and a lower stress response. The low Curie temperatures of PMN–PT and PZN–PT limit their operating range, where as PYN–PT has a higher Curie temperature. Research in relaxor ferroelectrics is still very promising due to the wide array of different compositions of the $PbZrO_3$–$PbTiO_3$–$Pb(B_1B_2)O_3$ ternary system.

Polyvinylidene fluoride: Polyvinylidene fluoride (PVDF) ($-CH_2-CF_2-$), a semi-crystalline homopolymer, is technologically very important due to its piezoelectric and pyroelectric properties. These characteristics make it useful in sensor and battery applications requiring the highest purity, strength, and resistance to solvents, acids, bases and heat and low smoke generation (Jiang et al., 1997). PVDF is very flexible, exhibits good stability over time and does not depolarize when subjected to very high alternating electric fields. Like other current conducting polymers, the application potential of PVDF can be enormously escalated if its piezoelectric properties are enhanced to produce high electromechanical coupling and large force generating capabilities.

3.5.3 Measurement Circuits

3.5.3.1 Equivalent circuit of piezoelectric sensors

A piezoelectric sensor could be seen as a charge generator, or a capacitor (parallel plates) with capacitance of

$$C = \frac{\varepsilon_\gamma \varepsilon_0 A}{d} \tag{3.5.5}$$

where A is the area of piezoelectric chip; d is the thickness of piezoelectric chip; ε_γ is the relative permittivity of the piezoelectric material.

Thus, a piezoelectric sensor could be equivalent to the voltage source in series with the capacitor, as shown in Fig. 3.5.4a, where

$$U = \frac{q}{C} \tag{3.5.6}$$

A piezoelectric sensor could also be equivalent to the charge source in parallel with the capacitance, as in Fig. 3.5.4b.

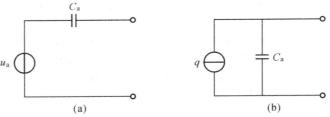

Fig. 3.5.4. Equivalent circuit of piezoelectric sensors: (a) Voltage source, (b) Charge source

In practical use, the equivalent capacitance of the cable (C_c), input resistance (R_i) and capacitance (C_i) of amplifiers, and leakage resistance (R_a) of the piezoelectric sensor are taken into consideration. Thus the actual equivalent circuit is shown as follows in Fig. 3.5.5.

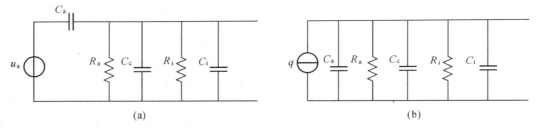

Fig. 3.5.5. Actual equivalent circuit of piezoelectric sensors: (a) Voltage source; (b) Charge source

3.5.3.2 Measurement circuit of piezoelectric sensors

Since the piezoelectric sensor has high inner resistance and low output energy, it is necessary to include the preamplifier circuit with high input resistance, so that

(1) High input resistance is replaced by low input resistance.

(2) The weak output of sensors is amplified.

According to the output of piezoelectric sensors, the preamplifiers could be classified into: Voltage amplifier and Charge amplifier.

- *Voltage amplifier (Impedance converter)*

Shown in Fig. 3.5.6 is the schematic of voltage amplifier circuit and its equivalent circuit.

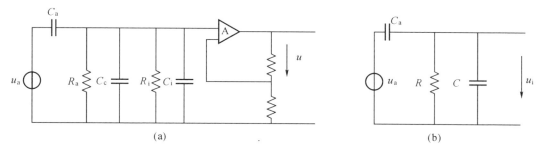

Fig. 3.5.6. Voltage amplifier circuit & its equivalent: (a) Voltage amplifier circuit; (b) Equivalent circuit

In Fig. 3.5.6b, $R = R_a R_i / (R_a + R_i)$, $C = C_c + C_i$. If the piezoelectric chip is applied with sine force $f = F_m \sin \omega t$, thus

$$U_a = \frac{dF_m \sin \omega t}{C_a} = U_m \sin \omega t \qquad (3.5.7)$$

where U_m is the output voltage, $U_m = dF_m / C_a$, d is the piezoelectric coefficient.

The input voltage of amplifier

$$\dot{U}_i = \frac{df j R \omega}{1 + jR\omega C + jR\omega C_a} \qquad (3.5.8)$$

Let U_{im} be the amplitude of $\dot{U}_i$,

$$U_{im} = \frac{dF_m R\omega}{\sqrt{1 + \omega^2 R^2 (C_a + C_c + C_i)}} \qquad (3.5.9)$$

The phase difference between the input voltage and the force applied:

$$\Phi = \frac{\pi}{2} - \arctan\left[\omega(C_a + C_c + C_i)R\right] \qquad (3.5.10)$$

In ideal condition, R_a and R_i are infinite, so U_{im} could be expressed by

$$U_{im} = \frac{dF_m}{C_a + C_c + C_i} \qquad (3.5.11)$$

The above formula indicates the input voltage of amplifier is irrelative to frequency. Usually, when $\omega/\omega_0 > 3$, ω and ω_0 could be taken as irrelative. ω is used to notify the inverse of time constant of the measurement circuit, where $\omega_0 = 1/\left[(C_a + C_c + C_i)R\right]$.

When piezoelectric elements are applied with static forces, the pre-amplifier will output zero voltage, because the charges will leak through the input resistance of the amplifier and the inherent leakage resistance of the sensor. Therefore, piezoelectric sensor is inapplicable for static force measurements.

- **Charge amplifier**

Charge amplifier is often used as input circuit of piezoelectric sensors, the equivalent circuit is shown in Fig. 3.5.7. A notifies the gain of the Operational Amplifier, and the output voltage U_0 could be given by

$$U_0 \approx U_{cf} = -\frac{q}{C_f} \qquad (3.5.12)$$

where U_0 is the output voltage. U_{cf} is the voltage across the feedback capacitor.

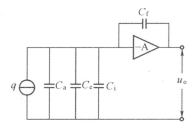

Fig. 3.5.7. Equivalent circuit of charge amplifier

With characteristics of amplifiers, we have the output voltage of charge amplifier

$$U_o = -\frac{Aq}{C_a + C_c + C_i + C_f + (1+A)C_f} \qquad (3.5.13)$$

usually, $A = 10^4 \sim 10^8$, so when $(1+A)C_f \gg C_a + C_c + C_i$, the above formula could be approximated by

$$U_0 \approx -\frac{q}{C_f} \qquad (3.5.14)$$

which indicates the characteristics of a charge amplifier: the output voltage of a charge amplifier depends only on the input charge and the feedback capacitor, having nothing to do with cable capacitance, and is in direct proportion to q.

3.5.4 Biomedical Applications

Piezoelectric sensors are configured as direct mechanical transducers or as resonators.

3.5.4.1 Sensors based on direct piezoelectric effect

A wide variety of mechanical sensors are based on the piezoelectric effect, which can be applied for detecting pressure impulse or movements.

Monitoring of breathing conditions during sleeping is one of the crucial tests for appropriate diagnosis of sleep disorders. The structure of a kind of piezoelectric flexible transducers (Yuu et al., 2008) is shown in Fig. 3.5.8.

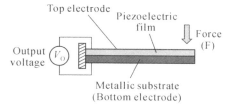

Fig. 3.5.8. A schematic configuration for airflow monitoring

The sensor consists of a stainless steel (SS) foil substrate, a piezoelectric ceramic film, a top electrode, and a protection film. The PZT composite piezoelectric film is fabricated onto the 40 μm thick SS foil using a sol-gel spray technique. The thickness of the piezoelectric film was 60 μm. The top electrode is constructed using a silver paste. The SS foil serves as a bottom electrode as well as the substrate. The dimension of the active transducer area is 4 mm by 20 mm. The sensor is covered with a polymeric film that protects the sensor from moisture and scratches. Due to the porosity inside the film and the thin substrate, the sensor has high flexibility. The transducer can be used as an ultrasonic probe and unimorph-type bending sensor which could be applied in various types of industrial applications such as nondestructive testing, medical diagnosis and physiological monitoring.

3.5.4.2 Sensors based on piezoelectric resonators

- *Quartz crystal microbalance*

Quartz crystal microbalance (QCM) have been in steady use for a number of years as a convenient tool to determine mass loading of material layers with layer thicknesses ranging well below the monolayer level. A QCM typically consists of a slab of thin, single-crystal, piezoelectric quartz, with very large lateral dimensions in comparison to its thickness, which is sandwiched between two metal electrodes. Fig. 3.5.9 shows a schematic representation of the quartz crystal microbalance.

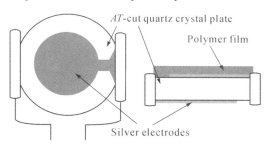

Fig. 3.5.9. Schematic representation of the quartz crystal microbalance

QCMs utilize the piezoelectric qualities of quartz crystals. Applying alternating current to the quartz crystal will induce oscillations. With an alternating current between the electrodes of an *AT*-cut crystal, a standing shear wave is generated. The resonance is so narrow that oscillators are highly stable and accurate in the determination of the resonance frequency. It has been shown that there is an explicit quantitative relationship between a shift in the resonant frequency and added mass on the silver electrode, Δm (Kurosawa et al., 2006).

$$\Delta F_N = -N \frac{2F_1^2}{A\sqrt{\mu\rho}} \Delta m \qquad (3.5.15)$$

where N is the order of the overtone (N=1, 3, 5, 7, . . .), ΔF_N is the change in the oscillation frequency of a quartz crystal of Nth mode, Δm represents the change in mass on crystal electrodes. The fundamental frequency of the quartz crystal is represented as F_1, A is the electrode area of the quartz, μ is the elastic modulus of the quartz, and ρ is the quartz density. The equation was examined experimentally by measuring changes in oscillation frequency when the deposit mass of polymer films affected the oscillating frequency of QCM.

The QCM can be used under vacuum, in gas phase (King, 1964) and more recently in liquid environments (Höök & Kasemo, 2001). It is useful for monitoring the rate of deposition in thin film deposition systems under vacuum. In liquid, it is highly effective at determining the affinity of molecules (proteins, in particular) to surfaces functionalized with recognition sites. Larger entities such as viruses or polymers can be investigated, as well.

- *Infrared-sensitive resonator*

Thermal infrared detectors are broadband detectors and can be operated at room temperature without cooling. These detectors can be designed to operate near the room temperature thermodynamic noise limit arising from the thermal conductance fluctuation between the sensing element and the supporting substrate.

More recently, thickness shear mode resonators from quartz have been proposed and demonstrated as sensitive infrared detectors. The unprecedented temperature sensitivity along with the low noise performance that can be achieved in quartz crystal oscillators is the principle of operation of several thermal sensors based on acoustic waves.

A kind of micro-machined Y-cut quartz-resonator-based thermal infrared detector array (Kao and Tadigadapa, 2009) is shown in Fig. 3.5.10. 1 mm diameter and 18 μm thick (90 MHz) inverted mesa configuration quartz resonator arrays with excellent resonance characteristics have been fabricated by RIE etching of quartz. The average resonance frequency of the array was 89.65 MHz with a maximum deviation in the resonance frequency of ±0.59%. The temperature sensitivity of the resonators was measured by placing the resonator array in an oven and varying the temperature in the range from 22 to 38 °C. The resonator array was allowed to equilibrate at each temperature for more than 30 min until a stable resonance frequency was achieved. As expected a linear dependence on temperature for all the three resonators was observed and the average temperature sensitivity was measured to be 7.2 kHz/°C. Fig. 3.5.11 shows the measured temperature dependence of the resonance frequency.

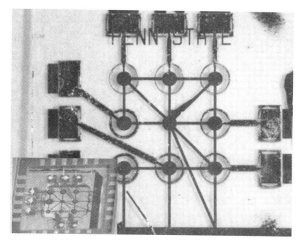

Fig. 3.5.10. Photograph of the fabricated quartz resonator infrared sensor array with eight resonators per chip (reprinted from (Kao and Tadigadapa, 2009), Copyright 2009, with permission from Elsevier)

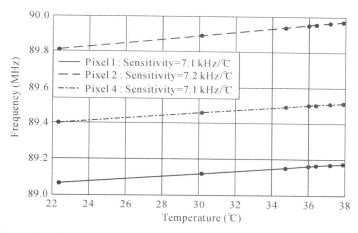

Fig. 3.5.11. Experimentally measured temperature sensitivity of the Y-cut quartz resonators

Modern imagers based upon thermal infrared detectors are typically implemented using micromachining techniques. Deposition of thin film ferroelectric materials has allowed the construction of better thermally isolated IR sensor structures. The microbridge structures offer better thermal isolation and lower cost products through batch fabrication techniques (Hanson et al., 2001). Fig. 3.5.12a shows the exploded view of the thermal infrared detectors pixel, Fig. 3.5.12b shows a SEM photograph of the Section of a 320×240 pixel array, and thermal infrared detectors, and Fig. 3.5.12c shows a thermal image obtained. Although the noise equivalent temperature difference of the detector was only 0.21 °C, the images show very high quality which has been attributed to the high modulation transfer function of these detectors.

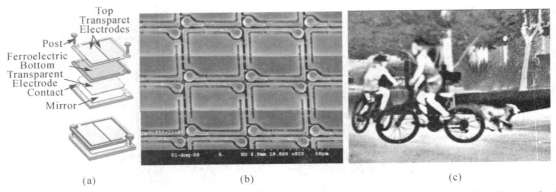

Fig. 3.5.12. Thermal infrared detectors: (a) Exploded view of a thermal infrared detectors pixel; (b) SEM photograph of a part of 330 × 240 pixel IR imaging array; (c) An IR image obtained (reprinted from (Hanson et al., 2001), Copyright 2001, with permission from SPIE)

3.6 Magnetoelectric Sensors and Measurement

A magnetoelectric sensor is one kind of sensor which is able to convert displacement, velocity, acceleration and other measuring quantities to electrical signals by magnetoelectric effect. They can be divided into magnetoelectric induction sensors and hall sensors. Magnetoelectric sensors with large output power, stable performance, as well as bandwidth working-frequency are widely used in medical, automation, mechanical engineering and other fields.

3.6.1 Magnetoelectric Induction Sensors

Magnetic induction sensors are active sensors which do not require auxiliary power.

3.6.1.1 Working principle

According to Faraday's law of electromagnetic induction, when a conductor is moving in a stable magnetic field perpendicularly to the magnetic field direction, the induced electromotive force of the conductor is

$$e = \frac{d\Phi}{dt} = Bl\frac{dx}{dt} = Blv \tag{3.6.1}$$

where Φ is the effective magnetic flux, B is the magnetic induction intensity of the stable magnetic field, l is the effective length of the conductor, v is the relative velocity of the conductor to the stable magnetic field.

When there are ω circles in the time-varying electromagnetic field, the induced electromotive force of the induction coil is

$$e = -\omega\frac{d\Phi}{dt} \tag{3.6.2}$$

For the different modes above, there are two different types of magnetoelectric induction sensors: variable magnetic flux mode (Fig. 3.6.1) and constant magnetic flux mode (Fig. 3.6.2).

Fig. 3.6.1 is a variable magnetic flux type sensor used to measure the rotation angular velocity of rotating objects. When the gear rotates, the teeth bumping causes the changes of magnetic flux, which results in, generating induced electromotive force. Fig. 3.6.2 is the typical structure of constant flux-type magnetic sensors, which consists of a permanent magnet, a coil, two springs, and a metal skeleton. The magnetic circuit system produces a constant DC magnetic field, the working air gap of the magnetic circuit is fixed, therefore the magnetic field between a air gap is constant. Its moving parts could be a coil (moving coil type) or a magnet (moving iron type). The principle of the moving-coil-type (Fig. 3.6.2a), and the moving-iron-type (Fig. 3.6.2b) are identical.

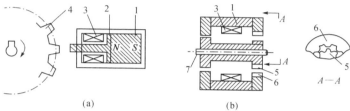

Fig. 3.6.1 Structure schematic diagraph of magnetoelectric induction sensor with variable magnetic field: (a) Open magnetic circuit; (b) Closed magnetic circuit. 1: permanent magnet, 2: soft magnet, 3: induced coil, 4: iron gear, 5: internal gear, 6: external gear

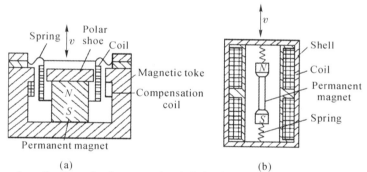

Fig. 3.6.2 Structure schematic diagraph of magnetoelectric induction sensor with constant magnetic flux: (a) Moving coil; (b) Moving iron

The induced electromotive force of permanent magnetic field is:

$$e = -\omega B_0 l v \quad (3.6.3)$$

where, B_0 is the magnetic induction intensity in working air gap, l is the average length of the coils, ω is the number of turns of the coils in working air gap, and v is the relative velocity.

3.6.1.2 Basic characteristics

In Fig. 3.6.3 the output current of magnetoelectric induction sensor I_0 is:

$$I_0 = \frac{E}{R + R_f} = \frac{B_0 l \omega v}{R + R_f} \quad (3.6.4)$$

where, R_f is the input resistance, R is the equivalent resistance of the coils. The current sensitivity of the sensor is:

$$S_I = \frac{I_0}{v} = \frac{B_0 l \omega}{R + R_f} \qquad (3.6.5)$$

When the working temperature varies or under external magnetic field's disturbing and mechanical vibrating, the sensitivity will change, and the relative error is

$$\gamma = \frac{dS_I}{S_I} = \frac{dB}{B} + \frac{dl}{l} - \frac{dR}{R} \qquad (3.6.6)$$

When the temperature changes, all the three items in Eq. (3.6.6) are nonzero values. Temperature compensation is necessary by using a thermo-magnetic splitter.

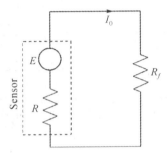

Fig. 3.6.3 The simplified model of the magnetoelectric induction sensor

In addition, Magnetic sensors also generate a certain degree of non-linear error. When the current I goes through the sensor's coils, the alternating magnetic flux Φ_1 exists. It will add to the working magnetic flux of the permanent magnet, and thus causes the change of the magnetic flux in air gap.

In Fig. 3.6.4, when the relative velocity of the coils to the permanent magnet increases, there will be a larger induced electromotive force and larger induced current. The additional magnetic field's direction is opposite to the working magnetic field, thus the working magnetic field and the sensitivity decrease. Otherwise, when the coils move in the opposite direction, the additional magnetic field's direction is the same with the working magnetic field, thus the sensitivity increases. The higher the sensor's sensitivity, the larger the current is in the coils, and thus the worse the nonlinear error. To eliminate the error, it is possible to add compensating coils to the sensor as Fig. 3.6.2a.

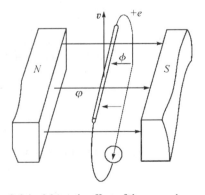

Fig. 3.6.4 Magnetic effect of the sensor's current

3.6.1.3 Measuring circuit for magnetoelectric induction sensor

Magnetoelectric induction sensors output the induced electromotive force directly, with high sensitivity, so gain amplifier is not necessary in the measuring circuits. This kind of sensors can measure speed directly. Integral and differential circuits are needed when measuring displacement and acceleration.

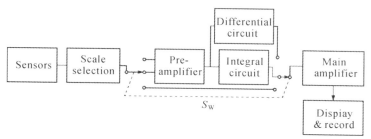

Fig. 3.6.5. Measuring circuit for magnetoelectric induction sensor

3.6.1.4 Biomedical applications

Magnetic sensors can be used in biomedical flow tests (see Section 2.1.3). At present, the magnetic flow meter has been used as a standard method for the measurement of intravascular blood flow. The sensitivity of magnetic measurement for the volume flow rate has nothing to do with the velocity distribution. The magnetic method can be widely applied from the thickest blood vessels to a 1 mm diameter in humans.

According to the previous description of the electromagnetic induction law, when a conductor moves in a magnetic field while cutting the magnetic field lines, it will generate induced electromotive force. As shown in Fig. 3.6.6a, the uniform magnetic flux density B, the electromotive force generated from diameter EE' is equal to

$$V = 2aBv \qquad (3.6.7)$$

where B is magnetic flux density (Gs), v is velocity (cm/s). Or it can be explained as

$$V = \frac{2QB}{\pi a} \qquad (3.6.8)$$

where Q is the volume flow rate (cm^3/s), $Q = V\pi a$

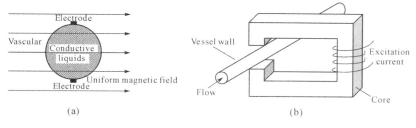

Fig. 3.6.6. Magnetic blood flow meter: (a) Cross-section diagram; (b) Principle

The advantage of the electromagnetic flow method is the electromotive force has nothing to do with the blood flow distribution. For a certain degree of blood vessel diameters and the magnetic induction intensity, the electromotive force only relates to the instantaneous volume flow rate.

3.6.2 Hall Magnetic Sensors

A Hall sensor is a kind of magnetic-electric sensors, which is based on the Hall Effect. This phenomenon is discovered by U.S. physicist Hall (Dr. Edwin Hall, 1855 – 1938) in 1879 while studying metal conductors. With the development of semiconductor technology, Hall components began to use semiconductor materials. At present, the Hall sensors are widely used in the measurements of electromagnetic, pressure, acceleration, vibration, and so on.

3.6.2.1 Hall device

Dr. Hall found that, when the direction of the applied magnet field is not the same with that of the electrons, a voltage difference (Hall voltage) will be generated across the electrical conductor, which is transverse to the current and the magnetic field perpendicular to the current. He also found that this voltage was proportional to the current flowing through the conductor, and the magnetic flux density or magnetic induction perpendicular to the conductor (Fig. 3.6.7a).

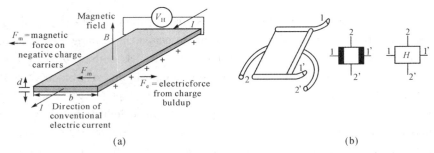

Fig. 3.6.7 Hall device: (a) Hall effect; (b) External structural schematic diagram. Graphic symbol 1,1': Exciting electrode, 2,2': Hall electrode

The current expressed in terms of the drift velocity is

$$I = nevbd \tag{3.6.9}$$

where n is the density of charge carriers.

The magnetic force is

$$F_m = eBv \tag{3.6.10}$$

where v is the drift velocity of the charge.

The Hall voltage is

$$U_H = E_H b \tag{3.6.11}$$

where E_H is the electric field intensity.

Since, $I = nevbd$

$$v = \frac{1}{nebd} \tag{3.6.12}$$

Since, $F_m = eBv$, $F_e = E_H e$, at equilibrium, $F_m = F_e$

$$E_H = Bv = \frac{BI}{nebd} \tag{3.6.13}$$

$$U_H = E_H b = \frac{BI}{ned} \tag{3.6.14}$$

The Hall coefficient is defined as $R_H = 1/ne$, so,

$$U_H = R_H \frac{IB}{d} = K_H IB \tag{3.6.15}$$

where $K_H = R_H/d$ is defined as the sensitivity of the Hall device.

As is seen from Eq. (3.6.15), the Hall voltage is in direct proportion to excitation and magnetic field intensity, the sensitivity is in direct proportion to the Hall coefficient R_H and in inverse proportion to the thickness of the Hall device. To improve the sensitivity, the Hall device is usually made into thin slice.

The resistance between the excitation electrodes is $R = \rho L/(bd)$, and $R = U/I = El/I = vl/(\mu nevbd)$ (because $\mu = v/E$, where μ is the electron mobility). So,

$$\frac{\rho L}{bd} = \frac{L}{\mu nebd}, \quad R_H = \mu \rho \tag{3.6.16}$$

From the above equation, only the semiconductor materials are suitable for the manufacture of the Hall device, because the semiconductor has a better Hall coefficient R_H.

The structure of the Hall device is shown in Fig. 3.6.7b. 1,1' are connected to an electrode with exciting current or voltage (exciting electrode). 2,2' are Hall output down-lead (Hall electrode).

The Hall element is the basic magnetic field sensor. It requires signal conditioning to make the output usable for most applications. The signal conditioning electronics needed are an amplifier stage and a temperature compensation. Voltage regulation is needed when operating from an unregulated supply. Fig. 3.6.8 illustrates a basic Hall Effect Sensor.

If the Hall voltage is measured when no magnetic field is present, the output is zero. However, if voltage at each output terminal is measured with respect to ground, a non-zero voltage will appear. This is the common mode voltage, and is the same at each output terminal. It is the potential difference that is zero. The amplifier shown in Fig. 3.6.8 must be a differential amplifier so as to amplify only the potential difference—the Hall voltage.

The Hall voltage is a low-level signal on the order of 30 mV in the presence of a one-gauss magnetic field. This low-level output requires an amplifier with low noise, high input impedance and moderate gain. A differential amplifier with these characteristics can be readily integrated with the Hall element using standard bipolar transistor technology. Temperature compensation is also easily integrated.

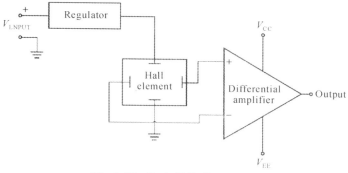

Fig. 3.6.8 Basic Hall effect sensor

3.6.2.2 Rated exciting current and maximum permitted exciting current

When the Hall element of the excitation current is zero, if the location of the magnetic flux density components is zero, then the hall electric potential should be zero. However, in the actual cases, this value is usually not zero. Then the measured electric potential is called the allelic potential. The reasons of this phenomenon are:

(1) Hall electrodes asymmetry, or not in the same equipotential surface;
(2) Non-uniform resistivity semiconductor material causing non-uniform resistance or geometry;
(3) The poor contact of the Incentive electrodes causes the unevenness of the electrode current.

Allelic resistance can also state the allelic potential,

$$r_0 = \frac{U_0}{I_H} \tag{3.6.17}$$

where U_0 is equipotential potential, r_0 is Allelic resistance, and I_H is excitation current.

3.6.2.3 Input resistance and output resistance

When the external magnetic field is zero and the Hall element is input with AC excitation, there is a DC potential, called the parasitic direct potential, in addition to the potential allelic exchange. The value of parasitic DC potential is generally below 1 mV, which affects the Hall element drift through temperature. It is caused by the following reasons:

Poor contact of the Incentive electrodes and Hall electrodes which forms the contact capacitance and causes a rectification effect;

Asymmetry in the size of two Hall electrodes causes the two different heat capacity points of electrodes, so that the thermal states are different, and causes formatting of the polar temperature potential.

3.6.2.4 Temperature coefficient for Hall voltage

In certain magnetic induction intensities with an excitation current, the temperature changes 1 °C, the Hall potential percentage changes in the temperature coefficient of electrical potential, and is called the temperature coefficient for Hall voltage. Usually, the temperature is an important factor of a Hall effect sensor.

- **Compensation for hall device's unequal electrical potential**

In ideal situations, A and B will be on the same equipotential surface, and unequal potential U_{AB} will be zero. However, it's not in this case in real situation. Some compensating circuits are shown above in Fig. 3.6.9.

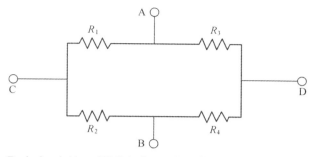

Fig. 3.6.9. Equivalent bridge of Hall device. A, B: Hall electrode; C, D: exciting electrode

- *Temperature compensating for hall device*

To reduce the temperature error of the Hall device, it is important to select suitable temperature coefficient devices and other constant temperature methods. For $U_H=K_H IB$, when choosing constant current supply, the Hall voltage will be more stable, by reducing the variation of the exciting current I caused by the variation of the input resistance as the temperature changes. The relationship between the sensitivity coefficient and temperature:

$$K_H = K_{H0}(1+\alpha\Delta T) \qquad (3.6.18)$$

where K_{H0} is the changed K_H under temperature T, ΔT is the change of temperature, α is the temperature coefficient of Hall voltage. α is generally a positive value. When the temperature increase, the Hall voltage will increase by $\alpha\Delta T$ times. If decreasing the exciting current to keep the K_H unchanged, the effect of K_H change could be avoid.

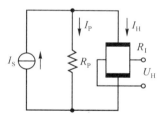

Fig. 3.6.10. Temperature compensating circuit

Fig. 3.6.10 shows the temperature compensating circuit. Suppose the initial temperature is T_0, the input resistance is R_{10}, the sensitivity coefficient is K_H, and the shunt resistance is R_{P0}. When temperature increase to T, the parameters in the circuit will be

$$R_1 = R_{10}(1+\delta\Delta T) \qquad (3.6.19)$$

$$R_P = R_{P0}(1+\beta\Delta T) \qquad (3.6.20)$$

where δ is the temperature coefficient of input resistance, β is the temperature coefficient of shunt resistance.

$$I_H = \frac{R_P I_S}{R_P + R_1} \qquad (3.6.21)$$

By arranging the equations above and omitting the high items as $\alpha, \beta, (\Delta T)^2$,

$$R_{P0} = \frac{(\delta - \alpha - \beta) R_{I0}}{\alpha} \tag{3.6.22}$$

When the devices are selected, the input resistance R_{I0}, temperature coefficient δ, and the temperature coefficient of Hall voltage are set values. So the shunt resistant R_{P0} and temperature coefficient β can be calculated.

3.6.2.5 Biomedical applications

- **Hall whisker sensor**

Hall effect devices produce a very low signal level and thus require amplification. While suitable for laboratory instruments, the vacuum tube amplifiers available in the first half of the 20th century were too expensive, power consuming, and unreliable for everyday applications. It was only with the development of the low cost integrated circuit that the Hall effect sensors became suitable for mass applications. Many devices now sold as Hall effect sensors in fact contain both the sensor described above and a high gain integrated circuit (IC) amplifier in a single package. Recent advances have resulted in the addition of ADC (Analog to Digital) converters and I²C (Inter-integrated circuit communication protocol) IC for direct connection to a microcontroller's I/O port being integrated into a single package. Reed switch electrical motors using the Hall Effect IC is another application. Here is an application of the Hall sensor (Zhong et al., 2009).

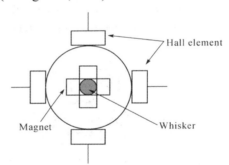

Fig. 3.6.11. Structure of whisker Hall sensor

The sensor is a two-tier board structure whose lower root fixes the antennae and the upper layer feels the displacement of the tentacle. The two plates are connected by a nylon column and maintain a spacing of 20 cm to 30 cm as shown in Fig. 3.6.11. The tentacle crosses the center of the upper hole (diameter 1,215 mm) and four small magnets fixed at the top of the tentacle at a 5 – 15 mm away from the hole. The small magnets around the hole correspond with the four Hall elements (UGN3503), these constitute a "follicle" role.

When the tentacles are bent, the magnets deviate from the original position, the change of magnetic fields cause the increases or decreases of the Hall element output voltage, and thus the signals in the x direction and y direction form a differential output respectively. The sensor has a two-dimensional awareness.

- **Hall respiration flow sensor**

There are various methods for designing Hall flow meters, but the general principle is the same: each actuation of the sensor, by a magnet or by shunting the magnetic field, corresponds to a measured quantity of water or air. In the example shown, the magnetic field is produced by magnets mounted on the impeller blade (Fig. 3.6.12). The impeller blade is turned by the air flow. The sensor produces two outputs per revolution. This kind of flow rate sensor needs two conversions; firstly, the flow rate is converted to the rotation speed of the blade turn. Secondly, the Hall sensor converts rotation speed to pulse signal.

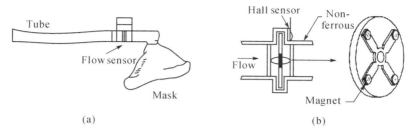

Fig. 3.6.12. Respiration flow sensor: (a) Mask of respiration flow meter; (b) Structure of the Hall respiration flow sensor

3.7 Photoelectric Sensors

A photoelectric sensor is a device used to detect the distance, absence, or presence of an object by using a light transmitter and a photoelectric receiver.

3.7.1 Photoelectric Element

A photoelectric element can convert the light signals into electrical signals using a Photoelectric Effect. It is the primary element that constitutes the photoelectric sensor. There are many advantages such as quick response, simple structure, convenient to use, noncontact detection and high reliability.

3.7.1.1 Photoresistor

A photoresistor or light dependent resistor or cadmium sulfide (CdS) cell is a resistor whose resistance decreases with increasing incident light intensity. It can also be referred to as a photoconductor. A photoresistor is made of a high resistance semiconductor. If light falling on the device is of high enough frequency, photons absorbed by the semiconductor will give bound electrons enough energy to jump into the conduction band. The resulting free electrons (and their hole partner) conduct electricity, thereby lowering resistance.

- *The structure and working principle of the photoresistor*

Photoresistor, also known as a light tube, is an optoelectronic device made of semiconductor materials without polarity. On condition of no illumination, the dark resistance is very large while dark current is very small. When it is exposed to the light with a certain range of wavelength, bright resistance will decrease; meanwhile the bright current will increase.

Hence, for a photoresistor, the greater the dark resistance, the smaller the light bright resistance, the higher the sensitivity of the photoresistor will be. Fig. 3.7.1 shows the structure of a photoresistor. It is applied on the glass substrate layer of semiconductor material with metal electrodes at both ends, and metal electrode connected to the circuit.

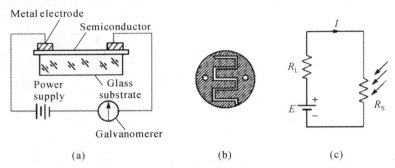

Fig. 3.7.1. The structure of the photoresistor: (a) The structure of the photoresistor; (b) The electrode of the photoresistor; (c) The schematic circuit diagram of the photoresistor

- *Some important parameters of photoresistor*

Dark resistance: the resistance value without the effect of the light. The corresponding current is called dark electricity.

Bright resistance: the resistance value within the effect of the light. The corresponding current is called bright electricity.

Photocurrent: the difference between dark resistance and the bright resistance.

- *Basic characteristics of the photoresistor*

Voltage-current characteristic: Under the certain illumination, the relationship between current and voltage of the photosensitive resistance is called volt-ampere characteristic. As is shown in Fig. 3.7.2, in a certain range, the *I-V* curve is a straight line which means that the value of the resistance is relevant to the illumination intensity, but has nothing to do with the current and voltage.

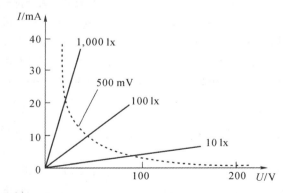

Fig. 3.7.2. The volt-ampere characteristic of sulfuration cadmium photosensitive resistance

Illumination characteristic: The illumination characteristic of the photosensitive resistance describes the relationship between photocurrent (I) and the illumination intensity. Most photosensitive resistance's illumination characteristic is nonlinear as shown in Fig. 3.7.3.

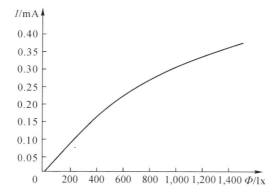

Fig. 3.7.3. The illumination characteristic of the photosensitive resistance

Spectrum characteristic: The relationship between the relative photosensitivity of the photosensitive resistance and the incidence wavelength is defined as spectrum characteristic of the photosensitive resistance or the spectrum response. Different materials have different spectrum responses. Even the same material will have different photosensitivity due to the change of the wavelength. Fig. 3.7.4 is the spectrum characteristic of the photosensitive resistance.

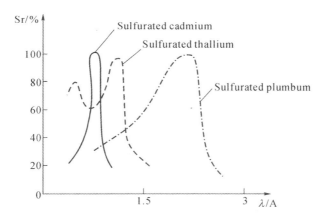

Fig. 3.7.4. Spectrum characteristic of the photosensitive resistance

Frequency characteristic: The relationship between the relative photosensitivity of the photosensitive resistance and the incidence frequency of light is defined. Different materials have different Frequency Characteristic. Fig. 3.7.5 is the spectrum characteristic of the photosensitive resistance.

Temperature characteristic: Similar with other semiconductor devices, the working temperature will produce the obvious affect to photosensitive resistance. The relationship between the relative photosensitivity of the photosensitive resistance and the working temperature is defined. And the different material will have different temperature Characteristic. Fig. 3.7.6 is the temperature Characteristic of the sulfuration lead photosensitive resistance.

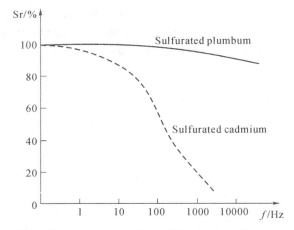

Fig. 3.7.5. Frequency characteristic of the photosensitive resistance

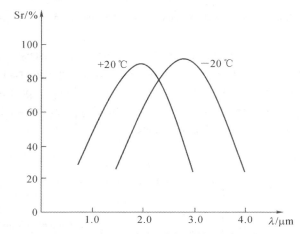

Fig. 3.7.6. Spectrum temperature characteristic of the sulfuration lead photosensitive resistance

3.7.1.2 *Photodiode and transistor*

A photodiode is a type of photodetector capable of converting light into either current or voltage, depending upon the mode of operation.

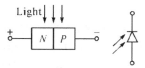

Fig. 3.7.7. Structure and symbol of the photodiode

    In the circuit, the photodiode always works in the reverse state. Without illumination, the reverse resistance is very large and the countercurrent (dark current) is very small; when PN junction is exposed to the light, photoelectron and photo-induced cavities will be produced near the PN junction, thus the photocurrent is formed. As a result, without illumination, the diode is under the cut-off state, while on the light condition, it is in the conduction state.

- *Structure*

Photodiodes are similar to regular semiconductor diodes except that they may be either exposed (to detect vacuum UV or X-rays) or packaged with a window or optical fiber connection to allow light to reach the sensitive part of the device.

Photosensitive transistors have two PN junctions. When the collecting electrode is connected to a higher electrical potential than the emitting electrode and leaving the base electrode unconnected, to the collecting electrode will have inverse biased potential. When light irradiates the collecting electrode, the Electron-Hole Pair will appear. The photo-induced electron will be pulled to the collecting electrode and the hole will be in base electrode, increasing the electrical potential between base electrode and the emitting electrode. Thus more electrons will flow to collecting electrode to form output current.

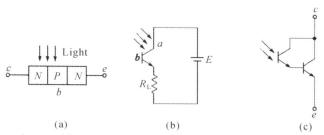

Fig. 3.7.8. The NPN type photosensitive transistor: (a) Structure schematic diagram; (b) Basic circuit; (c) Equivalent circuit of the Darlington photosensitive transistor

- *Basic characteristics*

Spectrum Characteristic: The relationship between the relative photosensitivity of the the photosensitive diode and transistor and the incidence wavelength is defined as spectrum characteristic of the photosensitive diode and transistor or the spectrum response. Different materials have different spectrum responses. And even the same material will have different photosensitivity due to the change of the wavelength. Fig. 3.7.9 is the spectrum characteristic of the photosensitive diode and transistor. From the curve, we ca see that silicon-based photosensitive diode and transistor has maximum sensitivity in 0.9 μm wavelength and germanium-based photosensitive diode and transistor has maximum sensitivity in 1.5 μm wavelength.

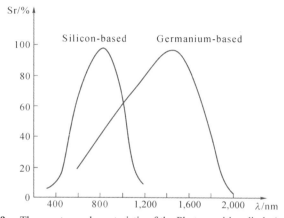

Fig. 3.7.9. The spectrum characteristic of the Photosensitive diode (transistor)

Voltage-current characteristic: Fig. 3.7.10 describes the voltage-current characteristic of the silicon photosensitive diode/transistor. Under the illumination, the countercurrent will increase if the illumination intensity increases. Under different illumination intensities, the voltage-current characteristic curves are almost parallel which means that unless it reaches saturate, the output will never influenced by the values of the different voltages.

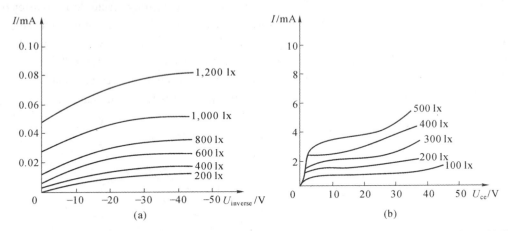

Fig. 3.7.10. Volt characteristic of the silicon photosensitive diode/transistor. (a) Silicon photosensitive diode; (b) Silicon photosensitive transistor

Frequency characteristic: The relationship between the relative photosensitivity of the photosensitive diode/ transistor and the incidence frequency of light is defined. Different materials have different frequency characteristic. Fig. 3.7.11 is the spectrum characteristic of the photosensitive diode/ transistor.

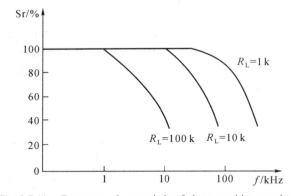

Fig. 3.7.11. Frequency characteristic of photo sensitive transistor

Temperature characteristic: The relationship between the relative photosensitivity of the photosensitive diode/transistor and the working temperature is defined as shown in Fig. 3.7.12. From the Fig. 3.7.12a, we can see that temperature have very heavy affects on the dark current of device, however has very little affects to light current of device as shown in Fig. 3.7.12b. So, it is necessary to compensate the dark current of device through electronic circuit, otherwise it will cause output errors.

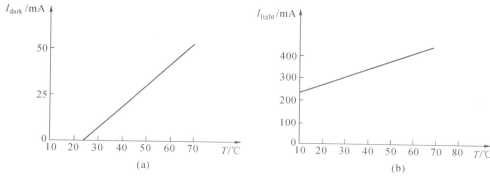

Fig. 3.7.12. Temperature characteristic of photo sensitive transistor

- *Applications*

P-N photodiodes are used in similar applications to other photodetectors, such as photoconductors, charge-coupled devices, and photomultiplier tubes. Photodiodes are used in consumer electronic devices such as compact disc players, smoke detectors, and the receivers for remote controls in VCRs and televisions. In other consumer items, such as camera light meters, clock radios (the ones that dim the display when it's dark) and street lights, photoconductors are often used rather than photodiodes, although in principle either could be used.

Photodiodes are often used for accurate measurement of light intensity in science and industry. They generally have a better, more linear response than photoconductors. They are also widely used in various medical applications, such as detectors for computed tomography (coupled with scintillators) or instruments to analyze samples (immunoassay). They are also used in pulse oximeters.

3.7.1.3 Photovoltaic sensors

The photovoltaic effect involves the creation of a voltage (or a corresponding electric current) in a material upon exposure to electro-magnetic radiation. Though the photovoltaic effect is directly related to the photoelectric effect, the two processes are different and should be distinguished. In the photoelectric effect, electrons are ejected from a material's surface upon exposure to radiation of sufficient energy. To the contrary in the photovoltaic effect is different, in that the generated electrons are transferred from different bands (i.e., from the valence to conduction bands) within the material, resulting in the buildup of a voltage between two electrodes.

- *Working principle—photo-induced volta effect*

Actually it is a large-area PN junction, when one surface of the PN junction is exposed to the light, for example the P surface, if the electron energy is larger than the prohibit bandwidth of the semiconductor material, then a pair of free electrons and a hole will be produced while the P surface absorbs a photon. Free electron-hole pairs are diffused inward from the surface quickly, and finally produce an electromotive force closely related with the illumination intensity under the effect of a junction electric field.

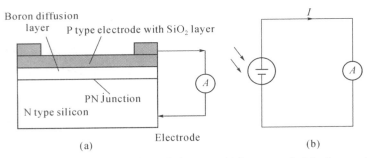

Fig. 3.7.13. The principle diagram of the silicon photovoltaic sensor: (a) Structure principle diagram; (b) Equivalent circuit

- **Basic characteristics**

Spectrum characteristic: the sensitivity of photovoltaic changes when the wavelength of light changes. As Fig. 3.7.14 shows, the peaks of spectral are corresponding to the incident light wavelength. Silicon photovoltaic is near 0.8 μm and selenium is 0.5 μm. The spectral response of Silicon photovoltaic is from 0.4–1.2 μm in the wavelength range, while the selenium is only 0.38–0.75 μm. Therefore, Silicon photovoltaic cells can be applied in a wide wavelength range.

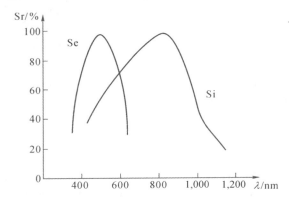

Fig. 3.7.14. The spectrum characteristic of the photovoltaic sensor

Illumination characteristic: Under different illuminations, the photocurrent or photo-emf is different; the relationship between the illumination and the photocurrent is called the illumination characteristic. Fig. 3.7.15 shows that in the wide range, the short-circuit current (Fig. 3.7.15a (Curve1)) and light intensity is a linear relationship However, the open-circuit (Fig. 3.7.15a (Curve2)) voltage is nonlinear while in the 2000 lx illumination, it tends to saturated. Therefore, as a measuring element, the photovoltaic must be treated as a current source instead of a voltage source.

Temperature characteristic: Temperature characteristic of photovoltaic describes the open circuit voltage and short circuit current with temperature changing. As it is related to the temperature drift of photovoltaic, this characteristic affects the precision of measurement. The open-circuit voltage decreases fast with the temperature increasing. However, the short circuit current increases slowly. For these reasons, it is best to ensure that the temperature stays constant or to take into account the temperature compensation.

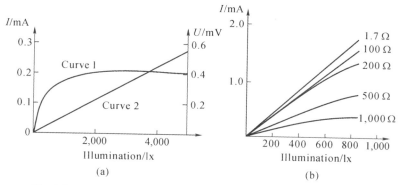

Fig. 3.7.15. The illumination characteristic of the Si photovoltaic sensor

3.7.2 Fiber Optic Sensors

A fiber optic sensor is a sensor that uses optical fiber either as the sensing element (intrinsic sensors), or as a means of relaying signals from a remote sensor to the electronics that process the signals (extrinsic sensors).

Optical fibers can be used as sensors to measure strain, temperature, pressure and other quantities by modifying a fiber so that the quantity to be measured modulates the intensity, phase, polarization wavelength or transit time of light in the fiber. Sensors that vary the intensity of light are the simplest, since only a simple source and detector are required. A particularly useful feature of such fiber optic sensors is that they can, if required, provide distributed sensing over distances of up to one meter.

3.7.2.1 *Optical fiber's structure and its principle of transmitting the light*

- *Optical fiber structure*

As shown in Fig. 3.7.16, fiber core is the central cylinder, envelop is the layer outside fiber core and safety layer is the layer outside of the envelop which is used to enhance the mechanic intensity. The refractive index of the fiber core is larger than that of the envelop. The conductivity of optic fiber is based on the characteristics of the fiber core and envelop.

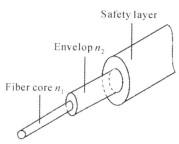

Fig. 3.7.16. Optical fiber structure

Biomedical Sensors and Measurement

- **Light transmission principle of optic fiber—total reflection**

As shown in Fig. 3.7.17 to satisfy the total reflection in optic fiber, the angle of incidence θ_i should satisfy

$$\theta_i \leq \theta_c = \arcsin \frac{\left(n_1^2 - n_2^2\right)^{1/2}}{n_0} \tag{3.7.1}$$

Generally, the optic fiber is set in atmosphere, where $n_0 = 1$. So the above equation can be represented by

$$\theta_i \leq \theta_0 = \arcsin\left(n_1^2 - n_2^2\right)^{1/2} \tag{3.7.2}$$

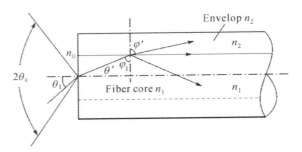

Fig. 3.7.17. Light transmission principle of optic fiber

3.7.2.2 Basic characteristic of optic fiber

Numerical aperture (NA):

$$NA = \arcsin\left(n_1^2 - n_2^2\right)^{1/2} \tag{3.7.3}$$

Numerical aperture represents the optic fiber's capability of the collecting light. The larger the value NA is, the more powerful the optic fiber will be. No matter how large the emitting power, only in the case when the angle of incidence is smaller than $2\theta_c$, will the optic fiber be conductive. When the angle of incidence is too large, the light will escape from the optic fiber causing the leaking of the light.

Mode of optic fiber: In order to provide light with a different angle of incidence, the amount of reflection on the interface different and the landscape intensity of the light interference will also be different. This will cause a different transmitting mode.

3.7.3 Applications of Photoelectric Sensors

Photoelectric sensors have many applications. Here, the detection of pulse oximeter and fiber optic temperature sensor will be introduced.

3.7.3.1 Detection of pulse oximeter

A pulse oximeter is a medical device that indirectly measures the oxygen saturation of a patient's blood (as opposed to measuring oxygen saturation directly through a blood sample) and changes in blood volume in the skin. It is often attached to a medical monitor so staff can see a patient's oxygenation at all times. Most monitors also display the heart rate. Portable, battery-operated pulse oximeters are also available for home blood-oxygen monitoring. A blood-oxygen monitor displays the percentage of arterial hemoglobin in the oxyhemoglobin configuration. Acceptable normal range is from 95% to 100%, although values down to 90% are common.

A pulse oximeter is a particularly convenient non-invasive measurement instrument. Typically it has a pair of small light-emitting diodes (LEDs) facing a photodiode through a translucent part of the patient's body, usually a fingertip or an earlobe (Fig. 3.7.18). One LED is red, with wavelengths of 660 nm, and the other is infrared, 905, 910, or 940 nm. Absorption of these wavelengths differs significantly between oxyhemoglobin and its deoxygenated form, therefore from the ratio of the absorption of the red and infrared light the oxy/deoxyhemoglobin ratio can be calculated. The absorbance of oxyhemoglobin and deoxyhemoglobin is the same (isosbestic point) for the wavelengths of 590 and 805 nm; earlier oximeters used these wavelengths for correction for hemoglobin concentration.

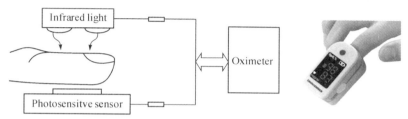

Fig. 3.7.18. Schematic diagram and picture of fingertip Pulse Oximeter—MD300C

3.7.3.2 Fiber optic temperature sensor

A fiber optic sensor, because of its unique properties and widespread attention, has rapidly developed. Among these, the most commonly used one is the fiber optic temperature sensor. Fiber not only has the electric insulation, but also is very little affected by the temperature. Therefore, it is widely used in body temperature measurement.

The liquid crystal optical fiber temperature sensor is the earlier development of a fiber optic sensor. For a certain wavelength, the color or reflectivity of liquid crystal changes with temperature.

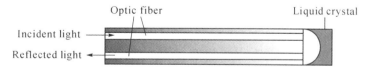

Fig. 3.7.19. Optical fiber sensor structure

The principle of this fiber optic sensor is shown in Fig. 3.7.19. An incident light is injected into the fiber, after being reflected by the liquid crystal, it is transmitted by the outgoing optical fiber and then detected by the receiver. As the crystal is affected by the temperature, the reflected light intensity changes. Therefore, the reflected light intensity is a function of temperature.

Fiber optic sensors are mainly characterized by high sensitivity, small size, anti-interference and no electric signal, so it can be safely used to check the life and the body, especially for heart-related measurements.

In addition, the special feature that fiber can be curved, optical fiber temperature sensors can be used to detect the body temperature changes in certain specific locations, such as hyperthermia in tumors, the sensor travels through hypodermic needles or a catheter into the cancer patient's body, where it monitors and controls the temperature of tumorous lesion (Fig. 3.7.20).

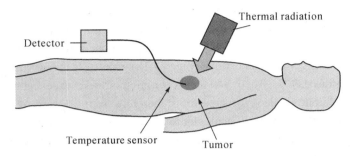

Fig. 3.7.20.   Optical fiber temperature sensor for monitoring the tumor tissue

3.8   Thermoelectric Sensors and Measurement

Temperature is one of the most important variables of the environment and the human body. Using the particular effect of thermosensitive elements or thermocouple sensors, temperature measurement (Wang and Ye, 2003) has long been a reality. The basics of temperature measurement are summarized by Fraden (Fraden, 1991). As a potent method for measuring temperature, thermoelectric sensors have been widely used not only in scientific and engineering applications, but also in the biomedical field for temporary and long-term monitoring of body temperature. To satisfy the rising demand for continuous and local measurement, integrated thermometer sensors need to be developed. Here we'll introduce the basis of thermosensitive elements, thermocouple sensors and newly developed integrated temperature sensors, and how they can be used for body temperature measurement.

3.8.1   Thermosensitive Elements

The electrical resistivities of conductor and semiconductor materials vary with changing temperature which is called thermoresistive effect. This kind of element is called thermosensitive resistor. The high-accuracy resistance-temperature relationship of some materials can be used to sense various non-electrical quantities just as temperature, velocity, concentration and density of medium. With the development of technology, thermometers based on thermoresistive effect have been used from the triple point of equilibrium hydrogen to the freezing point of silver.

3.8.1.1   *Resistance temperature detectors*

A resistance temperature detector (RTD) (Harsányi, 2000) is the commonly used term for temperature sensors, the operation of which is based on the positive temperature coefficient of metals. The resistance-temperature relationship of RTD can be derived from:

$$R(T) = R_0(1 + \alpha \Delta T) \qquad (3.8.1)$$

where R_0 is the resistance at a reference temperature T_0, α is the temperature coefficient of resistance at T_0, and ΔT is the actual temperature difference related to T_0. For the majority of metals, α is the function of temperature, but can be seen as a constant within a limited range.

Resistance thermometer elements must meet four conditions as follows:
(1) High temperature coefficient of resistance;
(2) Stable chemical and physical properties;
(3) High electrical resistivity;
(4) Excellent reproducibility.

Common resistance materials for RTDs include platinum, nickel and copper. Among these materials, platinum is used widely and in fact, as an interpolation standard. The resistance of platinum via temperature characteristic can be approximated as:

$$R_t = R_0[1 + AT + BT^2] \qquad (0\,°C < T < 640\,°C) \qquad (3.8.2)$$

$$R_t = R_0[1 + AT + BT^2 + C(T-100)T^3] \qquad (-240\,°C < T < 0\,°C) \qquad (3.8.3)$$

where the constants are $A = 3.96847 \times 10^{-3}$, $B = -5.847 \times 10^{-7}$, $C = -4.22 \times 10^{-12}$. Since the B and C coefficients are relatively small, the resistance changes almost linearly with the temperature.

Copper is used occasionally as an RTD element. Its low resistivity forces the element to be longer than a platinum element. However its linearity and low cost make it an economical alternative in industrial applications. Its upper temperature limit is only about 150 °C. The resistance of copper via temperature characteristic can be approximated as

$$R_t = R_0\left[1 + AT + BT^2 + CT^3\right] \qquad (-50\,°C < T < 150\,°C) \qquad (3.8.4)$$

where the constants are $A = 4.28899 \times 10^{-3}$, $B = -2.133 \times 10^{-7}$, $C = -1.233 \times 10^{-9}$.

As shown in Fig. 3.8.1, these elements almost always require insulated leads attached. At low temperatures, PVC, silicon rubber or PTFE insulators are common to 250 °C. Above this, glass fiber or ceramic are used. The measuring point and usually most of the leads require a housing or protection sleeve. This is often a metal alloy which is inserted into a particular process. Often more consideration is taken in selecting and designing protection sheaths than sensor itself, as this is the layer that must withstand chemical or physical attack and also offer convenient process attachment points. In order to minimize the effects of the lead resistances, a three or four-wire configuration can be used.

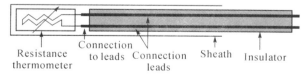

Fig. 3.8.1. Construction of resistance thermometer

3.8.1.2 Thermistors

Thermistors differ from RTDs in that the material used in a thermistor is generally ceramic or polymer, while RTDs use pure metals. The temperature response is also different; RTDs are useful over larger temperature ranges, while thermistors typically achieve a higher precision within a limited temperature range.

Thermistors can be classified into two types, depending on the sign of temperature coefficient of resistance. If the resistance increases with increasing temperature, the device is called a positive temperature coefficient (PTC) thermistor. If the resistance decreases with increasing temperature, the device is called a negative temperature coefficient (NTC) thermistor. Typical characteristics of two temperature-dependent thermistors are compared in Fig. 3.8.2.

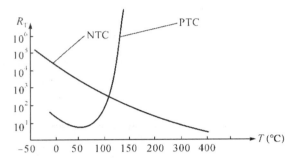

Fig. 3.8.2. Typical characteristics of two temperature-dependent thermistors

PTC-thermistors are made of a doped polycrystalline ceramic containing barium titanate ($BaTiO_3$) and other compounds. Steep increases of resistance at a certain critical temperature make them particularly useful as self-regulation heating elements, current limiting devices, etc.

NTC-thermistors are made from a pressed disc or cast chip of semiconductor oxides, such as the precisely controlled mixtures of the oxides of Mn, Co, Ni, Cu and Zn. NTC-thermistors can be used as inrush-current limiting devices in power supply circuits and are regularly used in automotive applications.

The main parameters of thermistors are:

(1) Nominal resistance R_0 at 25 °C;

(2) Temperature coefficient of resistance α, generally at 20 °C;

(3) Dissipation factor H, the measure of loss-rate of power;

(4) Specific heat capacity C, the measure of the heat energy required to increase the temperature of a unit quantity of a thermistor by a unit of temperature;

(5) Time constant r, the ratio of dissipation factor H and specific heat capacity C.

The high resistivity of the thermistor contributes to a distinct measurement advantage. The four-wire resistance measurement may not be required as it is with RTDs. A measurement lead resistance of 10 Ω produces only 0.05 °C error. This error is a factor of 500 times less than its RTD counterpart.

3.8.2 Thermocouple Sensors

Thermocouple sensors are temperature sensors that are easy to use and obtain. With the high sensitivity, linearity and functional long-term stability, they are widely used in industry.

3.8.2.1 Principle of thermocouples

Thermocouple sensors are inexpensive and interchangeable, and can measure a wide range of temperature. The main limitation is accuracy: system errors of less than one Kelvin (K) can be difficult to achieve.

The thermoelectric effect (also called Seebeck effect) is the theoretical basis for thermocouple sensors. The effect is that a voltage is created in the presence of a temperature difference between two different conducting (metal or semiconductor) materials, A and B. This causes a continuous current in the conductors if they form a complete loop.

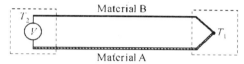

Fig. 3.8.3. Schematic representation of the thermocouple

As shown in Fig. 3.8.3, a simple thermocouple is made from a coupling or junction of two materials. In the circuit, T_1 is the DUT temperature of the "hot" point and T_2 is the stable reference temperature of the "cold" point.

Charge carriers in the materials (electrons in metals, electrons and holes in semiconductors, ions in ionic conductors) will diffuse when one end of a conductor is at a different temperature than the other. Hot carriers diffuse from the hot end to the cold end, since there is a lower density of hot carriers at the cold end of the conductor. Cold carriers diffuse from the cold end to the hot end for the same reason. The voltage caused by this phenomenon can be derived from:

$$E_T = -\int_{T_0}^{T}(\sigma_A - \sigma_B)dT \tag{3.8.5}$$

where σ_A and σ_B denote the Thomson coefficient of the materials A and B.

Charges also diffuse in the junction. The created electrical potential difference is defined as follows:

$$E_C = \frac{k(T_1 - T_0)\ln(N_A/N_B)}{e} \tag{3.8.6}$$

where k is Boltzmann constant, N_A and N_B are the free-charge densities of the materials A and B, and e is the electronic charge. Thus the total thermoelectromotive force is

$$E_{AB}(T_1, T_0) = E_C - E_T = \frac{k(T_1 - T_0)\ln(N_A/N_B)}{e} - \int_{T_0}^{T}(\sigma_A - \sigma_B)dT \tag{3.8.7}$$

3.8.2.2 Sensitivity of thermocouple sensors

The thermoelectromotive force can be approximated as follows:

$$E_{AB}(T_1, T_0) = \alpha(T_1 - T_0) + \beta(T_1^2 - T_0^2) \tag{3.8.8}$$

where α and β are constants. For small temperature differences, the sensitivity K of thermocouple sensors can be taken from derivation of Eq. (3.8.8):

$$K = \frac{\mathrm{d}E_{\mathrm{AB}}(T_1 - T_0)}{\mathrm{d}T} = \alpha + 2\beta T \tag{3.8.9}$$

Generally, K is between 6 and 80 μV/K.

3.8.2.3 Cold junction compensation

Having a junction of known temperature, while useful for laboratory calibration, is not convenient for most measurement and control applications. They incorporate an artificial cold junction using a thermally sensitive device such as a thermistor or diode to measure the temperature of the input connections at the instrument, with special care to minimize any temperature gradient between terminals. Hence, the voltage from a known cold junction can be simulated, and the appropriate correction applied. This is known as cold junction compensation.

3.8.3 Integrated Temperature Sensors

Traditional analog temperature sensor, including thermoresistive sensors and thermocouple sensors, may have poor linearity in certain scope of temperature, which means cold junction compensation or lead wire compensation is inevitable. By contrast, integrated temperature sensors have advantages of good sensitivity, high linearity and quick response. As the name implies, an integrated temperature sensor is the integration of a driving circuit, data processing circuit and a requisite logical control circuit on a single IC. With small size and convenient usage, integrated temperature sensors have become more widely used in recent years.

3.8.3.1 Diode temperature sensor

The ordinary semiconductor diode can be used as a temperature sensor. The forward biased voltage across a diode has a temperature coefficient of about 2 mV/°C and is reasonably linear. For an ideal diode, the diode voltage V_F can be expressed as follows:

$$V_F = \frac{E_{g_0} - kT \ln(\alpha T \gamma / I_F)}{q} \tag{3.8.10}$$

where E_{g_0} is the band gap of semiconductor at absolute zero, k is the Boltzmann constant, α is a constant temperature-independent and related with cross sectional area, γ is the constant dependent on electron mobility, I_F is the current in diode, and q is the electron charge.

With proper doping concentration, diode voltage V_F can be proportional to T within a certain temperature scope. To acquire larger linearity, two differential pair diodes with the same characteristics are applied (working currents are I_{F1} and I_{F2} respectively), and only the forward current will appear in the result:

$$\Delta V_F = \frac{kT \ln(I_{F1} / I_{F2})}{q} \tag{3.8.11}$$

Thus, the voltage difference ΔV_F is independent of the reverse current, and is absolutely linear to temperature T.

3.8.3.2 Triode temperature sensor

A triode temperature sensor is based on the principle that the base-emitter voltage V_{BE} is linear to temperature T if the current in collector I_C is constant. Take the case of NPN crystal triode, the voltage V_{BE} and current I_C have the following relation:

$$V_{BE} = \frac{E_{g_0}}{q} - \frac{kT\ln(\alpha T\gamma/I_C)}{q} \qquad (3.8.12)$$

where α is a constant temperature-independent and related with junction area and base width, γ, the constant dependent on electron mobility, E_{g_0}, the band gap of semiconductor at absolute zero. Similar to diode temperature sensors, V_{BE} has an approximate linear relation with T if I_C remains constant. In pursuit of a higher precision of measurement, the linear compensation method, like differential pair triodes, as shown in Fig. 3.8.4, is necessary.

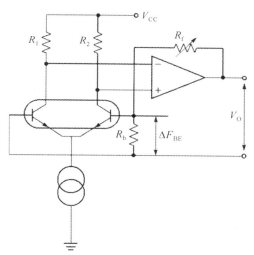

Fig. 3.8.4. Temperature measuring circuit using differential pair triodes

3.8.3.3 Integrated temperature sensor

A triode temperature sensor, together with peripheral detection circuits, will compose an integrated temperature sensor. Within a working range from −50 to 150 °C, it generates current or voltage generally proportional to temperature.

Integrated temperature sensors can be divided into voltage output types and current output types. Voltage output temperature sensors such as AD22100, AD22103, LM135/235/335, and so on, provide voltage output signals with relatively low output impedance. All require an excitation power source and are essentially linear. Current output sensors act as a high-impedance, constant current regulator and require a supply voltage. AD590, AD592, TMP17 and LM134/234/334 are all very common current output sensors. Provided by ADI, AD590 is a 2-terminal integrated temperature sensor and can be used in any temperature- sensing application below 150 °C. The inherent low cost of a monolithic integrated circuit combined with the elimination of support circuitry makes the AD590 an attractive alternative for many temperature measurement situations.

As is shown in Fig. 3.8.5, the measurement circuit for current output temperature sensor often uses proportional to absolute temperature (PTAT) and the output voltage is expressed as follows:

$$V_o = \frac{R_2}{R_1}\left(\frac{kT}{q}\right)\ln n \tag{3.8.13}$$

where n is area ratio of triode Q_1 and Q_2, k the Boltzmann constant, T the measurand temperature, q the electron charge.

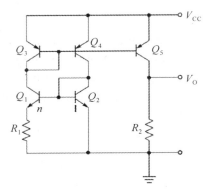

Fig. 3.8.5. PTAT circuit

3.8.4 Biomedical Applications

3.8.4.1 *Thermal sensor*

Thermometers have been used in clinical testing for the past few decades. Nowadays the demand of continuous monitoring devices is increasing, especially wearable instruments (Bonato, 2003). A skin thermography testing system is no exception. In 2007, Giansanti and Maccioni (Giansanti and Maccioni, 2007) made their research on a wearable integrated thermometer sensor for skin contact thermography known to the public, which now makes continuous monitoring in breast cancer investigation possible.

The thermal sensor unit is arranged through a 4 row-4 column matrix box, where each cell corresponds with an area of 4 mm×4 mm, and is monitored by one thermometer that carries the IM335 component, as shown in Fig. 3.8.6a. The face of the matrix box, made of a special permeable sponge, makes contact with the skin, and carries the 16 thermal sensor packages. As showing in Fig. 3.8.6b, a service unit comprising the multiplexing circuit, the processing and conditioning circuit, the power supply unit with the oscillation and stabilization circuitry, and a Pentium IV PC guarantees the device's normal operation.

A test was designed to measure hand skin temperature under conditions of gradual loading of physical activity, as is shown in Fig. 3.8.6c. The result indicates a thermal resolution less than 0.03 °C, and a spatial resolution equal to 1.6×10^{-5} m^2. In clinical testing, the maximum rate of thermal skin variation equals to 3.1 °C/0.25 h. The usage of this type of wearable thermometer paves the way for the testing of breast cancer thermography.

Chapter 3 Physical Sensors and Measurement

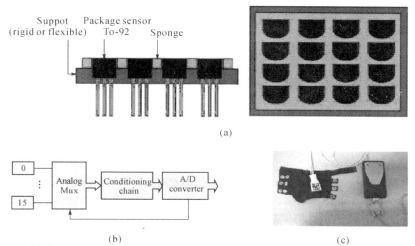

(a)

(b)   (c)

Fig. 3.8.6. A wearable integrated thermometer sensor: (a) The 3D design of the integrated thermometer; (b) Block diagram of the device; (c) Details of the integrated thermometer, the elastic bandage and the service unit (reprinted from (Giansanti and Maccioni, 2007), Copyright 2007, with permission from Elsevier)

3.8.4.2 *Multiple-sensor micro-system for pulmonary function diagnostics*

Asthma and COPD (chronic obstructive pulmonary disease) affect 10% – 20% of the population world wide, and this number is still increasing. The development of a portable pocket sized electronic multiple-sensor micro-system for low cost, high volume equipment for improved diagnosis of pulmonary diseases and diagnostic functions in general have been implemented by van Putten et al. (2000). The microsystem can measure peak expiratory flow, relative humidity, pressure and temperature. Different operating modes allow the measurement of pressure, flow velocity and temperature using the same sensor configuration. The core part is a single chip multiple-sensor (Fig. 3.8.7) which is obtained by applying silicon and MEMS technology. The sensing elements adopt P-type doped resistive elements, which guarantees the best reproducibility and accuracy. Two separate temperature- sensing elements are integrated with a value of about 60 kΩ at room temperature. The temperature measurements reflect an almost linear relationship between the temperature coefficient and the doping concentration.

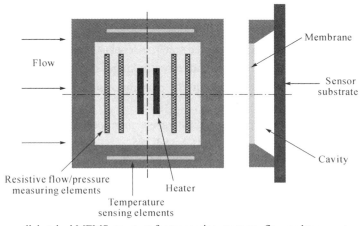

Fig. 3.8.7. Principle parallel etched MEMS structure for measuring pressure, flow and temperature. Not on scale. Actual sensor size is about 4 mm^2

133

References

Ba A. & Sawan M., 2003. Integrated programmable neurostimulator to recuperate the bladder functions. *IEEE CCECE Conference*, Montreal. 1, 147-150.

Bakhoum E.G. & Cheng M.H.M., 2010. Novel capacitive pressure sensor. *Journal of Microelectromechanical Systems*. 19, 443-450.

Bonato P., 2003. Wearable sensors/systems and their impact on biomedical engineering. *IEEE Engineering in Medicine and Biology Magazine*. 22, 18-20.

Bottom V.E., 1982. *Introduction to Quartz Crystal Unit Design*. Van Nostrand-Reinhold, New York, 4.

Chen Y., Wang L. & Ko W., 1990. A piezoplymer finger pulse and breathing wave sensor. *Sensors and Actuators A-Physical*. 21-23, 879-882.

Chiang C.C., Lin C.C.K. & Ju M.S., 2007. An implantable capacitive pressure sensor for biomedical applications. *Sensors and Actuators A-Physical*. 382-388.

Ding M., Wang X.Z., Yang L.G. & Zhong Y.N., 2008. Method for improving the precision of inductance-transducer. *China Measurement & Testing Technology*. 34, 17-19.

Fraden J., 1991. Noncontact temperature measurement in medicine. In: *Bioinstrumentation & Biosensors*. Marcel Dekker, New York, 511-549.

Gaunt R.A. & Prochazka A., 2006. Control of urinary bladder function with devices: successes and failures. *Progress in Brain Research*. 152, 163-194.

Giansanti D. & Maccioni G., 2007. Development and testing of a wearable integrated thermometer sensor for skin contact thermography. *Medical Engineering & Physics*. 29, 556-565.

Gowrishetty U., Walsh K.M., Aebersold J., Jackson D., Millar H. & Roussel T., 2008. Development of ultra-miniaturized piezoresistive pressure sensors for biomedical applications. *University/Government/Industry Micro/Nano Symposium, UGIM 2008, 17th Biennial*, 89-92.

Goyal A., Zhang Y. & Tadigadapa S., 2005. Y-cut quartz resonator based calorimetric sensor. *Proceedings of IEEE Sensors Conference*. 1241-1244.

Hanson, C.M., Beratan H.R. & Belcher J.F., 2001. Uncooled infrared imaging using thin-film ferroelectrics. *Proceedings of SPIE*. 4288, 298-303.

Harsányi G., 2000. *Sensors in Biomedical Applications*. Technomic Publishing Company, Inc., 65-68.

He F., Huang Q. & Qin M., 2007. A silicon directly bonded capacitive absolute pressure sensor. *Sensors and Actuators A-Physical*. 135, 507-514.

Höök F. & Kasemo B., 2001.Variations in coupled water, viscoelastic properties, and film thickness of a Mefp-1 protein film during adsorption and cross-linking: a quartz crystal microbalance with dissipation monitoring, ellipsometry, and surface plasmon resonance study. *Analytical Chemistry*. 73, 5796-5804.

Hu X. & Wang C., 2008. Research progress on invasive monitoring of blood pressure for patients. *Chinese Nursing Research* (in Chinese). 22(1C), 193-195.

Huang, C.T., Shen C.L., Tang C.F. & Chang S.H., 2008. A wearable yarn-based piezo-resistive sensor. *Sensors and Actuators A-Physical*. 141(2), 396-403.

Jeong J.W., Jang Y.W., Lee I., Shin S. & Kim S., 2009. Wearable respiratory rate monitoring using piezo-resistive fabric sensor. *World Congress on Medical Physics and Biomedical Engineering*, Munich, Germany, 282-284.

Jiang Z., Carroll B. & Abraham K.M., 1997. Studies of some poly (vinylidene fluoride) electrolytes. *Electrochimica Acta*. 42, 2667-2677.

Jobbágy Á., Csordás P. & Mersich A., 2007. Blood pressure measurement at home. *IFMBE Proceedings*. 5, 3453-3456.

Kao P. & Tadigadapa S., 2009. Micromachined quartz resonator based infrared detector array. *Sensors and Actuators A-Physical*. 149, 189-192.

Kensall D.W., Pamela T.B., Jianbai W. & Craig R.F., 2008. High-density cochlear implants with position sensing and control. *Hearing Research*. 242, 22-30.

Kim J.H., Park K.T., Kim H.C. & Chun K., 2009. Fabrication of a piezoresistive pressure sensor for enhancing sensitivity using silicon nanowire. *Solid-State Sensors, Actuators and Microsystems Conference, Transducers 2009, International*, 1936-1939.

King W.H., 1964. Piezoelectric sorption detector. *Analytic Chemistry*. 36, 1735-1739.

Kittel C., 1981. Introduction to solid state physics. *John Wiley & Sons*, New York.

Kurosawa S., Parka J.W., Aizawa H., Wakida S.I., Taoa H. & Ishiharab K., 2006. Quartz crystal microbalance immunosensors for environmental monitoring. *Sensors & Actuators B-Chemical*. 22, 473-481.

Marco S., Samitier J., Ruiz O., Morante J.R. & Esteve J., 1996. High-performance piezoresistive pressure sensors for biomedical applications using very thin structured membranes. *Measurement Science & Technology*. 7, 1195-1203.

Martineza F., Obietaa G., Uribea I., Sikorab T. & Ochoteco E., 2009. Polymer-based flexible strain sensor. *Procedia Chemistry*. 1(1), 915-918.

Mazeika G.G. & Swanson R., 2007. *Respiratory Inductance Plethysmography an Introduction*. Pro-Tech Services, Inc., 5-8.

Moreau-Gaudry A., Sabil A., Benchetrit G. & Franco A., 2006. Use of respiratory inductance plethysmography for the detection of swallowing in the elderly. *Dysphagia*. 20, 297-302.

Peng C., 2000. *Principle and Application of Biomedical Sensors* (in Chinese). Higher Education Press, Beijing, China, 157-160.

Poole K.A., Thompson J.R., Hallinan H.M. & Beardsmore C.S., 2000. Respiratory inductance plethysmography in healthy infants: a comparison of three calibration methods. *European Respiratory Journal*. 16, 1084-1090.

Pramanik C. & Saha H., 2006. Low pressure piezoresistive sensors for medical electronics applications. *Materials and Manufacturing Processes*. 21, 233-238.

Rajagopalan S., Sawan M., Ghafar-Zadeh E., Savadogo O. & Chodavarapu V.P., 2008. A polypyrrole-based strain sensor dedicated to measure bladder volume in patients with urinary dysfunction. *Sensors*. 8, 5081-5095.

Rey P., Harvet P.C & Delaye M.T., 1997. A high density capacitive pressure sensor array for fingerprint sensor application. *Proceedings of Transducers '97*. 1453-1456.

Sander C.S., Knutti J.W. & Meindl J.D., 1980. A monolithic capacitive pressure sensor with pulse-period output. *IEEE Transactions on Electron Devices*. 17, 927-930.

Sato N., Shigematsu S., Morimura H. & Yano M., 2005. Novel surface structure and its fabrication process for MEMS fingerprint sensor. *IEEE Transactions on Electron Devices*. 52, 1026-1032.

Schenk M., Amrein M. & Reichelt R., 1996. An electret microphone as a force sensor for combined scanning probe microscopy. *Ultramicroscopy*. 65, 109-118.

Schwartz R.W., Ballato J. & Haertling G.H., 2004. Piezoelectric and electro- optic ceramics. Ceramic Materials for Electronics. Dekker. New York, 207-315.

Sessler G.M. & West J.E., 1962. Self-biased condenser microphone with high capacitance. *Journal of the Acoustical Society of America*. 34, 1787-1788.

Shaw A.J., Davis B.A., Collins M.J. & Carney L.G., 2009. A technique to measure eyelid pressure using piezoresistive sensors. *IEEE Transactions on Biomedical Engineering*. 56, 2512-2517.

Shung K.K., Smith M.B. & Tsui B.W.M., 1992. *Principles of Medical Imaging*. Academic Press, San Diego, 134-135.

Tadigadapa S. & Mateti K., 2009. Piezoelectric MEMS sensors: state-of-the-art and perspectives. *Measurement Science & Technology*. 20, 092001.

Valma J.R. & Kerry G.B., 2001. A review of therapeutic ultrasound: effectiveness studies. *Physical Therapy*. 81, 1339-1350.

van Putten A.F.P., van Putten M.H.P.M., Eichler R., Dankwart F., Pellet C., Laville C., Puers B., De Bruijker D., Correia C.M.B.A., Morgado M., Rien ten Wolde Berends, J., van Putten P.F.A.M. & van Putten M.J.A.M., 2000. Multiple-sensor micro-system for pulmonary function diagnostics for COPD and asthma patients. *1st Annual International IEEE-EMBS Special Topic Confercnce on Microtechnoloyies in Medicine & Biology*. 2000, 567-571.

Wang J., Gulari M. & Wise K.D., 2005. An integrated position-sensing system for a mems-based cochlear implant. *Electron Devices Meeting, IEDM Technical Digest, IEEE International*. 121-124.

Wang P. & Ye X., 2003. *Modern Biomedical Sensing Technology* (in Chinese). Zhejiang University Press, Hangzhou, China, 181-196.

Xu Y., 1991. *Ferroelectric Materials and Their Applications*, Amsterdam: North-Holland, 61-63, 226-229.

Yuu O., Mohamed D., Kobayashi M. & Jen C.K., 2008. Piezoelectric membrane sensor and technique for breathing monitoring. *Ultrasonics Symposium*. 795-798.

Zhang Z. & Wu T., 2003. The techniques of non-invasive blood pressure measurement and its development. *Chinese Journal of Medical Instrumentation*. 27(3).

Zhong L., Yang Z. & Fan Q., 2009. Design and analysis of biominetic whisker sensor. *Transducer & Microsystem Technologies*. 28, 80-83.

Chapter 4

Chemical Sensors and Measurement

Chemical sensors have been widely used in the biomedical field. With the rapid development of microelectronics and microprocessing technology, chemical sensors have grown to be more and more miniaturized and integrated. Combined with new information processing technology, intelligent chemical sensor arrays such as e-Nose and e-Tongue have been developed. Meanwhile, microfluidic chips enable continuous monitoring of chemical substances in living organisms.

4.1 Introduction

This chapter introduces the principles and characteristics of some typical chemical sensors including ion sensors, gas sensors and humidity sensors. Furthermore, e-Nose, e-Tongue, microfluidic chips and wireless sensor networks are also presented.

4.1.1 History

The history of chemical sensors can be traced back to 1906. Cremer, the pioneer of chemical sensors, discovered the phenomenon that glass thin films could react with hydrogen ions in solution and invented the glass electrode for measuring pH which promotes the development of chemical sensors. With further studies, glass-based thin-film pH sensors entered a practical stage in 1930. However, the research on chemical sensors progressed slowly before the 1960s and only a study on humidity sensors employing lithium chloride was reported in 1938.

Since the 1960s, numerous phenomena, like the ion-selective response of the silver halide film and the selective response of zinc oxide to flammable gases, have been discovered. Along with the application of new materials and principles, the research on chemical sensors has entered a new era and developed very rapidly. Chemical sensors such as gas sensors, ion sensors and humidity sensors obtained the preliminary application especially electrochemical sensors and ion-selective electrodes.

In the late 1980s, with the measuring methods and fabrication technologies of chemical sensors constantly expanded by microelectronic technology, chemical sensors based on optical, thermal and mass signals were fully developed. This enriched the research contents of chemical sensors including electrochemical sensors, optical chemical sensors, mass sensors, and thermochemical sensors. So electrochemical sensors lost their edge, and the modern history of chemical sensors began.

Chemical sensors with advantages of high selectivity, high sensitivity, fast response, wide measuring range, etc., have caught people's attention and have served in many different fields such as environmental protection and monitoring, industrial and agricultural production, food testing, weather forecast, health care, and diagnosis of diseases (Fig. 4.1.1). They have already become one of the main development trends in contemporary analytical chemistry.

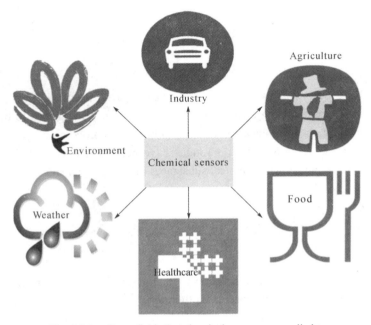

Fig. 4.1.1. Some fields that chemical sensors are applied to

The 1st International Meeting on Chemical Sensors was held in Fukuoka, Japan in 1983. Subsequently, it has been held every two years since the 3rd meeting in 1990 and there have been a total of 13 sessions to the present day. The meetings were held in various locations in Europe, the United States, Japan, China and other areas in the world. At the same time, some other international academic conferences associated with chemical sensors such as Biosensors, Eurosensors and the Asia Conference on Chemical Sensors were held one after another. Chemical sensors also played an important role in the Pure and Applied Chemistry International Conference. All of these show that the research and development of chemical sensors are very active and eye-catching throughout the world.

Along with the rapid development of modern science and technology and the mutual penetration between the disciplines, basic research on chemical sensors has become more and more active. The emergence of new technologies such as microprocessing, molecular imprinting, functional membrane, pattern recognition, micromachining, enable chemical sensors to be functioned, arrayed and integrated with neural network and pattern recognition chips. Chemical sensor networks' great vitality improved the measurement performance and remote testing capabilities of chemical sensor. In a word, chemical sensors will become more miniaturized, integrated, multifunctional, intelligent and network capable in the future.

4.1.2 Definition and Principle

The definition of chemical sensors by Wolfbeis (Wolfbeis, 1990) in 1990 is as follows:

Chemical sensors are small-sized devices comprising a recognition element, a transduction element, and a signal processor capable of continuously and reversibly reporting a chemical concentration.

The description above is pragmatic while the definition by the IUPAC (International Union of Pure and Applied Chemistry) in 1991 is general:

A chemical sensor is a device that transforms chemical information, ranging from concentration of a specific sample component to total composition analysis, into an analytically useful signal.

As a kind of analytical device, chemical sensors are so effective that they can detect the object molecules in the presence of interfering substances. This sensing principle is shown in Fig. 4.1.2.

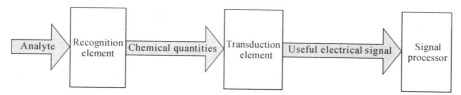

Fig. 4.1.2. Sensing principle of chemical sensors

4.1.3 Classification and Characteristics

There are millions of chemical substances with different compositions and properties existing in the natural world. A certain chemical substances can be detected by one or more kinds of chemical sensors, and one sensor may detect several substances. So the species of chemical sensors are multitudinous and the classification of chemical sensors has been accomplished in several different ways. Following the principles of signal transduction was made by IUPAC in 1991, chemical sensors are classified into ion sensors, gas sensors, and humidity sensors according to the property of analytes (see Fig. 4.1.3). In combination with computer information processing technology, intelligent chemical sensor arrays like electronic nose (e-Nose), electronic tongue (e-Tongue) were developed with characteristic of miniaturization, integration and portability, micro total analysis system (μTAS) has emerged in the last several decades.

The characteristics of chemical sensors, as follows, are generally accepted. Chemical sensors should:
- Transform chemical quantities into electrical (or optical) signals;
- Respond rapidly;
- Maintain their activity over a long time period;
- Be small;
- Be cheap;
- Be specific, i.e., they should respond exclusively to one analyte, or at least be selective to a group of analytes.

The above list could be extended with, e.g., the postulation of a low detection limit, or a high sensitivity. These mean that low concentration values should be detected (Gründler, 2006).

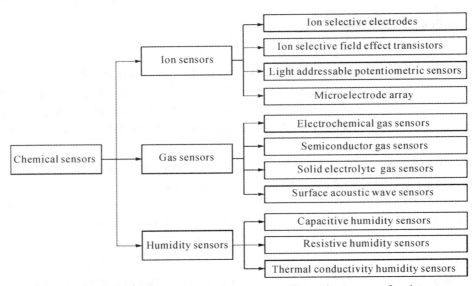

Fig. 4.1.3. Classification of chemical sensors according to the property of analytes

4.2 Electrochemical Fundamental

There are three kinds of electrochemical measurement systems which are primary cells, electrolytic cells and conductivity cells. The electrochemical measurement is related to the interactions between miscellaneous molecules and ions in the solution, which are introduced in detail as the following sections.

4.2.1 Measurement System

Electrochemical sensor is an important category in the field of chemical sensor. It reflects the electrochemical reactions on the electrode-electrolyte interface, which convert chemical signals into electrical signals. The basic electrochemical measurement system consists of the electrolyte, electrodes and a measurement circuit, where the electrolyte is the analyte and ion conductor (Fig. 4.2.1).

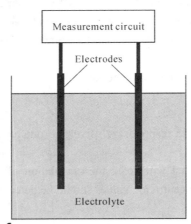

Fig. 4.2.1. Basic measurement system of electrochemical sensor

An electrode is a sensitive device made of metal or other materials. At the electrode-electrolyte interface, charges transfer by the movements of ions and make the measurement circuit working. These processes are regarded as energy conversion. As the ion concentration is converted into electric signal, electrochemical process is formed.

Then we are going to introduce three kinds of electrochemical measurement systems.

Measurement circuit of the primary cell. As shown in Fig. 4.2.2, when K switches to the standard battery E_s and the potentiometer is at the position of D, modulate R to make $U_{AD}=E_s$. Thus the galvanometer will point to zero under this condition. While K switches to the electrodes, E_x will not equal to E_s due to the chemical reaction. Thus by moving the potentiometer from D to D' to make galvanometer point to zero again, E_x can be expressed as follows:

$$E_x = AD' \cdot E_s / (AD) \tag{4.2.1}$$

Because the resistance between electrodes is pretty high, a high input impedance galvanometer or other impedance circuit is indispensable.

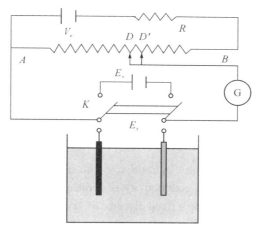

Fig. 4.2.2. Basic measurement system of the primary cell

Measurement circuit of the electrolytic cell. If the chemical pool is an electrolytic cell, chemical reaction will result from external power. Fig. 4.2.3 shows this process. V_e is the power supply. Then oxidation occurs on anode is called oxidizing-electrode while reduction occurs on cathode is called reducing electrode.

Measurement circuit of the conductivity cell. The electrolyte is an ion conductor, thus ion concentration can be detected by measuring the conductance. Fig. 4.2.4 shows the basic measurement system. It is a Wheatstone bridge with two electrodes in the electrolyte. The impedance Z_x between the two electrodes is one arm of the circuit bridge. In order to eliminate polarization, the power supply is AC. By modulating the potentiometer R to make the output of bridge to be zero, the position of B can express the output.

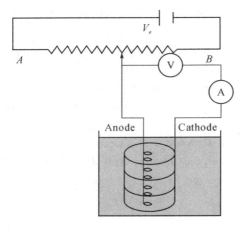

Fig. 4.2.3. Basic measurement system of the electrolytic cell

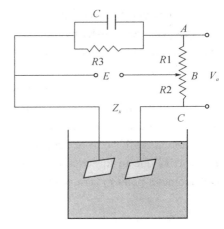

Fig. 4.2.4. Basic measurement system of the conductivity cell

4.2.2 Basic Conception

The electrolyte refers to aqueous solution consists of acid, alkali or salt which will lead to the generation of cations and anions in solution. The main properties of electrolyte are as followings.

4.2.2.1 Solution conductivity

Solution conductance is relevant to the electrode size, the distance between the electrodes, and the electrolyte concentration. The solution conductivity is defined as the conductance of 1mol electrolyte between two flat electrodes separated by a fixed distance of 1cm as the following equation:

$$\lambda = V k \qquad (4.2.2)$$

Here V is the solution volume which contains 1 mol solute. If c is the molarity of solute in 1000 mL solution, the solution volume containing 1 mol solute is:

$$V = \frac{1000}{c} \qquad (4.2.3)$$

Integrating Eq. (4.2.2) and Eq. (4.2.3) we get conductance G:

$$G = k\frac{A}{l} = \frac{\lambda}{V}\frac{A}{l} = \frac{\lambda A}{1000 l} c = Kc \qquad (4.2.4)$$

where A/l is conductivity cell constant, k is the conductivity, K is the coefficient.

Fig. 4.2.5 shows that electrolytes can be categorized as strong electrolytes and weak electrolytes according to the conductivity. From the curves we can see the conductivity increases with the increase of solution concentration in the beginning and decreases with the concentration when it exceeds a certain value. It is because the gravitation between cations and anions raises with the increase of solution concentration and leads to the restraint of ion movement. Nevertheless this phenomenon is not significant for weak electrolytes.

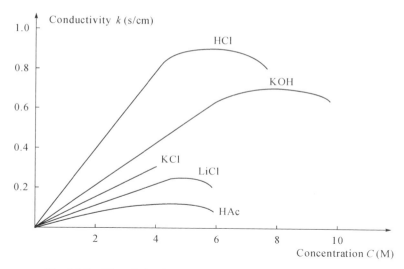

Fig. 4.2.5. The relationship between conductivity and concentration

4.2.2.2 *Ionization constant*

In weak electrolytes, there is a dynamic equilibrium between non-ionized molecule and ion, which is called ionization equilibrium. Ionicity is defined to express the ionization degree of equilibrium. It is the ratio of ionized molecule to total solution molecule, which is marked as α. Ionization constant K is the parameter to show the ionization capability of electrolyte.

If electrolyte $[BA]$ ionized as A^- and B^+ in solution. In equilibrium:

$$BA \rightleftarrows B^+ + A^- \tag{4.2.5}$$

Then the ionization constant is:

$$K = \frac{[A^-][B^+]}{[BA]} \tag{4.2.6}$$

where $[BA]$ represents the concentration of BA. The relationship between the ionization constant K and α is:

$$K = \frac{(c\alpha)^2}{c(1-\alpha)} \tag{4.2.7}$$

where c is the molarity of $[BA]$.

4.2.2.3 *Activity and activity coefficient*

There are interactions between ions and solvent molecules in electrolyte solution. Only concentration itself can hardly represent the "effective concentration". So activity is introduced to represent this conception. When solute is diluted infinitely, ion activity equals its concentration.

The relationship between activity α and concentration c is as follows:

$$\alpha = \gamma c \tag{4.2.8}$$

where γ is activity coefficient which represents the deviation of the electrolyte concentration and the effective concentration. Infinite activity is equal to the solution concentration only when solution is diluted, that is $\gamma \leq 1$.

4.2.2.4 The generation of electrode potential

Interface reaction. When an electrode is submerged in the electrolyte, there will be chemical reactions at the interface of electrode and electrolyte. If the chemical potential of zinc in electrode is larger than that in electrolyte, the zinc ion in electrode will depart the crystal lattice into the solution. Thus double electrode layer is formed and generates potential which obstruct the motion of zinc ions. In the equilibrium there will be a pretty narrow potential restrict on the interface and this potential is called equilibrium electrode potential. As Fig. 4.2.6 represents, the electronic interface migration is tiny enough to be ignored. But for inert electrode, the ion-exchange reaction is notable. The magnitude and polarity of this potential is depended on the category of electrode and the concentration of metal ions.

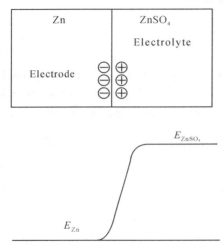

Fig. 4.2.6. Relationship between electrode and interface potential

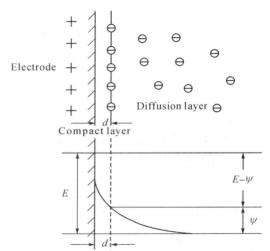

Fig. 4.2.7. Potential distribution of electrode and interface potential

Interface potential distribution. Due to the interaction of ion thermal motion and static electricity, if the electrode is a good conductor, the leftover charge will attach closely to the interface of electrode and solution. As Fig. 4.2.7 represents double electrode layer is formed. It contains two parts. One is attached to electrode (compact layer d), its thickness is the ionic radius of the hydration layer. Its potential is $E-\psi$ as the figure shows. The other is diffusion layer of potential ψ. It changes exponentially.

The magnitude and polarity of ψ have eminent effects on the electrode reactions. It depends on the solution character, ion concentration, and active substance adhesion at the interface.

4.2.2.5 Electrode potential and Nernst equation

The electrode potential cannot be measured by a single electrode. A reference electrode must be introduced. If the reference electrode potential is defined to be zero, thus the measured electrode potential can be determined. When the activities of all the reactants equal 1 and the partial pressure is 1

Chapter 4 Chemical Sensors and Measurement

atmosphere if there is any gas, the potential vs. the standard hydrogen electrode at 25 ℃ is the standard electrode potential E^0.

However, in practice, they are not always at the standard state. With the variety of ion activities, electrode potential will deviate from the standard state. Therefore we need the Nernst equation to calculate the electrode potential. If the electrode reaction is reversible:

$$\text{Oxidation state} + Ze \rightleftarrows \text{Reduced state} \quad (4.2.9)$$

So the electrode potential expressed by the Nernst equation is:

$$E = E^0 + \frac{RT}{ZF} \ln \frac{\alpha_{\text{oxidation state}}}{\alpha_{\text{reduced state}}} \quad (4.2.10)$$

where E^0 is standard potential relative to the standard hydrogen electrode. R is the universal gas constant. F is the Faraday constant. Z is the number of transferred electrons. T is the absolute temperature.

Electrochemical battery electromotive force. Fig. 4.2.8 shows the Daniel cell. The left is zinc electrode (negative) while the right is copper electrode (positive). The whole electrode reaction is:

$$Zn + Cu^{2+} \rightleftarrows Zn^{2+} + Cu \quad (4.2.11)$$

The reaction can be expressed as:

$$-Zn \mid ZnSO_4 \parallel CuSO_4 \mid Cu+ \quad (4.2.12)$$

here "|" represents the boundary of two phase, and "||" represents the salt bridge, so the potential of cell is:

$$E = E_1 + E_{lj} - E_2 \quad (4.2.13)$$

here E_1 is the potential of the copper electrode. E_2 is the potential of the zinc electrode. E_{lj} is liquid junction potential.

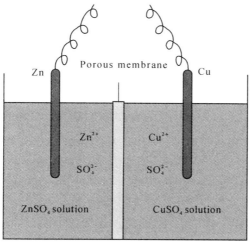

Fig. 4.2.8.   Structure of Daniel cell

The function of porous membrane is to isolate the two solutions whereas ions can go through it. If the mobilities of the ions in both sides are the same, the liquid junction potential will be zero.

$$E = E_1 - E_2 = (E_{Cu}^0 + \frac{RT}{ZF} \ln \frac{\alpha_{Cu^{2+}}}{\alpha_{Cu}}) - (E_{Zn}^0 + \frac{RT}{ZF} \ln \frac{\alpha_{Zn^{2+}}}{\alpha_{Zn}}) \quad (4.2.14)$$

145

For general chemical reactions:

$$aA + bB \rightarrow cC + dD \quad (4.2.15)$$

$$E = E^0 - \frac{RT}{ZF} \ln \frac{\alpha_c \alpha_d}{\alpha_a \alpha_b} \quad (4.2.16)$$

4.2.2.6 Liquid junction potentials and salt bridge

Liquid junction potential is generated by ion diffusing from one side of the porous membrane to the other. Different ions have distinct diffusion speed and an electrical double lager of positive and negative charges will be produced at the junction of the two solutions. The potential difference at the junction caused by the ionic transfer is called liquid junction potential.

The major causes of the liquid junction potential are as follows.

Same solution with different concentrations. As Fig. 4.2.9 shows, diffusion from high concentration to low concentration is due to the difference of hydrogen ion and chloride mobility. The hydrogen ion with higher speed will be accumulated more on the right side. Thus the right side possesses a positive potential positive while the left is negative. When this phenomenon appears, the transfer speed of hydrogen ion will slow down while that of chloride will speed up. As a result a balance is obtained. The liquid junction potential can be described as follows:

$$E = -\frac{RT}{ZF} \cdot \frac{U_+ + U_-}{U_+ - U_-} \ln \frac{C_1}{C_2} \quad (4.2.17)$$

where C_1, C_2 represents the two concentrations, U_+ and U_- is the ionic mobility of cations and anions.

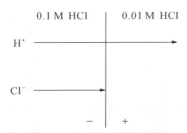

Fig. 4.2.9. Schematic diagram of ion diffusion at the liquid junction interface (category 1)

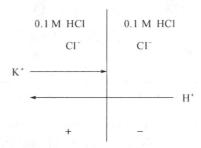

Fig. 4.2.10. Schematic diagram of ion diffusion at the liquid junction interface (category 2)

Same concentration with different electrolytes. As Fig. 4.2.10 shows the condition of same concentration with different electrolytes except a common ion. Due to the similar electrolyte concentration of both sides, chloride is regarded motionless while hydrogen ions diffuse to the left and potassium ions diffuse to the right. The speed of hydrogen is larger than that of potassium. So the left obtains positive charge while the right obtains negative ones.

To eliminate the liquid junction potential, salt bridge is inserted between the two solutions as shown in Fig. 4.2.11. The solution in salt bridge normally is saturated potassium chloride. Due to the high concentration, ions in the salt bridge will diffuse to both sides. And there is no much difference between the speeds of potassium and chloride ions, thus the liquid junction potential is pretty small, which is only several millivolts. The requirements of the solution in salt bridge are high concentration, the speeds of anode and cathode are similar, and it has no chemical reaction with the electrolyte.

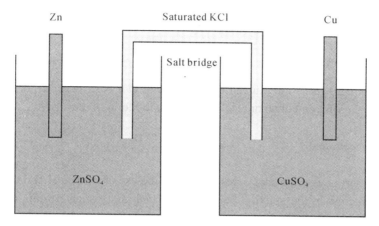

Fig. 4.2.11. Schematic diagram of salt bridge

4.2.3 Classification of Electrodes

According to the role of electrodes playing in the electrochemical measurement, they can be classified as indicator electrode, working electrode, reference electrode, and auxiliary electrode.

4.2.3.1 Indicator electrode

Indicator electrode serves as a transducer responding to the excitation signal and to the composition of the solution being investigated, but that does not affect an appreciable change of bulk composition within the ordinary duration of a measurement. The most commonly used indicator electrode includes ion-selective electrode, gas-sensing electrode, and bioelectrode.

The simplest type of direct indicator electrode is a metal, M, in contact with a solution containing its own cation, M^+. At the metal-solution interface, a potential develops that is proportional to the activity of the metal ion in solution. The potential can be measured directly with respect to a reference electrode using the simple arrangement shown in Fig. 4.2.12. The voltage measured is simply the difference between the potential at each electrode:

$$E_{cell} = E_{indicator} - E_{reference} \qquad (4.2.18)$$

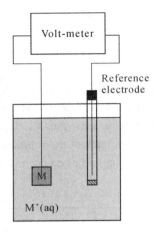

Fig. 4.2.12. Electrochemical cell for a potentiometric measurement with a metallic indicator electrode

Inert metal electrodes like Pt or Au can be used as indicator electrodes for ions involved in redox reactions that occur in solution but do not include the metallic form of the analyte.

4.2.3.2 Working electrode

Working electrode serves as a transducer responding to the excitation signal and the concentration of the substance of interest in the solution being investigated, and which permits the flow of current sufficiently large to effect appreciable changes of bulk composition within the ordinary duration of a measurement. The working electrode represents the most important component of an electrochemical cell. It is at the interface between the working electrode and the solution that electron transfers of greatest interest occur. The most commonly used working electrode includes hanging mercury electrode, mercury thin-film electrode, glassy carbon electrode, graphite electrode and metal electrode such as gold electrode, platinum electrode and copper electrode.

4.2.3.3 Reference electrode

A reference electrode is used in measuring the working electrode potential of an electrochemical cell. It should be constructed using half-cell components that are stable over time and with changing temperature, present at well-defined values of activity and should possess fixed, reproducible electrode potentials. The most common reference electrode is standard hydrogen electrode, which compose of an inert solid like platinum on which hydrogen gas is adsorbed, immersed in a solution containing hydrogen ions at unit activity.

Practical application of the standard hydrogen electrode is limited by the difficulties in preparing solutions containing H^+ at unit activity and maintaining unit activity for $H_2 (g)$ in the half-cell. Most experiments carried out in aqueous solutions utilize one of two other common reference half-cells: saturated calomel electrode or silver-silver chloride electrode (Ag/AgCl) (Fig. 4.2.13).

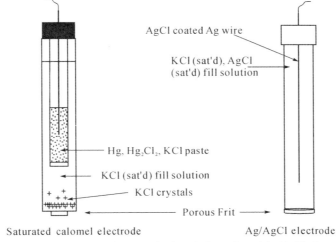

Fig. 4.2.13. Schematic of saturated calomel electrode and Ag/AgCl electrode

4.2.3.4 Auxiliary electrode (Counter electrode)

The purpose of the auxiliary electrode is to provide a pathway for current to flow in the electrochemical cell without passing significant current through the reference electrode. There are no specific material requirements for the electrode beyond it not adversely influencing reactions occurring at the working electrode. Remember that if a reduction occurs at the working electrode, there must be an oxidation that takes place at the auxiliary electrode. Care should be taken that electrode products formed at the auxiliary electrode do not interfere with the working electrode reaction. The auxiliary electrode can be physically separated from the working electrode compartment using a fritted tube, but one should be aware that under certain circumstances this can have a deleterious effect.

The most commonly used material for the auxiliary electrode is platinum, due to its inertness and the speed with which most electrode reactions occur at its surface. Other less expensive materials may also be used as auxiliary electrodes. Choices include carbon, copper, or stainless steel if corrosion is not an issue for a particular electrolyte solution or reaction.

4.3 Ion Sensors

There are several sensors that can be used in the determination of ions such as ion-selective electrode sensors (ISE), ion-selective field-effect transistor sensors (ISFET), light addressable potentiometric sensors (LAPS) and microelectrode array sensors (MEA). We will describe each of these ion sensors in detail in the following sections.

4.3.1 Ion-Selective Electrodes

An ion-selective electrode (ISE) is defined as an electroanalytical sensor with a membrane whose potential indicates the activity of the ion to be determined (the determinand) in a solution (the analyte). The membranes of ISEs consist either of liquid electrolyte solutions or of solid or glassy electrolytes that usually have negligible electron conductivity under the conditions of measurement. ISEs have certain undoubted advantages: (1) they do not affect the test solution; (2) they are portable; (3) they are suitable for direct determinations and can be used as sensors for titrations; and (4) they are not expensive (Koryta, 1986).

4.3.1.1 Principle

In potentiometry with ISEs, the electromotive force (EMF) of the cell with general type is measured (Fig. 4.3.1 and Fig. 4.3.2).

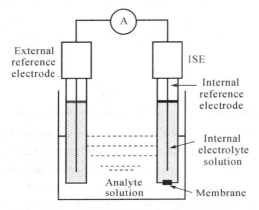

Fig. 4.3.1. The schematic illustration of an ISE measurement

| External reference electrode | External electrolyte | Analyte solution | Membrane | Internal electrolyte | Internal reference electrolyte |

Fig. 4.3.2. Schematic diagram of reference electrode, solution and membrane

The analyte contains the ion to be determined. The actual ISE system consists of internal reference electrode, internal electrolyte solution, the analyte solution, and the membrane. The liquid junction, denoted by double vertical bars, between the electrolyte solutions, the reference electrodes and the analyte solution, respectively, can unfavorably influence the EMF, as the liquid junction potential (a kind of diffusion potential), $\Delta\varphi_L$, formed at these phase boundaries depends on the concentrations and the mobilities of the ions present in the solutions in contact. The liquid junction potential can be nearly eliminated if: (1) the solutions in the reference electrodes contain a salt with high concentrations which has approximately equal cation and anion mobilities; (2) the solutions of the reference electrodes are separated from the analyte solution, respectively, by a salt bridge; (3) both the solution of the reference electrode and the solution in contact with an identical excess (indifferent) electrolyte.

If the liquid junction potentials are eliminated, the EMF of the cell is:

$$E = E_2 - \Delta\varphi_M - E_1 \tag{4.3.1}$$

where E_1 and E_2 are the electrode potentials of the reference electrodes and $\Delta\varphi_M$ is the membrane potential. If only the determinand ion x with charge number z influences the membrane potential, its value is:

$$\Delta\varphi_M = \left(\frac{RT}{zF}\right)\ln\left[\frac{\alpha_x}{\alpha_{x0}}\right] \tag{4.3.2}$$

where R is the gas constant, T is the absolute temperature, F is the Faraday constant, and α_x and α_{x0} are the activities of x in the analyte solution and the internal reference solution, respectively. The ISE potential then is:

$$E_{ISE} = E_{0,ISE} + \left(\frac{RT}{zF}\right)\ln\left(\frac{\alpha_x}{\alpha_{x0}}\right) \tag{4.3.3}$$

where $E_{0,ISE}$ is the "standard" ISE potential.

This is the case when only the determinand ion interacts with the membrane without changing its composition. An interferent there is another type of ion present in the analyte that can also interact with the membrane by an exchange reaction with the determinand present in the membrane. When both the determinand x and the interferent K have the same charge number z, this behavior is described by the Nikolsky equation:

$$E_{ISE} = E_{0,ISE} + \left(\frac{RT}{zF}\right)\ln[\alpha_x + K_{x,K}^{pot}\alpha_k] \tag{4.3.4}$$

where $K_{x,K}^{pot}$ is the selectivity coefficient for the ion K with respect to the ion x. In fixed site (solid or glassy) membranes, the selectivity coefficient is a function of the equilibrium constant of the exchange reaction or of the ratio of solubility products. In mobile site (liquid) membranes, $K_{x,K}^{pot}$ depends on the ratio of partition coefficients and on ion-pairing constants or on the ratio of stability constants of complexes formed in the membrane phase.

4.3.1.2 Characterization

The properties of an ISE are characterized by parameters like:

- **Detection limit**

According to the IUPAC recommendation, the detection limit is defined by the cross-section of the two extrapolated linear parts on the ion-selective calibration curve. As shown in Fig. 4.3.3, when the ionic activity of sample solution gets smaller, the linear part of the calibration curve CD gradually bends into another linear part EF. The detection limit is the ionic activity A corresponding to the potential where CD and EF intersect.

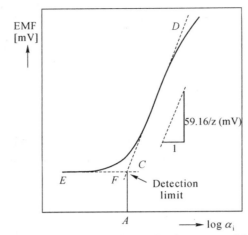

Fig. 4.3.3. The schematic illustration of the calibration curve and the detection limit

- *Impedance*

The resistance of an ISE is determined by the electrode materials, for example the resistance of the glass membrane electrode is several hundred megohm while it is only a few kilo-ohm for the crystal membrane electrode.

In practice, we usually use the resistance of an ISE to describe the impedance of the electrolytic cell that consists of ISE, sample solution and reference solution. We can calculate the resistance of the ISE by measuring the potential difference E_x of the electrolytic cell first, and then the potential V of a resistance R_e paralleled with the cell is obtained, so the resistance of the cell is:

$$R_x = \frac{E_x - V}{V} R_e \qquad (4.3.5)$$

- *Response time*

In earlier IUPAC recommendations, it was defined as the time between the instant at which the ISE and a reference electrode are dipped in the sample solution (or the time at which the ion concentration in a solution is changed on contact with ISE and a reference electrode) and the first instant at which the potential of the cell becomes equal to its steady-state value within 1 mV or has reached 90% of the final value (in certain cases also 63% or 95%). Usually the response time is less than 1 s, or even only a few milliseconds.

- *Temperature coefficient, drift, and lifetime*

If $E_{ISE} = E_{0,ISE} + (RT/(zF)) \ln \alpha_x$ describes the E_{ISE} behavior, then the influence of temperature is mainly reflected in the change of the Nernst coefficient, $RT/(zF)$. The temperature effect on the liquid-junction potential, which is never completely eliminated, and on the activity coefficient of the determinand also cannot be neglected. If, however, E_{ISE} is also influenced by the interferent, temperature coefficients of all factors contributing to the selectivity coefficient must be accounted for.

Both the drift of E_{ISE} and the lifetime of ISE mainly depend on dissolution of membrane components in the analyte and on the changes in the surface structure of the membrane. The lifetime of an accurate liquid-membrane ISE can exceed one year, while the shelf life of an all-solid-state ISE is practically unlimited.

4.3.1.3 Applications

Among various classes of chemical sensors, ISEs are one of the most frequently used potentiometric sensors during laboratory analysis as well as in industry, process control, physiological measurements, and environmental monitoring. The most commonly used ISE is the pH glass electrode, which contains a thin glass membrane that responds to the H^+ concentration in a solution. Other ions that can be measured include fluoride, bromide, cadmium and gases in solutions such as ammonia, carbon dioxide and nitrogen oxide.

As shown in Fig. 4.3.4, a typical commercial electrode is made of a glass tube ended with small glass bubble. Inside the electrode is usually filled with a buffered solution of chlorides (for pH probe is usually 0.1 mol/L HCl) in which silver wire covered with silver chloride is immersed. The active part of the electrode is the glass bubble with a typical wall thickness of 0.05 – 0.2 mm. When the glass membrane is exposed to the solution, a thick hydrated layer is formed (5 – 100 nm), which exhibits improved mobility of the ions.

Besides, the glass electrodes can also be applied to the detection of sodium, potassium and ammonium ions that depends mainly on the component of the glass materials. The normal glass membrane is composed of $Na_2O/Al_2O_3/SiO_2$, and the selectivity for different ions is available while the proportion of these three components changes.

Nowadays intracellular environmental monitoring has been given increasing attention. It can be classified to monitoring of ions (Ca^{2+}, H^+, K^+, Na^+, etc.), small molecules (O_2, CO_2, NH_3, etc.) and a variety of macromolecules. Apparently, Ca^{2+} is a regulator of physiological functions which plays an important role in the nerve conduction, muscle contraction and second messenger regulation. So it is crucial to monitor the calcium ion.

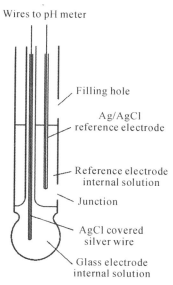

Fig. 4.3.4. Schematics of typical combination glass electrode, which is made of a glass tube ended with small glass bubble

Ion selective microelectrodes can be applied to monitor the intracellular calcium ion, for example the transient releasing of extracellular Ca^{2+} stimulated by light in cardiac myocytes can be measured by microelectrodes. As the ion-selective microelectrode shown in Fig. 4.3.5, the diameter of the tip is less than 1 μm, and a liquid calcium ionophore (ETH129) is utilized as the electrode- sensitive material.

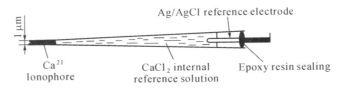

Fig. 4.3.5. The structure chart of calcium ion-selective microelectrode

The microelectrode must be calibrated before and after use. The calibration device is shown in Fig. 4.3.6, and it is carried out in a solution and the pCa of the standard solution is 2 – 7.

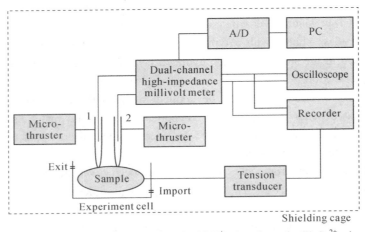

Fig. 4.3.6. Experimental setup of microelectrode: (1) K^+ microelectrode; (2) Ca^{2+} microelectrode

The myocardium whose diameter was 0.3 – 0.4 mm and whose length was about 0.5 mm used in the experiment was obtained from living frogs and stored in a none-calcium solution. First, the myocardium was moved into the physiological cell as shown in Fig. 4.3.6. Then K^+ and Ca^{2+} microelectrodes were inserted into the myocardium using a micro-thruster. The signals of the two microelectrodes obtained by a high-impedance millivolt meter were shown through an oscilloscope. An electrical pulse was used to stimulate the myocardium whose action potential signals and tension changes were recorded. At last, ultraviolet light pulse (wavelength 350 nm, pulse width 100 μs, energy about 100 mJ) was added to the back of the myocardium.

In order to investigate the effect of Ca^{2+} on the myocardial action potential, DM-nitro-phenol calcium was added into the solution which can release Ca^{2+} under the light pulse. Fig. 4.3.7 shows the results of this experiment, which briefly demonstrates the effect of extracellular Ca^{2+} on the cardiac myocytes calcium channel.

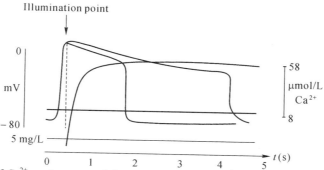

Fig. 4.3.7. The effect of Ca^{2+} on the myocardial action potential: (1) Ca^{2+} concentration before illumination; (2) The increase of Ca^{2+} concentration after illumination; (3) The action potential before illumination; (4) The action potential after illumination

4.3.2 Ion-Selective Field-Effect Transistors

In 1970, Bergveld replaced the metal plate in an IGFET (insulated-gate field-effect transistor) with a glass electrode membrane and obtained the first ISFET (ion-selective field-effect transistor) (Dzyadevych et al., 2006). In this device, the drain current of the field-effect transistor, which is the measured quantity, depends on the field in the insulator (SiO_2 or Si_3N_4) separating the ion-selective membrane from the p-type silicon wafer of the transistor. So the field gate is a function of membrane potential. During the recent 40 years, ISFETs for determination of H^+, halide ions, K^+, Na^+, Mg^{2+}, Ag^+, Ca^{2+}, CN^- and other ions, have been reported.

4.3.2.1 Principles

Structurally, the ISFET is very similar to the IGFET, and a typical construction of an n-channel IGFET is shown in Fig. 4.3.8.

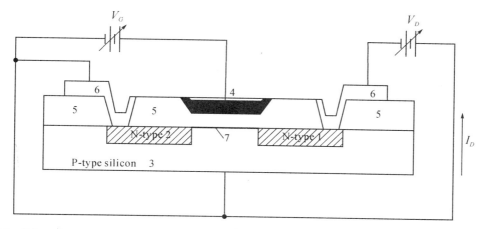

Fig. 4.3.8. Schematic diagram of an IGFET. (1) Drain; (2) Source; (3) Substrate; (4) Gate; (5) Insulator; (6) Metal contacts; (7) Conducting channel

It consists of a p-type silicon substrate with source and drain diffusions separated by a channel which is overlain by SiO₂ as insulator and a metal gate. The polarity and magnitude of the gate voltage difference (V_G) applied between the substrate and the gate are chosen so that an n-type inversion layer is formed in the channel between the source and drain regions. The magnitude of the drain current (I_D) is determined by the effective electrical resistance of the surface inversion layer and the voltage difference (V_D) between the source and the drain.

The ISFET differs from the IGFET in several respects: (1) the solution (analyte) is in direct contact with the gate insulator layer and a reference electrode in the solution replaces the metal gate; (2) silicon nitride, Si_3N_4, overlying the SiO_2 is used to provide a charge blocking interface and an improved pH response; (3) other membranes, such as poly containing valinomycin can be added to confer other ion selectivities to the ISFET; (4) the successful encapsulation of all regions of the device other than the gate region to be exposed to the analyte solutions is mandatory (Fig. 4.3.9).

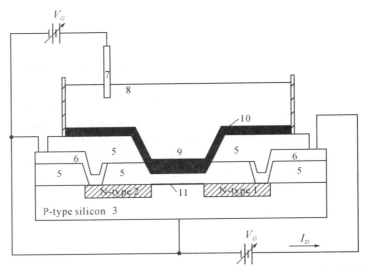

Fig. 4.3.9. Schematic diagram of a composite gate, dual dielectric ISFET. (1) Drain; (2) Source; (3) Substrate; (5) Insulator; (6) Metal contacts; (7) Reference electrode; (8) Solution; (9) Electroactive membrane; (10) Encapsulant; (11) Conducting channel

A diagram of the complete electrochemical system, together with the relevant electrical potentials (i.e. differences in inner potentials between the bulk phases), is shown in Fig. 4.3.10, where V_{ref} is the potential of the reference electrode, E_I is the potential difference between the sample solution and the membrane, V_G' is the potential of the depletion layer. By analogy with the IGFET gate voltage difference (V_G), it is possible to define an equivalent ISFET gate voltage difference (V_G^*) as the electrical potential difference between the bulk phases of the semiconductor and gate material:

$$V_G^* = E_I + V_{ref} + V_G' \tag{4.3.6}$$

$$E_I = E_{0,I} + \left(\frac{RT}{zF}\right)\ln\alpha_x \quad \text{or} \quad E_I = E_{0,I} + \left(\frac{RT}{zF}\right)\ln[\alpha_x + K_{x,K}^{pot}\alpha_k] \tag{4.3.7}$$

Therefore, the interface charge will alter while the ionic activity of the solution changes, leading to the variety of membrane potential. In theory, changes of pH value and redox potential can be measured by ISFETs.

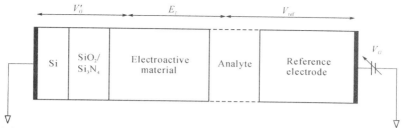

Fig. 4.3.10. Schematic diagram of composite gate, dual dielectric ISFET showing potential difference contributions

4.3.2.2 Characteristics

ISFETs are used to measure the ionic activity in the electrolyte solution with both electrochemical and transistor characteristics. Compared to the traditional ISE, they have the following advantages:
- High sensitivity, fast response time, high input impedance and low output impedance, with both impedance conversion and signal amplification functions which can be used to avoid interference from external sensors and secondary circuit.
- Small size, especially applicable for biodynamic monitoring.
- They are suitable for mass production and easy to be miniaturized and integrated by the integrated circuit technology and micro-processing technology.
- All solid-state structure makes the high mechanical strength available.
- Easy to realize on-line control and real-time monitoring.
- The sensitive materials can be conductive or insulated.

4.3.2.3 Ion sensitive membranes

The key component of an ISFET is the sensitive membrane which is primarily fabricated by insulating material. The first ISFET gate material utilized was silicon dioxide, obtained in the conventional MOSFET technology by heating silicon up to 1,100 °C in a dry oxygen atmosphere. Accompanied with this type of structure, there are many disadvantages such as poor insulativity, low sensitivity and bad linearity. Therefore, we often use a double-gate structure (double-layer or multi-layer), such as a redeposited layer of Al_2O_3 on the insulating layer to achieve a good response to pH values.

Solid film: This is a film with high ion selectivity. Sodium ion-sensitive film is formed by aluminosilicate or sodium silicate materials while potassium and calcium ions-sensitive film is fabricated by organic polymer membrane materials.

Liquid film: The polyvinyl chloride (PVC) film is commonly used by putting the ion activity solution and plasticizer together with the PVC to form a layer of liquid film.

4.3.2.4 Applications

Initially ISFETs were serving as new probes for electrophysiological experiments, but this challenge has not been taken up by the field. Recently, publications paid more attention on the monitoring of cell metabolism in which electrophysiological signals are not measured but are physiological. It mainly focused on the extracellular acidification rate of a cell culture. The pH in the cellular microenvironment (pH_M) is an important regulator of cell-to-cell and cell-to-host interactions. This is, for example, of particular importance in the field of tumor biology and intercellular signaling. The pH_M is reduced significantly in the interstitium of solid tumors in comparison to the values of normal interstitial fluid. Additionally, the extracellular acidification rate of a cell culture is an important indicator of global cellular metabolism (Lehmann et al., 2000).

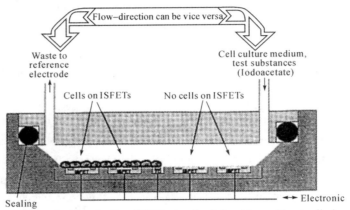

Fig. 4.3.11. Measurement setup showing the four ISFETs, two loaded and two without cells (reprinted from (Lehmann et al., 2000), Copyright 2000, with permission from Elsevier Science B.V.)

Lehmann et al. (2000) developed a method measuring the pH_M on line and in real time in the immediate vicinity (10 – 100 nm) of the cell plasma membranes. As shown in Fig. 4.3.11, in a flow through chamber, adherent tumor cells (LS174T) were cultured on specially developed pH-ISFET arrays to elucidate how the pH of cell-covered ISFETs differs from the pH of ISFETs in cell-free regions. The pH-sensitivity of the Al_2O_3-ISFETs is 56.119±2.12 mV/pH. The output signal of the ISFETs as the measure for the pH value is given by the source voltage V_{GS} relative to the reference potential. The perfusion rate of the cell culture medium was increasing between 1.3 and 4.3 mL/h in a stop and flow mode, and the effect of Triton X-100 on the pH of the cells was studied then.

As the results shown in Fig. 4.3.12, the pH of cell-covered ISFETs is lower than that of the ISFETs in cell-free regions, and immediately after Triton X-100 containing medium reached the cell culture, the sensor with the cells showed a characteristic acidification peak. The sensor without cells did not show that peak. The acidification peak of the cell-ISFET was followed by an increase of 29.294 mV relative to the constant pumping signal which is equivalent to a pH-increase of 0.529 pH-units.

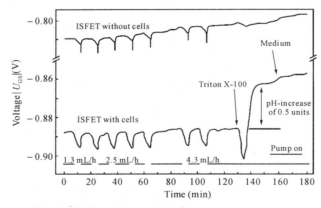

Fig. 4.3.12. The whole measurement showing the difference between cell-covered ISFETs and ISFETs in cell-free regions, and the effect of Triton X-100 addition (reprinted from (Lehmann et al., 2000), Copyright 2000, with permission from Elsevier Science B.V.)

4.3.3 Light Addressable Potentiometric Sensors

As a result of the development of semiconductors, the light addressable potentiometric sensor (LAPS) has gradually become the hot item in the late 1980s, and it's the most popular ion sensitive sensor at present.

4.3.3.1 Principle and fundamental

LAPS is a semiconductor based on the potential sensitive device that usually consists of the metal-insulator-semiconductor (MIS) or electrolyte-insulator-semiconductor (EIS) structure. LAPS with EIS structure is schematically shown in Fig. 4.3.13a. The LAPS consists of the heterostructure of $Si/SiO_2/Si_3N_4$. An external DC bias voltage is applied to the EIS structure to form accumulation, depletion and inversion layer at the interface of the insulator (SiO_2) and semiconductor (Si) in different working status. When a certain light pointer illuminates the LAPS chip, the semiconductor absorbs energy and leads to energy band transition (i.e. produces electron–hole pairs). Usually, electron and hole would compound soon and current is unable to be detected by peripheral circuit. If LAPS is biased in depletion, an internal electric filed exists across the depletion layer, an internal electric field appears across the depletion layer, and the width of the depletion layer is a function of the local value of the surface potential. When a modulated light pointer illuminates the bulk silicon, light induced charge carriers are separated by the internal electric field and thus photocurrent can be detected by the peripheral circuit. The amplitude of the photocurrent depends on the local surface potential. Therefore, by detecting the photocurrent of LAPS, localized surface potential can be obtained (Hafeman et al., 1988).

The characteristic sigmoid-shape current-voltage curve of n-type LAPS is shown in Fig. 4.3.13b. Three regions can be identified as the cutoff region, the working region, and the saturated region.

In the cutoff region, there is no photocurrent. In the working region, the photocurrent rises almost linearly with the decreasing voltage. The bias voltage is set to the point of inflection of this curve and kept constant during the long time detection process of LAPS. At this point the photocurrent is most sensitive to the change of the surface potential. In the saturated region, the photocurrent is saturated, which means that the change of bias voltage within this range will not chang the photocurrent. If the bias voltage applied is kept constant, external potential changes coupled to the bias voltage can be determined by detecting the change of the photocurrent.

By illuminating parts of the surface of the device with a modulated light-pointer, additional charge carriers are generated and ac-photocurrent flows. This photocurrent is due to a rearrangement of charge carriers in the depletion layer of the semiconductor. The arrangement of charge carriers is voltage dependent. If an additional potential is applied, the characteristic photocurrent-voltage curve of the LAPS shifts along the voltage axis. The most common functions of LAPS are detecting pH and extracellular potential signals.

For pH detection, a layer of Si_3N_4 is fabricated on the surface of LAPS. According to the site-binding theory (Siu et al., 1979; Bousse et al., 1982), a potential difference which is related to the concentration of H^+ in the electrolyte, forms at the insulator (Si_3N_4/SiO_2) and solution. This potential is coupled to the bias voltage applied to the sensor. A larger concentration of H^+ provides a larger value of this potential difference, causing the I-V curve to shift along the axis of bias voltage (Fig. 4.3.13b). When the bias voltage is kept constant in the middle region, change of the photocurrent indicates the pH change of the electrolyte.

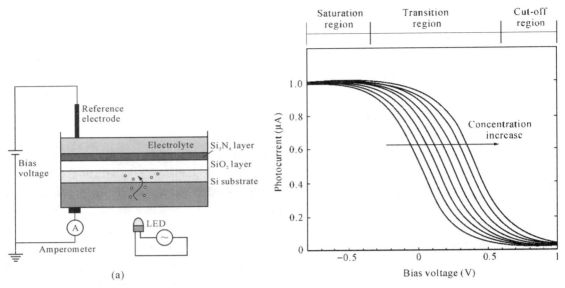

Fig. 4.3.13. (a) Working principle of the LAPS. (b) Characteristics *I-V* curve of n-type LAPS

4.3.3.2 Numeric analysis

Numeric analysis helps in further understanding the working principle of LAPS, as well as the optimization of the sensor and system. In this section, the electric circuit model of LAPS is introduced in a simple way. The equivalent circuit model of the interface of semiconductor presents a direct view of the way how LAPS works.

Modeling of LAPS focused mainly on the EIS system. Here we will briefly introduce the circuit model most commonly used, as shown in Fig. 4.3.14a.

Fig. 4.3.14a shows the qualitative view of charge distribution within the EIS system. Charge is mainly distributed in four region: charge in the semiconductor Q_s, the interfacial charge Q_o, charge in the electrolyte space-charge region Q_d, and counterion charge at the inner Helmoltz plane (IHP) Q_β, which is negligible here. Since the system is electric neuter, sum of the charge is zero.

$$Q_s + Q_o + Q_d = 0 \qquad (4.3.8)$$

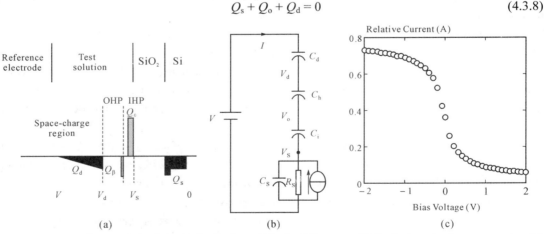

Fig. 4.3.14. (a) Qualitative view of charge distribution within the EIS system; (b) Equivalent circuit representation; (c) Simulation of *I-V* curve of LAPS

According to the so-called site-binding theory (Siu and Cobbold, 1979), and with reference to a silicon nitride (Si_3N_4) insulator, the interfacial charge is:

$$Q_0 = e\left(\frac{[H^+]_s^2 - K_+ K_-}{[H^+]_s^2 + K_+[H^+]_s + K_+ K_-}\right)N_{sil} + e\left(\frac{[H^+]_s}{[H^+]_s + K_{N_+}}\right)N_{nit} \quad (4.3.9)$$

where N_{sil} and N_{nit} are the silanol and amine binding site density, respectively; K_+, K_- and K_{N+} represent the dissociation constants of the surface chemical reactions.

The charge density in the electrolyte space-charge region can be estimated by the Gouy-Chapman-Stern theory.

$$Q_d = \sqrt{8\varepsilon_e \varepsilon_0 KTC_0} \sinh\left[\frac{e(V - V_d)}{2KT}\right] \quad (4.3.10)$$

Q_s can be estimated from the Piosson's equation:

$$Q_s = \pm\sqrt{2\varepsilon_e \varepsilon_0 KT}\sqrt{\left[p_0(e^{-\frac{eV_s}{KT}} + \frac{eV_s}{KT} - 1) + n_0(e^{\frac{eV_s}{KT}} - \frac{eV_s}{KT} - 1)\right]} \quad (4.3.11)$$

An equivalent circuit for the LAPS is shown in Fig. 4.3.14b. Where V is the applied bias voltage, C_d is the Gouy-Chapman layer capacitance, C_h is the Helmoltz layer capacitance, C_i is the insulator capacitance, C_s is the depletion region capacitance and R_s is the resistance that models the recombination process. From Fig. 4.3.14b, we can easily get these two equations:

$$V_d - V_o = \frac{Q_d}{C_h} \quad (4.3.12)$$

$$V_o - V_s = \frac{Q_o}{C_i} \quad (4.3.13)$$

In Eq. (4.3.9) – (4.3.10), C_i and C_h are determined from experiment, thus, we can get Q_d, Q_o, Q_s, V_d, V_s and V_o from these six equations. The relationship between the bias voltage and photo-induced current can be simulated, shown in Fig. 4.3.14c, which is consistent with *I-V* curve from the experiment, shown in Fig. 4.3.14b.

4.3.3.3 Applications

Since the introduction of LAPS in 1988, Hafeman et al. proposed a measurement device for biological applications, the first LAPS was mainly developed for biological investigations, e.g., a phospholipids bilayer membrane-based LAPS, a sandwich immunoassay for human chorionic gonadotropin (HCG) and an enzyme-based (urease) microchamber-LAPS device. Recently, concerns about the contamination of water by heavy metals such as Pb^{2+}, Cu^{2+}, Cd^{2+} and Hg^{2+} has been proposed because of the toxicity of such metals on a broad spectrum of organisms, including humans.

When LAPS is applied to the pH measurements, dissociation groups of SiN-OH are on the surface of the sensitive membrane Si_3N_4, which reacts with the H^+ ions to maintain the electrochemical dissociation equilibrium (Zhang et al., 1999). Therefore, a net charge exists in the sensitive membrane surface apart from point-of-zero-charge (PZC). The net charge will attract free ions with opposite charge in the solution to form electric double layer. The ionization equilibrium is as follows:

$$SiN\text{-}OH \rightleftarrows SiN\text{-}O^- + H_a^+ \quad (4.3.14)$$

$$\text{SiN-OH} + \text{H}_a^+ \rightleftarrows \text{SiN-OH}_2^+ \qquad (4.3.15)$$

where H_a^+ is the hydrogen ions located in the interface between the sensitive membrane and the solution, the relationship between the concentration of H_a^+ and the bulk H^+ obeys the Boltzmann rule:

$$[\text{H}_a^+] = [\text{H}^+] e^{\frac{qE}{kT}} \qquad (4.3.16)$$

where E is the interfacial potential between the sensitive membrane and the solution, q is the quantity of electric charge, k is the Boltzmann constant, and T is the absolute temperature.

Finally we can obtain the relationship between the pH value of the solution and the interfacial potential:

$$E = 2.303 \frac{kT}{q} ([\text{pH}_{pzc}] - [\text{pH}]) \qquad (4.3.17)$$

where $[\text{pH}_{pzc}]$ is the pH value of the zero charge point, $[\text{pH}]$ is the pH value of the solution.

LAPS also can be applied to various ions detection when deposited by different transducer materials on the sensor surface. Mourzina et al. (2001) described a novel chalcogenide glass ion-sensitive membrane LAPS device for the detection of Pb^{2+}. In this study, the Pb-Ag-As-I-S chalcogenide glass is deposited on the LAPS structure by a pulsed laser deposition (PLD) technique as a Pb-ion-selective transducer material for the first time. Fig. 4.3.16 shows the scheme of pulsed laser deposition technique.

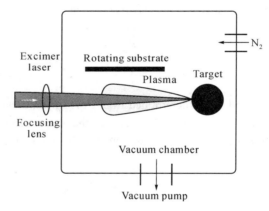

Fig. 4.3.16. Pulsed laser deposition technique

The main potential-determining process, which takes place at the interface between the chalcogenide glass membrane and the solution, is the exchange of primary ions between the solution and the exchange sites at the modified surface layer of the glass. Fig. 4.3.17 shows the dependence of the AC photocurrent I, measured in the external circuit on the applied bias potential V, for different concentrations of Pb^{2+} in the solution. The current-voltage curve moves reproducibly along the voltage axis depending on the Pb^{2+} concentrations.

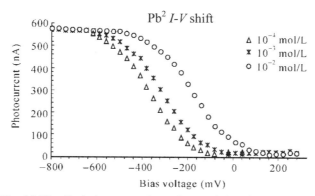

Fig. 4.3.17. Typical current-voltage characteristics of the Pb-LAPS

4.3.4 Microelectrode Array

The importance of microelectrodes attracted the attention of electrochemists first at the beginning of 1980s (Swan, 1980; Wightman, 1981). However, the benefits of microelectrodes have been widely recognized only when the developments in microelectronic technologies made it possible to reliably measure very low currents and to construct reproducible microstructures (Wightman, 1988). Studies concerning the theory, fabrication, and application of microelectrodes have been well documented over the past two decades. Microelectrodes, including single- and composite-microelectrodes can be produced in diverse geometries, with different electrode materials, and using various fabrication techniques. Thanks to the advantageous properties of microelectrodes, they have opened new possibilities in the research fields of electrochemistry, biotechnology, medicine, and environmental sciences (Cammann et al., 1991; Heinze, 1993; Ryan et al., 1994; Wollenberger et al., 1995; Anderson and Bowden, 1996).

4.3.4.1 Structure and principle

Microelectrodes are commonly known also as ultramicroelectrodes (UME). The definition of microelectrodes is ambiguous. Nevertheless, electrodes with dimensions of 10 s of micrometers or less, down to submicrometer are often called microelectrodes (Stulik et al., 2000). In general, microelectrodes can be classified as single microelectrodes and composite microelectrodes. For single microelectrodes, there are different electrode types, such as disc, cylinder, band, ring, sphere, hemisphere, etc. As for composite microelectrodes, it can be generally divided into array electrodes and ensemble electrodes (Fig. 4.3.18), depending on whether the surface of the composite electrodes consists of uniform (array) or random (ensemble) dispersions of a conductor region within a continuous insulating matrix (Tallman and Petersen, 1990).

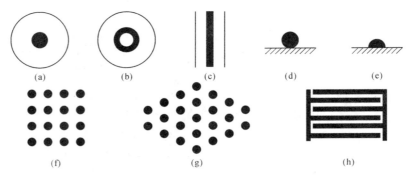

Fig. 4.3.18. Common types of microelectrodes (a) Disc; (b) Ring; (c) Band; (d) Sphere; (e) Hemisphere; (f) Square array; (g) Hexagonal array; (h) Interdigitated microband

The electrolytic process of microelectrode and the conventional electrode is the same in nature. When redox reaction occurs in the electrode system, concentration gradients are formed on the electrode surface, leading to the diffusion effects of the electro-active substance transfer from the bulk solution towards to the electrode surface. To disc electrode, for example, the diffusion equation is

$$\frac{1}{D}\frac{\partial c}{\partial t} = \frac{\partial^2 c}{\partial r^2} + \frac{1}{r}\frac{\partial c}{\partial r} + \frac{\partial^2 c}{\partial z^2} \tag{4.3.18}$$

where D is the diffusion coefficient, c is the bulk concentration of the solution, r is the electrode radius and z is the direction perpendicular to the surface of the electrode.

As shown on the right side of Eq. (4.3.18), the first two items show the radial diffusion, known as nonlinear diffusion, and the third item stands for the diffusion perpendicular to the direction of the electrode surface, called linear diffusion. For traditional electrodes, linear diffusion plays a leading role, but for the microelectrodes, nonlinear diffusion is the main component as shown in Fig. 4.3.19.

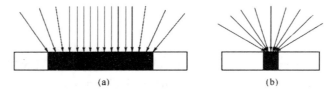

Fig. 4.3.19. The diffusion cross-sections of (a) the traditional electrode and (b) the microelectrode

For the steady-state, the mass transfer rate M for microelectrode is

$$M = \frac{4D}{\pi r} \tag{4.3.19}$$

It can be seen that the mass transfer rate is faster when the radius of the microelectrode gets smaller.

In electrolytic cell, if the electrode potential step occurs, the relationship between the charging current i_c caused by the electric double layer (Fig. 4.3.20) and time t is as follows:

$$i_c \propto \frac{\Delta E}{R} \exp\left(-\frac{t}{RC_u}\right) \tag{4.3.20}$$

where ΔE is the amplitude of the step potential, R is internal impedance of the electrolytic cell, C_u is the capacitance of the double electric layer (Fig. 4.3.20) and t is the duration of step potential.

The charging current i_c is an exponential decay with the index t, and it is also an exponential relationship between i_c and the electrode surface area, for C_u is in direct proportion to the electrode surface area. The smaller the electrode radius, the faster the charging current i_c decreases. Therefore, a microelectrode is able to achieve steady state in a short time and can respond faster, so it can be used in the transient electrochemical methods including voltammetry.

The current on the electrode consists of the Faraday component and the charge current. The Faraday current density of a micro-electrode is large and the charge current decays quickly, leading to an increasing signal to noise ratio, improved sensitivity and lower detection limit. So microelectrodes are applicable to the determination of trace substances.

Because of its small radius, the current density of microelectrode is significant, but the current intensity is very small for the small electrode surface area, only $1 \times 10^{-9} - 1 \times 10^{-12}$ A, so the ohmic drop iR caused by the electrolytic cell system is negligible. It can be applied to the detection of high-impedance solution without supporting electrolyte.

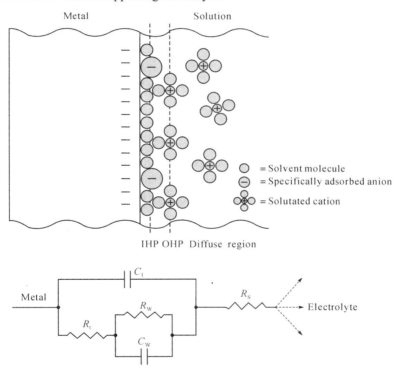

Fig. 4.3.20. Electrical double layer and its equivalent circuit diagram

The current of a single microelectrode is very small, which is at the pA – nA level. Microelectrode array consisting of a large number of microelectrodes could enhance the current signal without losing the characteristics of microelectrodes. The distance between the microelectrodes must be large enough to ensure that diffusion layers of microelectrodes do not overlap to get increased mass transfer capability. But the current density decreases when the spacing between electrodes increases. Empirically, microelectrode is ideal when the electrode spacing is 10 times the diameter of the electrode. The limited diffusion current for the disc microelectrode array is:

$$I = 4mnFDrc \qquad (4.3.21)$$

where m is the number of the electrodes, n is the number of electrons transferred, r is the radius of the single electrode and c is the concentration of the electro-active substance, F is the Faraday constant, D is the diffusion coefficient.

4.3.4.2 Characteristics

Compared to macroelectrodes, the attractive features of microelectrodes can be summarized as follows:
- High mass transport rate (or mass flux) due to radial diffusion;
- Rapid attainment of steady state, showing sigmoidal cyclic voltammogram and limiting current;
- Capability of performing high-speed experiments, because of its significantly reduced charging current and small RC time constant;
- Possibility of measurements in high resistance solutions, due to its extremely small current (i.e. the immunity to ohmic drop);
- High sensitivity thanks to the improved ratio of faradaic-to-charging current (IF/IC) and signal-to-noise ratio (S/N);
- Independent of convection;
- No necessity of deoxygenation;
- Increased current response with composite microelectrodes.

4.3.4.3 Heavy metals-sensitive MEA

Cai W. et al. (Cai et al., 2011), at Zhejiang University designed an Au-MEA for trace heavy metals detection. As shown in Fig. 4.3.21a the Au-MEA consisted of 30×30 Au microdisks of 10 μm diameter separated by 150 μm from each other. In Fig. 4.3.21b, a Pt foil as the counter electrode (CE) and an Ag/AgCl foil as the reference electrode (RE) were attached on the other side of printed circuit board and also encapsulated using epoxy resin. After mercury deposition was carried out on the Au, the MEA was ready to detect heavy metals such as Zn(II), Cd(II), Pb(II) and Cu(II).

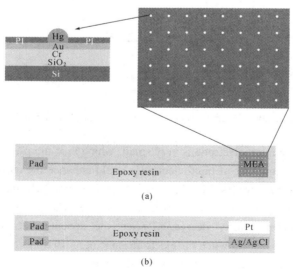

Fig. 4.3.21 Structure of the MEA sensor: (a) On one side is silicon-based Hg-coated Au microelectrodes array; (b) On the other side is Pt and Ag/AgCl electrodes

Then the analytical performance of mercury-coated gold MEA was studied using differential pulse anodic stripping voltammetry (DPASV) for determination of Zn(II), Cd(II), Pb(II) and Cu(II) in the acetate buffer with pH 4.5 (Fig. 4.3.22). The detect sample consisted of Zn(II), Cd(II), Pb(II) and Cu(II) whose concentrations were 80, 3, 3 and 10 μg/L, respectively. After four additions, voltammograms for Zn(II), Cd(II), Pb(II) and Cu(II) were obtained and shown good linearity with their linear ranges separately in 10 – 600, 1 – 100, 1 – 200 and 2 – 300 μg/L, respectively.

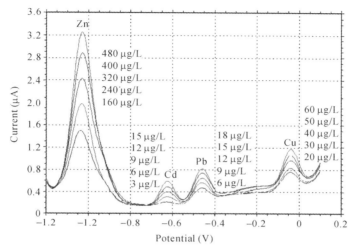

Fig. 4.3.22. The DPASV for standard additions of Zn(II), Cd(II), Pb(II) and Cu(II)

4.4 Gas Sensors

Gas sensors are an important category in the family of chemical sensors. There are a variety of classification criteria. According to the gas sensitive materials and the mechanism of the interaction between gases and the sensitive materials, gas sensors can be divided into semiconductor gas sensors, solid electrolyte gas sensors, electrochemical gas sensors, optical gas sensors, surface acoustic wave gas sensors, infrared gas sensors and so on. This Section focuses on the widely used electrochemical gas sensors, semiconductor gas sensors, solid electrolyte gas sensors and surface acoustic wave gas sensors.

4.4.1 Electrochemical Gas Sensors

Electrochemical gas sensors are used to detect and monitor low levels of toxic gases and oxygen levels in both domestic and industrial situations where it is essential to ensure that the air is safe to breathe.

4.4.1.1 *Structure and principle*

The most common type of electrochemical sensor is the 3-electrode fuel cell as shown in Fig. 4.4.1.
Electrochemical gas sensors contain two or three electrodes, occasionally four, in contact with an electrolyte. The electrodes are typically fabricated by fixing a high surface area precious metal on to the

porous hydrophobic membrane. The working electrode contacts both the electrolyte and the ambient air to be monitored usually via a porous membrane. The electrolyte most commonly used is a mineral acid, but organic electrolytes are also used for some sensors. The electrodes and housing are usually in a plastic housing which contains a gas entry hole for the gas and electrical contacts.

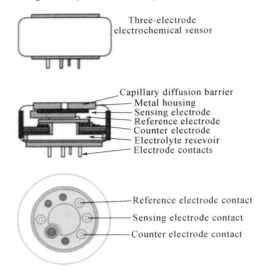

Fig. 4.4.1. Typical electrochemical sensor layout

The air being measured diffuses into the cell through the diffusion barrier (capillary) and filters. When it comes into contact with the sensing electrode, the toxic gas present in the sample undergoes an electrochemical reaction. In the case of carbon monoxide, for example, the reaction is:

$$CO + H_2O \rightarrow CO_2 + 2H^+ + 2e^- \qquad (4.4.1)$$

The carbon dioxide generated diffuses away into the air, while the positively charged hydrogen ions (H^+) migrate into the electrolyte. The electrons generated charge the electrode but are removed as a small electric current by the external measuring circuit.

This oxidation reaction is balanced by a corresponding reduction reaction at the counter electrode:

$$O_2 + 4H^+ + 4e^- \rightarrow 2H_2O \qquad (4.4.2)$$

So at one electrode, water is consumed while electrons are generated, and at the other, water is recreated and electrons are consumed. Neither reaction can occur if no carbon monoxide is present. By connecting the two electrodes, the small electric current generated between them is measured as directly proportional to the concentration of carbon monoxide in the air.

The reference electrode controls the whole process. It remains totally immersed in electrolyte, sees no gas and is not allowed to pass any current. The reference electrode always remains at the same electrochemical potential (known as "rest-air potential", dependent on the material the electrode is made from, and the electrolyte used). The sensing electrode is electrically tied to the reference electrode ensuring its potential will not change even when it is exposed to its determinand and generating current. Usually the potential of the sensing electrode is maintained at exactly the same value as the reference electrode, but for some gases and some applications, performance benefits are gained by maintaining the potential of the sensing electrode at a fixed level above or below the potential of the reference. This is known as "biased" operation.

4.4.1.2 Applications

Reliable and accurate blood pressure and oxygenation measurements within the cardiovascular system are important clinical applications. A method of electrochemical combined with PDMS was adopted by Goutam Koley et al. (2009) for oxygen content measurements within the heart and blood vessels.

The blood oxygen sensing was performed based on the change in current flowing between a Pt electrode and an Ag/AgCl electrode kept in contact with KCl solution soaked filter paper. The current flowing between the electrodes, which were maintained at a potential difference equal to the reduction potential of dissolved oxygen, can respond to any change in the dissolved oxygen content in the KCl solution with high sensitivity. For estimating the oxygen content of a given test liquid, the sensor (and KCl soaked filter paper) can be separated from the liquid using a PDMS thin film as the intervening membrane. Due to high oxygen permeability of the PDMS membrane, the dissolved oxygen in the KCl solution will track the dissolved oxygen content in the test liquid quite accurately.

The fabricated sensor consisted of three layers that are a gas-permeable membrane (PDMS, film thickness: 30 μm), a membrane filter with the dimension of 20 mm×12 mm (Isopore VMTP4700, Millipore Corp., USA) containing electrolytic solution (KCl 0.1 mol/L). The schematic diagram of the sensor is shown in Fig. 4.4.2, Pt working electrode and Ag/AgCl reference electrode are fabricated on the top layer with electron beam deposition (electrode thickness: 100 nm). The electrode is a simple stripe design that the width of the Pt electrode is 10 mm and that of the Ag/AgCl electrode is 5 mm. The sensor is fabricated by stacking the electrodes, gas-permeable membrane and solution-permeable filter together. The chemical reactions are the following:

$$\text{Cathode (Pt)}: O_2 + 2H_2O + 4e^- \rightarrow 4OH^- \qquad (4.4.3)$$

$$\text{Anode (Ag/AgCl)}: 4Ag + 4Cl^- \rightarrow 4AgCl + 4e^- \qquad (4.4.4)$$

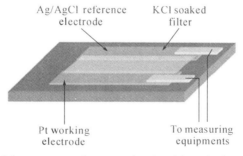

Fig. 4.4.2. Schematic diagram of the oxygen sensing set up (reprinted from (Koley et al., 2009), Copyright 2009, with permission from Elsevier Science B.V.)

The sensor was subject to pure Ar, 10% and 30% O_2 and the responses are shown in Fig. 4.4.3, the output current was significantly reduced by 30 μA, when 10% O_2 gas (10% O_2 and 90% Ar) was flown into the air-filled chamber. The response time to reach a steady current level was approximately 40 s. The current began to recover right after the gas flow was stopped, as air started to flow in the chamber. The recovery time to return to the original level of current was almost four times higher than the decay response time, which, however, can be reduced by flowing fresh air into the chamber at a high flow rate. As shown in Fig. 4.4.3a, the sensor output shows repeatable current response in the presence of 10% O_2, and the current was reduced by about 30 μA for each cycle. In contrast to the 10% O_2, exposure to 30%

O_2 made the output current changes in the reverse direction, and increase sharply by about 25 µA. This is because the oxygen current content in the sensor ambient was increased by 20% compared to the baseline value. The response time for the current to reach a steady value in this case was also about 40 s, similar to decay response time observed for 10% O_2. However, the recovery time observed was even longer than the first case.

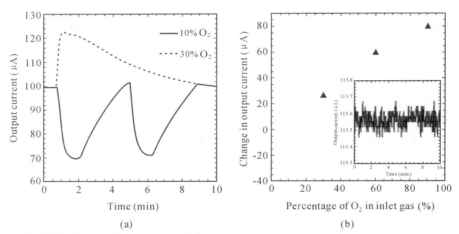

Fig. 4.4.3. Output current of the sensor to oxygen: (a) Time dependent electrode current as the ambient air is replaced by oxygen argon mixture; (b) Variation of maximum electrode current with oxygen concentration. Inset shows the rms noise plotted as a function of time (reprinted from (Koley et al., 2009), Copyright 2009, with permission from Elsevier Science B.V.)

To determine the sensor performance over a large range of oxygen concentration, it was further exposed to 60% and 90% O_2. We observed that the change in output current is +60 µA for 60% O_2, and +75 µA for 90% O_2. Fig. 4.4.3b shows the change in output current as the sensor is exposed to different oxygen composition from the baseline air environment. We observe that the output current changes much faster with change in oxygen composition for lower oxygen concentration, but gradually tends to saturate for higher oxygen concentration. This is possible because the oxygen generated current starts to get affected by the diffusion-limitation of dissolved oxygen at the Pt electrode.

4.4.2 Semiconductor Gas Sensors

Since 1962 it has been known that absorption or desorption of a gas on the surface of a metal oxide changes the conductivity of the material, this phenomenon was first demonstrated using zinc oxide thin film layers. The sensitivity of a surface to a gas can be as low as parts per billion (ppb). It is complicated to describe the sensitive mechanism of semiconductor gas sensors, while the fact of conductivity variation is distinct when the surface of the device absorbs the special gas molecular.

4.4.2.1 Structure and principle

- **Band theory**

Band theory states that within a lattice there exists a valence band and a conduction band. The separation between these two bands is a function of energy, particularly the Fermi level, defined as the highest available electron energy levels at a temperature. There are three main classes of material in band theory (see Fig. 4.4.4).

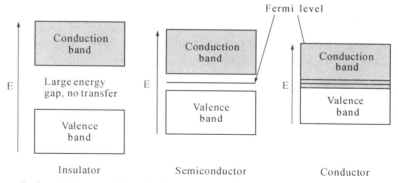

Fig. 4.4.4. Schematic band diagrams of an insulator, semi-conductor and conductor. Note the small gap in the semiconductor, where electrons with sufficient energy can cross and the overlapping of the bands in the conductor

Insulators have a large gap between the valence and conduction band (typically taken to be 10 eV or more), as such a lot of energy is required to promote the electron into the conduction band and so electronic conduction does not occur. The Fermi level is the highest occupied state at $T=0$. Semiconductors have a sufficiently large energy gap (in the region of 0.5 – 5.0 eV) so that at energies below the Fermi level, conduction is not observed. Above the Fermi level, electrons can begin to occupy the conduction band, resulting in an increase in conductivity. Conductors have the Fermi level lying within the conduction band.

- *Band theory applied to sensors*

Band theory as applied to gas sensors has been the subject of intense study for a number of years. The target gas interacts with the surface of the metal oxide film (generally through surface adsorbed oxygen ions), which results in a change in charge carrier concentration of the material. This change in charge carrier concentration serves to alter the conductivity (or resistivity) of the material. An n-type semiconductor is one where the majority charge carriers are electrons, and upon interaction with a reducing gas an increase in conductivity occurs. Conversely, an oxidising gas serves to deplete the sensing layer of charge carrying electrons, resulting in a decrease in conductivity. A p-type semiconductor is a material that conducts with positive holes being the majority charge carriers; hence, the opposite effects are observed with the material and shows an increase in conductivity in the presence of an oxidising gas (where the gas has increased the number of positive holes). A resistance increase with a reducing gas is observed, where the negative charge introduced into the material reduces the positive (hole) charge carrier concentration. A summary of the response is provided in Table 4.4.1.

Table 4.4.1 Sign of resistance change (increase or decrease) to change in gas atmosphere

Classification	Oxidising Gases	Reducing Gases
n-type	Resistance increase	Resistance decrease
p-type	Resistance decrease	Resistance increase

- *A p-Type sensor model for gas interaction*

A simple model for the response of a p-type sensor is demonstrated by Eq. (4.4.5) and Eq. (4.4.6), which shows the adsorption of an oxygen atom to the surface of the material, causing ionisation of the atom and yielding a positive hole (p^+). The positive hole and the ion can then react with a reducing gas such as carbon monoxide, forming carbon dioxide (k_2) or being removed through interaction with each other (k_{-1}). The difference in charge carrier concentration (in this case the positive hole) is manifest in a resistance change between the sensor's electrodes and read by the measurement circuitry.

$$\frac{1}{2}O_2 \underset{k_{-1}}{\overset{k_1}{\rightleftarrows}} O^- + p^+ \qquad (4.4.5)$$

$$O^- + p^+ + CO \xrightarrow{k_2} CO_2 \qquad (4.4.6)$$

- *The equivalent circuit model*

The response model, which includes the influence of sensor microstructure is a refinement of the conventional response model. The conventional model is found to apply to over a limited range of examples only, and assumes that the absorbed species on the surface of the metal oxide is the O^{2-} which is unlikely to be included as it is energetically unfavourable. Eq. (4.4.7) describes a relationship where the change in resistance is proportional to the concentration of the gas (in this example carbon monoxide, CO) and a sensitivity parameter A (the sensitivity parameter is constant for a given material at a given temperature), and R is the resistance after exposure to analyte gas and R_0 the baseline resistance Eq. (4.4.8).

$$\frac{1}{2}O_2 \underset{k_{-4}}{\overset{k_4}{\rightleftarrows}} O^{2-} + 2p^+ \qquad (4.4.7)$$

$$\frac{R}{R_0} = 1 + A[CO] \qquad (4.4.8)$$

Instead, Naisbitt et al. proposed that Eq. (4.4.5) is more probable, but to account for non-linear responses, there must be other factors influencing the response. First, the assumption is made that the only parts of the material that exhibit a response to the target gas, are the areas where the gas can reside and interact at the surface.

Thus, the material is split into three regions (Fig. 4.4.5): (1) the surface, (2) the bulk (inaccessible to the target gas), and (3) the neck or particle boundary (below this boundary, the material is no longer defined as a surface). The distance between the surface and the particle boundary is called the Debye length: the distance at which charge separation can occur. The Equation for the response in this model is found to be Eq. (4.4.9):

$$\frac{R_T}{R_{T,0}} = \gamma PB(1+A[CO]) = \frac{1}{\frac{1}{\gamma B} + \frac{1}{\gamma S(1+A[CO])}} \qquad (4.4.9)$$

where R_T is the total sensor resistance, $R_{T,0}$ is the baseline resistance in clean, dry air; $\gamma x = R_{x,0}/R_{T,0}$, ($x$ denotes particle boundary-PB; bulk-B; surface-S), so γ is the ratio of the baseline of x to the total baseline of sensor resistance.

Chapter 4 Chemical Sensors and Measurement

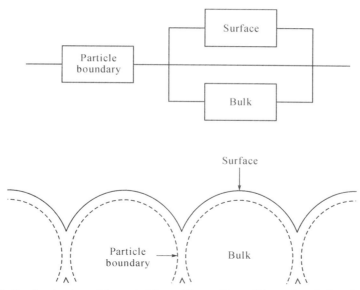

Fig. 4.4.5. Demonstrating the structure of the materials, and the positions of the surface, bulk and particle boundary. The model assumes the gas sensitivities of the surface and particle boundary are the same.

This work shows that the response time is directly related to the grain size and the size of the particle boundary in the material. The response model is different for n-type and p-type semiconductors. In forming the baseline resistance, oxygen is adsorbed on to the surface and abstracts electrons from the material; hence, this process will determine R_0. The resistivity of the p-type decreases relative to the bulk, and increase for the n-type. The relative contributions from the three resistors in the model then differ. For very small grain sizes, the grain can be considered to contain no bulk area at all (so the whole grain is considered to contribute to the surface area) in this instance, the simpler model and Eq. (4.4.9) is an adequate model for the response. If one considers the other extreme, where the grains are so large the contribution to the resistance or conductivity is negligible, the surface can be deemed to have a constant resistance. This model is expected to be generally applicable to both p and n-type sensors.

4.4.2.2 Applications

- *Carbon monoxide sensors*

Carbon monoxide (CO) is a colourless gas, with no odour, making it undetectable to humans. The gas has been shown to bind irreversibly to the iron centre of haemoglobin, the oxygen transport molecule in blood. The irreversible binding means that oxygen can no longer be absorbed, and at high levels of exposure this results in death. The maximum time weighted average exposure value ascribed by the United States National Institute of Occupational Safety and Health is 35 ppm over an 8 h period. The gas is mainly a product of poorly combusted organic material, such as petrol, oil or gas. Carbon monoxide is constantly in the public eye, largely because the home is such a susceptible place for carbon monoxide poisoning, usually resulting from a faulty gas powered boilers that lead to annual deaths. Carbon monoxide concentrations are particularly high in areas of industry, where fossil fuels are combusted for energy purposes, and in cities where high levels of traffic exist. Existing sensors are used in homes as a warning system to the otherwise undetectable carbon monoxide. Sensors fall in to two main types: 'blob' sensors, and electrical sensors. 'Blob' sensors are essentially a patch of metal oxide salts that interact with the monoxide and reduce the salt, forming carbon dioxide. The salt turns black when it is reduced, and the colour change alerts the observer. Though cheap, the alerting system requires the vigilance of the observer in order to recognise the change in concentration. Given that at high concentration carbon monoxide causes dizziness and confusion, the occupant may be in no condition to readily observe this change. Electronic carbon monoxide sensors come in two main types: thermistor type metal-oxide detectors which detect a change in heat when carbon monoxide lands on the oxide and reacts, (the change in temperature raising the alarm), and an electrolytic detector that works by sensing the change in charge carriers in an electrolyte solution when carbon monoxide interacts with an electrode of the device.

Carbon monoxide gas sensors have a myriad of applications, for not only home safety, but also in measuring atmospheric concentrations, in the exhaust of cars, and for process monitoring in industrial plants.

- *Carbon dioxide sensors*

Carbon dioxide is present in air at concentrations of 388 ppm in the Earth's atmosphere. Carbon dioxide has many uses: carbonating drinks, pneumatic applications, fire extinguishers, photosynthesis, it is an essential ingredient for the process to take place, lasers and refrigerants are just some examples where carbon dioxide has been utilized, by nature and by humans. Carbon dioxide is also a greenhouse gas. In the atmosphere, it absorbs infra-red energy, and vibrates, and then passes heat to its surroundings, increasing the ambient temperature. If the concentration of carbon dioxide increases in the atmosphere, the temperature increases too. Thus, accurate, reliable sensors with an appropriate temperature range could be used to measure the change in concentration. The recent increase in the concentration of carbon dioxide through human activity (the combustion of hydrocarbons and other carbon containing fuels such as coal, methane, petrol and kerosene with an appropriate amount of oxygen yields carbon dioxide and water) is leading to the concern about whether humans are causing global warming. Sensors can measure the output of CO_2 during combustion, thus giving important real-time information as to the amount of carbon dioxide being produced by the activity. Atmospheric carbon dioxide concentrations have risen steadily since 1958, and are expected to carry on rising as long as man satisfies its thirst for combusting fossil fuels. Carbon dioxide can cause substantial negative health affects to humans (Table 4.4.2) including drowsiness and at high enough concentrations suffocation.

Table 4.4.2 Detailing the effects of concentration and exposure time of CO_2 on human health

CO_2 concentration and exposure time	Effect on health (symptoms)
0.035%	Approximate atmospheric concentration, no noticeable effect
3.3% – 5.4% for 15 mins	Increased depth of breathing
7.5% for 15 mins	Feeling of an inability to breath, increased pulse rate, headache, dizziness, sweating, restlessness, disorientation, and visual distortion
3% for over 15 hours	Decreased night vision, colour sensitivity
10% for 1.5 mins	Eye flickering, increased muscle activity, twitching
>10%	Difficulty in breathing, impaired hearing, nausea, vomiting, a strangling sensation, sweating, after 15 minutes a loss of consceousness
30%	Unconsciousness, convulsions. Several deaths attributed to CO_2 at concentrations of more than 20%.

- *Nitrogen oxide (NOx) sensors*

NO_x are a varied group of gases, including the simplest form of nitric oxide (NO), nitrogen dioxide (NO_2), nitrous oxide (N_2O), dinitrogen trioxide (N_2O_3), dinitrogen tetroxide (N_2O_4) and dinitrogen pentoxide (N_2O_5). The gases are principally formed from the combustion of fossil fuels in internal combustion engines, where the energy of the combustion reaction helps combine nitrogen gas (N_2) and oxygen gas (O_2). Nitric oxide is a known component of photochemical smog, combining with hydrocarbons and oxygen to form a thick cloud over heavily industrialised areas. Photochemical smog, apart from the haze it produces, is irritating to the eyes, and also damages plant life in the affected areas. The monitoring of the concentrations of NO in the air can be particularly useful, as environmental agencies can use it to predict how likely smog is to form, thus being able to alert the public at times of significant risk. Nitrogen dioxide is toxic upon inhalation, but unlike carbon monoxide, it is easily detected by smell, one complicating factor of this is that exposure to the gas at 4 ppm anaesthetises the nose creating the possibility that increased concentrations in an environment may go unnoticed, causing potential health risks. The main risks of nitrogen dioxide appear to be their effect on the lungs. People with bronchitis or asthma are particularly sensitive to the gas, and lungs may become inflamed, leading to breathing difficulties. Animal subjects subjected to long-term exposure of NO_2 were found to have damaged lungs. Studies have linked the concentration of NO_2 and SO_2 to the exacerbation of conditions such as chronic bronchitis and emphysema. The monitoring of the gas in environments where NO_2 is at particularly high risk concentrations is desirable as one would like to reduce the risks to human health as much as possible.

- *Ammonia sensors*

Ammonia, NH_3 is a caustic and hazardous colourless gas with a characteristic pungent odour. The worldwide production of ammonia in 2007 was estimated at almost 140 million tonnes, a rise from approximately 2 million tonnes in 1945. 83% of the world's ammonia was used in the fertilization of crops in 2003 (global production was around 115 million tonnes).

Ammonia is a source of nitrogen to living organisms, and nitrogen presents in vital biological molecules such as amino acids (the building blocks of proteins). Ammonia is present in animal waste, and in particularly high concentrations in chicken faeces. In agricultural environments, where animals can live in unventilated environments, ammonia concentrations can become high enough to kill the animals. An ability to monitor and control these environments is highly desirable.

- *Other application*

This part takes the SnO_2 semiconductor gas sensor as an example to introduce a typical sensor for gas detection and its application information. This type of sensor device is developing fast, from sintered to thick and thin film, and has become the most widely used sensor in certain applications. SnO_2 is a kind of white powder with relative density 6.16 – 7.02 g/m^3, melting point 1,127 °C and boiling point over 1,900 °C. It does dissolve in heated strong acid or alkali solution, but not the same as in water.

There are three main factors contributing to the gas sensitive effect. The first one is its structure, generally, the more oxygen vacancy, the more evidence for sensitive effect. The next is that additives can also affect the sensitive process. Table 4.4.3 shows that, to some degree, different additives can make some new specialties. The third one is about temperature during the sintering and heating process.

Table 4.4.3 The additives to SnO_2 sensors

Additives	Detection gas	Working temperature (°C)
PdO, Pd	CO, C_3H_3, Alcohol	200 – 300
$PdCl_2$, $SbCl_3$	CH_4, CO, C_3H_3	200 – 300
Sb_2O_2, TiO_2, TlO_2	LPG, CO, Alcohol	200 – 300
V_2O_5, Cu	Alcohol, Acetone	250 – 400
Sb_2O_3, Pi_2O_3	Reducing gas	500 – 800

A number of sensor-based instruments on the market can measure the concentrations of reducing gases or vapors in the air. Examples include breath-alcohols analyzers used by police departments, carbon monoxide (CO) analyzers used in performing emission control measurements on vehicles, and methane detectors used to protect against explosions and other dangers from natural gas. All these applications have three common grounds: relatively low cost, could be operated by ordinary people rather than scientists and engineers, and be manufactured using similar technology.

The Figaro TGS gas sensors are based on a technology that uses powdered tin dioxide (SnO_2) sintered onto a semiconductor substrate in Fig. 4.4.6a. In normal operation the sensor element is heated to approximately 400 °C. Oxygen is adsorbed onto the surface of the SnO_2, where the oxygen molecules accept electrons. These electrons create a relatively high electrical potential barrier that is difficult for free electrons to cross. As a result, the electrical resistance is high and is a function of the partial pressure of oxygen (P_{O_2}) which is shown in Fig. 4.4.6b. When a kind of reducing gas or vapor (e.g., CO, methane, methanol) is adsorbed onto the surface and reacts with the oxygen, the resistance of the device drops.

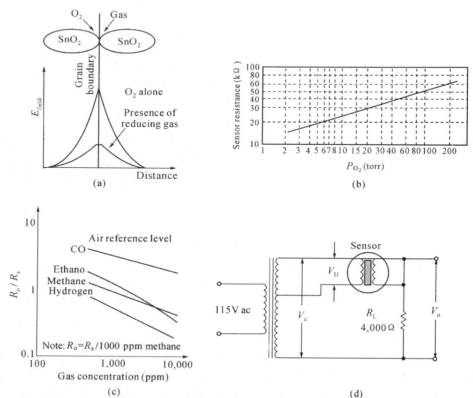

Fig. 4.4.6. Principle and application of the Figaro gas sensors: (a) Figaro gas/vapor sensor uses sintered; (b) Sensor resistance vs. partial pressure of oxygen; (c) Ratios for various gases and vapors; (d) Typical circuit diagram

Fig. 4.4.6c shows the ratio of the actual sensor resistance R_S of TGS2442 to a standard resistance R_0 for several different elements. The standard resistance R_0 is the value of R_S in an atmosphere of 1,000 parts per million (ppm) methane gases.

A typical circuit for the TGS sensors is shown in Fig. 4.4.6d. The heater voltage V_H heats the sensor element to the required temperature, while the operating voltage V_C provides excitation to the sensor element. A load resistance R_L is used to convert current flowing in the sensor to an output voltage V_O. The values of V_C and V_H vary from one sensor to another, but are typically in the range of 0.5 to 12 V. Some Figaro gas sensors for the detection of toxic gas, including TGS2442, are shown in Fig. 4.4.7.

Fig. 4.4.7. Figaro gas sensors for the detection of toxic gas

4.4.3 Solid Electrolyte Gas Sensors

A solid electrolyte is one of the types of solid state materials with the same ionic conduction characteristic as the electrolyte solution, and the solid electrolyte gas sensor is one kind of chemical cell taking the ionic conductor as the electrolyte. It does not need to make the gas pass through the breather membrane and dissolve in the electrolyte, this can avoid such problems as solution evaporation and electrode waste. Because of the high conductivity, sensitivity and selectivity, these types of sensors are widely used in the fields of petrochemical, environmental protection, mining industry, and food industry and so on.

4.4.3.1 *Structure and principle*

The solid electrolyte will have the obvious electrical conductivity only under a high temperature. Zirconia (ZrO_2) is a typical material for solid electrolyte gas sensors. The pure zirconia has clinohedral structure under normal temperature. When the temperature rises to about 1,000 °C, the allomorphism transformation will happen. Then the clinohedral structure turns into the polytropism structure, and follows the volume contraction and endothermic reaction, therefore it is an unstable structure.

Mixing ZrO_2 with stabilizer such as alkali soils calcium oxide CaO or rare earth yttrium oxide Y_2O_3, the ZrO_2 will become the stable fluorine cubic crystal. The stable degree is related to the density of stabilizer. The ZrO_2 is sintered under 1,800°C after being mixed with stabilizer, a part of zirconium ion will be substituted by the calcium ion, producing ($ZrO \cdot CaO$). Because Ca^{2+} is divalent ion, Zr^{4+} is quadrivalence ion, to maintain the electric neutrality, the oxygen ion O^{2-} hole will be generated in the crystal. This is why ($ZrO \cdot CaO$) transfers oxygen ions at high temperature, and ($ZrO \cdot CaO$) becomes oxygen ion conductor at 300 – 800 °C. But in order to pass oxygen ions actually, there must also be different partial pressure of oxygen (oxygen potentiometer) on the two sides of the solid electrolyte to form the so-called concentration cell. The structural principle is shown in Fig. 4.4.8, the precious metal electrodes are on both sides, forming sandwich structure with the intermediate dense (ZrO·CaO).

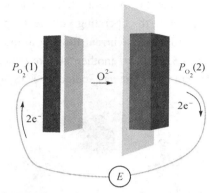

Fig. 4.4.8. The structural principle of concentration cell

Set the partial pressure of oxygen on both sides of the electrodes are $P_{O_2}(1)$ and $P_{O_2}(2)$ respectively, in the two electrode reactions occur as follows:

$$(+)\text{Pole: } P_{O_2}(2), \quad 2O^{2-} \rightarrow O_2 + 4e^- \tag{4.4.10}$$

$$(-)\text{Pole: } P_{O_2}(1), \quad O_2 + 4e^- \rightarrow 2O^{2-} \tag{4.4.11}$$

The electromotive force (EMF) of the reaction expressed by the Nernst equation:

$$E = \frac{RT}{nF} \ln \frac{P_{O_2}(1)}{P_{O_2}(2)} \quad \text{or} \quad E = 0.0496 T \ln \frac{P_{O_2}(1)}{P_{O_2}(2)} \tag{4.4.12}$$

As the above equation, fixing $P_{O_2}(1)$ at a certain temperature, the oxygen concentration of the sensor's positive pole can be equated by the above formula.

In addition to measuring oxygen, the application of β-Al_2O_3, carbonate, NASICON solid electrolyte sensors, can also be used to measure CO, SO_2, NH_4, CO_2 and other gases. New gas sensors have emerged in recent years, using antimony acids, La_3F, etc., can be used in low temperature and can be used to detect positive ions.

4.4.3.2 Applications

Recently, accurate measurement of CO_2 concentration in offices and houses has become widespread, as CO_2 is a good indicator of air quality. A practical CO_2 gas sensor for air quality control is developed using a combination of a $Na_3Zr_2Si_2PO_{12}$ (NASICON) as a solid electrolyte and Li_2CO_3 as a carbonate phase by Kaneyasu et al. (2000).

The construction of the CO_2 sensor element is shown in Fig. 4.4.9. The solid electrolyte sinter of NASICON—Na conductor, about 4 mm in diameter and about 0.7 mm in thickness—was used. A pair of gold electrodes was attached to both surfaces of the solid electrolyte by screen printing. A working electrode was pasted with lithium carbonate on one side of the electrode by screen printing and baked at 600 °C. A built-in Pt heater screen printed on an alumina plate was laminated on a reference electrode and sealed with glass. The sensor element was heated at 450 °C and EMF was measured by a high-impedance voltage meter.

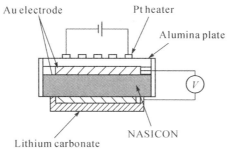

Fig. 4.4.9. Construction of the sensor element (reprinted from (Kaneyasu et al., 2000), Copyright 2000, with permission from Elsevier Science B.V.)

The construction of the CO_2 sensor is shown in Fig. 4.4.10. The sensor element was mounted on a resin base and the gas entrance was covered with a filter consisting of zeolite powder (Na/Y type, about 1 g) sandwiched between two non-woven fabrics. The size of the sensor was 24 mm in diameter and 31 mm in height.

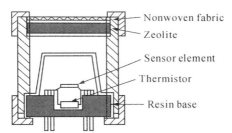

Fig. 4.4.10. Construction of the CO2 sensor (reprinted from (Kaneyasu et al., 2000), Copyright 2000, with permission from Elsevier Science B.V.)

The sensitivity of various gases is shown in Fig. 4.4.11. In this figure, change in EMF (ΔEMF) is calculated according to the expression as follows:

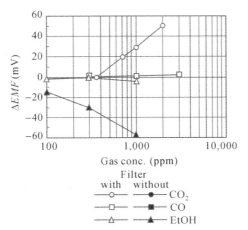

Fig. 4.4.11. Sensitivity of various gases (reprinted from (Kaneyasu et al., 2000), Copyright 2000, with permission from Elsevier Science B.V.)

$$\Delta EMF = EMF\ (CO_2=350\ ppm) - EMF\ (\text{measuring atmosphere}) \qquad (4.4.13)$$

ΔEMF of the sensor showed a linear relationship with the logarithm of CO_2 concentration and was slightly affected by interfering gases, such as carbon monoxide and ethyl alcohol, because of the zeolite filter. The EMF of the sensor increased as the surrounding temperature rose, necessitating a correction in the temperature dependence using a thermistor.

The heating condition stability of the EMF and ΔEMF in indoor atmospheres is shown in Fig. 4.4.12. Both the EMF and ΔEMF indicated excellent stability over 2 years. On the other hand, when the sensor was exposed to a high humidity atmosphere, the EMF decreased but ΔEMF stayed fairly stable. It is therefore possible to measure CO_2 concentration by calculating ΔEMF.

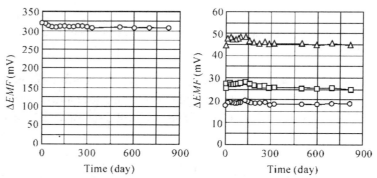

Fig. 4.4.12. Heating condition stability of the EMF and ΔEMF in indoor atmospheres (reprinted from (Kaneyasu et al., 2000), Copyright 2000, with permission from Elsevier Science B.V.)

4.4.4 Surface Acoustic Wave Sensors

Acoustic wave (AW) devices have received increasing interest in recent years in a wide range of applications where they are currently used as resonators, filters, sensors, and actuators. The AW family of devices includes the surface acoustic wave (SAW), the shear horizontal surface acoustic wave (SH SAW), the shear horizontal acoustic plate mode (SH APM), the flexural plate wave (FPW) or Lamb wave mode and thickness shear mode (TSM) devices. Although AW devices are already fabricated on a large scale for telecommunication systems such as the case in the mobile telephone industry, their development in the sensors field is still in the early stages. AW devices are highly sensitive to surface perturbation and changes. Thus, they are investigated as sensors for measurement in both gas and liquid environments. Measurements of parameters include temperature, pressure, liquid density, liquid viscosity, electrical conductivity, mass and visco-elastic changes of thin films and so on. However, their most common use is for chemical sensors (usually gas sensors).

4.4.4.1 *Structure and principle*

The basic structure of SAW is showed in Fig. 4.4.13. The output transducer and input transducer both are interdigital transducers (IDT), which are the core components of SAW. It is made up of interval distributed electrodes. These devices are fabricated by photolithography, the process used in the manufacture of silicon integrated circuits.

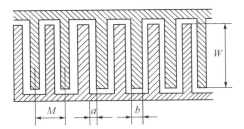

Fig. 4.4.13. The parameters of SAW

N is the number of electrodes, a is the interval of IDT, b is the width of IDT, w is the overlap length of electrodes and h is the thickness of IDT (Fig. 4.4.13). Here a, b and material factor of piezoelectric substrate collaboratively determined center frequency of the transducer. It`s relationship is:

$$f = \frac{v_R}{2(a+b)} \qquad (4.4.14)$$

where v_R is the speed of SAW. Because only in this length, SAW can be generated. Moreover w determined the intensity of SAW. The larger w is, the more powerful SAW will be.

When input transducer gets an AD input. This electrical signal will be converted into mechanical vibration signal due to the inverse piezoelectric effect. The mechanical vibration signal will spread to both sides of the substrate. One side signal is noise which should be eliminated by sound-absorbing material while the other side signal will be converted into electric signal again in output transducer due to the piezoelectric effect (Fig. 4.4.14). When the environment of the transducer changes like substance absorbed in the delay line, it will lead to the change of vibration frequency. Then through the variation of input transducer and output transducer, the weight information of the substance can be obtained.

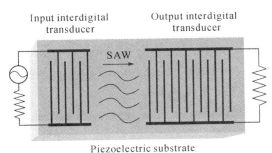

Fig. 4.4.14. Schematic picture of a typical SAW device design

4.4.4.2 Applications

In recent years, many types of renewable energy are receiving increasing attention. In particular, hydrogen energy may become a new clean energy for daily use. But any leak of hydrogen over a wide range of concentration (4% – 75%) will result in an explosion, and if humans are exposed to it in a closed space, it could cause asphyxiation. Therefore, a method for precisely detecting the content of hydrogen at room temperature is very much needed in the development of a hydrogen energy economy. A SAW sensor with Pt coated ZnO nanorods as the selective layer has been investigated for hydrogen detection by Huang et al. (2009).

The SAW sensor was fabricated based on a 128° YX-LiNbO$_3$ substrate with an operating frequency of 145 MHz, the SAW resonator was then connected to an amplifier to configure an oscillator. A dual delay line system as shown in Fig. 4.4.15, which consisted of two counterparts in the oscillator (one is coated with the selective material and the other is bare to execute common mode rejection), was realized to eliminate external environmental fluctuations. To function as an active element, the coated one contributes to a frequency shift by the interaction between the sensing material and the target gas. By comparison, the reference one, which has a bare surface, gives the signal of the environmental effects.

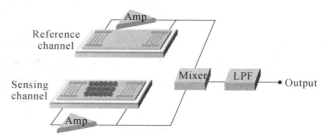

Fig. 4.4.15. Schematic diagram of a dual delay line configuration

Pt coated ZnO nanorods were chosen as the selective layer due to the advantages of simple fabrication, high sensitivity to hydrogen at room temperature, and no reaction to moisture. First of all, a thin ZnO film was deposited on the SAW delay line, the as-prepared substrate was immersed into an aqueous solution of zinc nitrate hydrate and methenamine at 95 °C for 5 h. Then, the substrate was rinsed with deionized water. Fig. 4.4.16. is a scanning electron microscope (SEM) image of the ZnO nanorods. Finally, a Pt film was coated over the ZnO nanorods as a catalyst by electron beam evaporation.

(a)   (b)

Fig. 4.4.16. SEM images of ZnO nanorods with growing time of 5 h: (a) Top view; (b) Side view (reprinted from (Huang et al., 2009), Copyright 2009, with permission from IOP Publishing)

The real-time responses of the dual-channel sensor to different H$_2$ concentrations are shown in Fig. 4.4.17. At the initial stage, the steady state of the base frequency was reached, and then nitrogen or hydrogen was led into the PDMS chamber. Testing cycles were implemented with constant exposure time and purge time to reach a new steady state or return to the baseline. The sensor was then exposed to different concentrations of hydrogen: 200, 500, 1,500, 2,500, and 6,000 ppm at room temperature. The responses were 8.36, 12.66, 17.47, 20, and 26.2 kHz respectively. It took less than 15 min to reach about 90% of the steady state, and the recovery time was about 2–3 min. The frequency shifts for different H$_2$ concentrations are shown in Fig. 4.4.17.

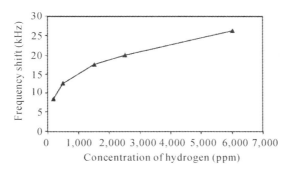

Fig. 4.4.17. Changes in frequency with H_2 concentration (reprinted from (Huang et al., 2009), Copyright 2009, with permission from IOP Publishing)

The results show that the Pt coated nanorod based SAW hydrogen sensor provides high sensitivity, fast response, and good repeatability while operating at room temperature and it also can avoid the influence of humidity.

4.5 Humidity Sensors

Humidity and temperature are the most frequently measured physical quantities in measurement science. Whereas the measurement of temperature can nowadays be done with a satisfactory accuracy, measurement of the water vapour content of gaseous atmosphere, i.e. hygrometry, appears much more complex. Water vapour is a natural component of air, and it plays an important role in a wide and various ranges of practical measurement situations. Today, different humidity sensors are developed for miscellaneous application. The requirements that humidity sensors must meet in order to satisfy a wide range of application are: (1) a good sensitivity over a wide range of both humidity and temperature, (2) a short response time, (3) a good reproducibility and a small hysteresis, (4) a good durability and long life, (5) resistance against contaminants, (6) negligible temperature dependence and (7) low cost.

In hygrometry, it is common to measure the relative humidity r_h along with the temperature T. The r_h is defined as:

$$r_h = \frac{p_w}{p_s} \times 100\% \qquad (4.5.1)$$

where p_w is the vapour pressure and p_s is the saturation pressure.

A parameter that is frequently measured is the dew point T_d (°C), i.e. the temperature and pressure at which gas begins to condense into liquid. Other ways to define the humidity are the absolute humidity (g/m³), the ratio with respect to the dry air, i.e. mixing ratio m_r (ppmv), and the deficit with respect to the saturation, i.e. saturation deficit p_{sd} (mabar). The vapour saturation pressure of water is plotted versus temperature (Fig. 4.5.1).

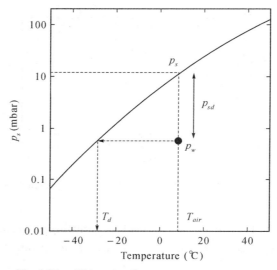

Fig. 4.5.1. Water saturation pressure vs. temperature

4.5.1 Capacitive Humidity Sensors

4.5.1.1 Principle

Capacitive humidity sensors are based on dielectric changes caused by the water vapour uptake on thin films. The properties of these sensors are determined by the hygroscopic material and the electrode geometry. Korvink et al. have considered four electrode geometries of capacitive humidity sensor. These designs, which differ in the ways they allow for a uniform electrical field distributing in the dielectric. Electrode geometry allows the vapour to diffuse freely into the dielectric, e.g., interdigitated electrodes (IDE) are favourable for rapid responses, but a thin top electrode (typically 10 – 80 nm Au), will result in a more uniform field distribution. An intermediate solution is to provide the top electrode with small meshes. The properties of IDE capacitors depend on the area of the electrode and gap spacing.

A capacitive humidity sensor is actually a plate capacitor. A polymer layer is placed between a metal electrode and a coated glass substrate (Fig. 4.5.2). The dielectric permittivity of the polymer depends on its water content which can be written in complex number form:

$$\varepsilon_{rh} = \varepsilon'_{rh} - j\varepsilon''_{rh} = \varepsilon'_{rh} - j\frac{\sigma}{\omega} \qquad (4.5.2)$$

where ε'_{rh} is the real part and ε''_{rh} is the dielectric loss factor. The latter equals the conductivity σ divides by the frequency ω of the applied electrical field. In a capacitive humidity sensor, change in dielectric permittivity is almost directly proportional to relative humidity in the environment. Typical change in capacitance is 0.2 – 0.5 pF for 1% relative humidity change.

For an optimal humidity exchange between the polymer layer and the surrounding air, the metal electrode is produced by a special process. The absence of additional insulation layer leads to a high sensitivity. The sensor capacity at relative humidity r_h can be described as:

$$C_{r_h} = \frac{\varepsilon_{r_h}\varepsilon_0 A}{d} \tag{4.5.3}$$

where ε_0 is the permittivity of vacuum, A is the area of the electrodes, d is the distance between the electrodes.

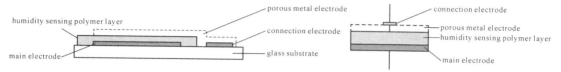

Fig. 4.5.2. Schematic construction of an capacitive humidity sensor and it's equivalent diagram

4.5.1.2 Characteristics

Capacitive technique is the most widely used technique for humidity sensors, where the relative humidity change is detected by the humidity-induced dielectric constant change of thin films. The most widely used materials as humidity-sensitive dielectrics are polyimide films, which provide a high sensitivity, a linear response, a low response time, and a low power consumption. The use of polyimide films allows the implementation of the humidity sensors together with integrated circuits. However, polyimide films have long-term reliability and chemical durability problems, especially in harsh environments. Furthermore, polyimide-based humidity sensors fail to operate properly when water condensation occurs on the sensor's surface at high RH levels, which can be eliminated by using additional elements such as on-chip or off-chip heaters.

4.5.2 Resistive Humidity Sensors

Resistive humidity sensors measure the change in electrical impedance of a hygroscopic medium such as a conductive polymer, salt, or a treated substrate. The impedance change is typically an inverse exponential relationship to humidity. Resistive sensors usually consist of noble metal electrodes either deposited on a substrate by photoresist techniques or wire-wound electrodes on a plastic or glass cylinder. The substrate is coated with a salt or conductive polymer (Fig. 4.5.3). The sensor absorbs water vapor and ionic functional groups are dissociated, resulting in an increase in electrical conductivity. There are three groups of materials: (1) ceramics, (2) polymers and (3) electrolytes.

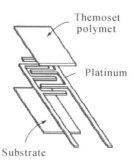

Fig. 4.5.3. Typical resistive humidity sensor configuration consisting of polymer coated electrodes on a suitable substrate

Ceramics based humidity sensors are popular and common-used resistive humidity sensors. It is usually made up of polymer film capacitors. When the ambient humidity changes, the dielectric constant of the humidity capacitor will change. It will lead to the change of the impedance of the capacitor. The merit of humidity capacitor is high sensitivity, high response speed, small hysteresis, and easy to be manufactured.

Various kinds of ceramics have been investigated as humidity sensitive materials, such as TiO_2, $LiZnVO_4$, $MnWO_4$, C_2O, and Al_2O_3. In general, ceramics have good chemical stability, high mechanical strength, and resistance to high temperature. However, they have nonlinear humidity-resistance characteristics and are not compatible with standard IC fabrication technologies.

Polymer based humidity sensors are another type of resistive humidity sensors that are present in the literature. Some researchers have examined the impedance changes of PVA (polyvinyl alcohol), TA (phthalocyaninosilicon), and Nafion with relative humidity. It has been reported that the humidity response and the stability of the polymer films are dependent on different chemical properties such as hydrophilicity, molecular, and ionic forms. Polymers also have nonlinear humidity-resistance characteristics.

An example of the electrolyte based humidity sensors is the LiCl dew point hygrometer fabricated with a composite of porous polymer and the salt. The humidity measurement is achieved by using the conductivity change of the LiCl polymer composition. The film is formed on an alumina substrate over interdigitated platinum electrodes. The device is integrated with a heater and a temperature sensor, which are connected to a control circuit to keep the vapor pressure of the saturated LiCl solution equal to the vapor pressure of the atmosphere. The humidity measurement is achieved by the measurement of the dew point temperature. Fig. 4.5.4 presents a prototype of the dew point hygrometer.

The major advantage of this sensor is that it is suitable for batch fabrication. However, it requires regular maintenance of the LiCl solution to guarantee the stable operation of the device, which is an important disadvantage.

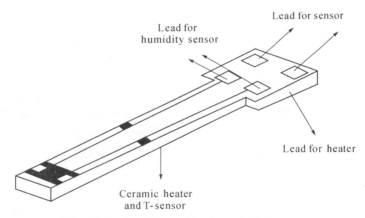

Fig. 4.5.4. A prototype of the dew point hygrometer

4.5.3 Thermal Conductivity Humidity Sensors

Thermal conductivity humidity sensors, also known as absolute humidity sensors, measure absolute humidity by calculating the difference between the thermal conductivity of dry air and air containing water vapor.

Thermal conductivity humidity sensors can be accomplished by a thermistor-based sensor. Two tiny thermistors (R_{t1} and R_{t2}) are supported by thin wire to minimize thermal conductivity loss to the housing. The left thermistor is hermetically sealed in dry air. Both thermistors are connected into a bridge circuit (R_1 and R_2), which is powered by voltage E. The bridge output is given by:

$$V_{out} = E \left(\frac{R_1}{R_1 + R_{t1}} - \frac{R_2}{R_2 + R_{t2}} \right) \qquad (4.5.4)$$

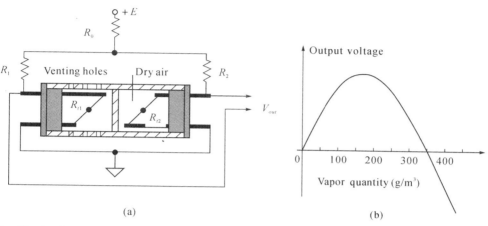

Fig. 4.5.5. Absolute humidity sensor with self-heating thermistors: (A) design and electrical connection; (B) output voltage

The thermistors develop self heating due to the passage of electric current (Fig. 4.5.5). Their temperature rises up to 170 °C over the ambient temperature. Initially, the bridge is balanced in dry air to establish a zero reference point. The output of this sensor gradually increases as absolute humidity rises from zero. At about 150 g/m³, it reaches the saturation and then decreases with a polarity change at about 345 g/m³.

4.5.4 Application

Relative humidity is sometimes measured with a device called a sling psychrometer. This instrument contains two thermometers, one of which is directly open to the air, while the other is immersed in a wet cloth, wick, or sponge. The psychrometer is rotated in the air being measured for several minutes, enough to permit a difference in the two temperatures to register. The wet bulb will show a lower ("suppressed") temperature that is a function of the relative humidity. Fig. 4.5.6 shows the relationship among the suppressed temperature, the air temperature, and the relative humidity. The RH can be determined from the intersection between the two temperatures.

Fig. 4.5.6 shows two relative humidity sensors that are based on the psychrometer principle. The sensor shown in Fig. 4.5.6a is the Shibaura HS-5. It consists of two thermistors (thermal resistors), one inside a sealed (dry) chamber and the other open to the air. Both thermistors are operated close to the self-heating point so that relatively small temperature changes create large resistance changes. When operated in a Wheatstone bridge configuration, the HS-5 will produce a voltage (V_0) that can be used to determine the relative humidity.

Fig. 4.5.6b shows the Figaro Engineering NH-series humidity sensor. It consists of a single assembly containing a thermistor RT_1 and a special absorptive sensor resistor element R_s. The sensor element consists of a high-polymer electrolyte deposited onto a porous ceramic bed mounted on an alumina substrate. A porous rubidium oxide (RbO_2) electrode covers the polymer. The humidity of the air causes the resistance of the ceramic/polymer to change in an exponential manner (Fig. 4.5.6c).

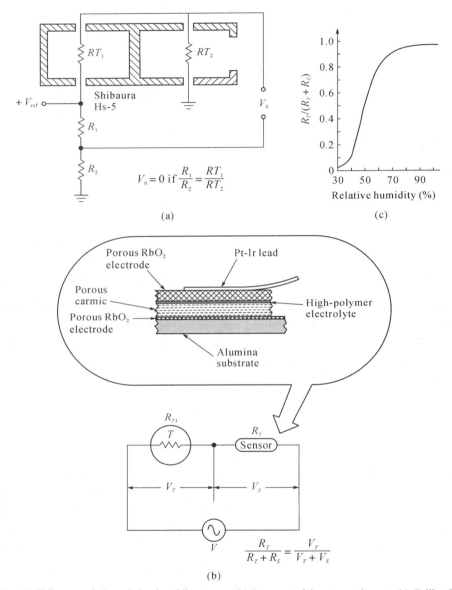

Fig. 4.5.6. (a) Shibaura resistive relative humidity sensor; (b) Structure of the sensor element; (c) Calibration curve

Shibaura Electronics is famous for thermal resistors in Japan. The latest in HS-series is the Shibaura HS-13 (Fig. 4.5.7), which is able to work out the cooking time according to the humidity when applied to the microwave oven. Its features are as follows:
• Remarkable downsizing: 22.8mm×22.8mm×12.7mm for the sensor part,

- Fast response to the humidity: as fast as 16 s,
- Short stabilization time: as short as 10 s,
- High durability: storage in 90% humidity and 200 °C for 2000 h.

Fig. 4.5.7. Shibaura humidity sensor HS-13

4.6 Intelligent Chemical Sensor Arrays

The intelligent chemical sensors generally include the sensor arrays and pattern recognition function. At present, the electronic or artificial nose (e-Nose) and electronic or artificial tongue (e-Tongue) have achieved great development.

4.6.1 e-Nose

e-Nose is an instrument, which comprises a sampling system, an array of chemical gas sensors with differing selectivity, and a computer with an appropriate pattern-classification algorithm, capable of qualitative and/or quantitative analysis of simple or complex gases, vapors, or odors.

4.6.1.1 Structure and principle

One cannot discuss the electronic nose without first comparing with the biological nose. Fig. 4.6.1 illustrates a biological nose and points out the important features of this "instrument". Fig. 4.6.2 illustrates the artificial electronic nose. Comparing the two is instructive. The human nose uses the lungs to bring the odor to the epithelium layer; the e-Nose has a pump. The human nose has mucous, hairs, and membranes to act as filters and concentrators, while the e-Nose has an inlet sampling system that provides sample filtration and conditioning to protect the sensors and enhance selectivity. The human epithelium contains the olfactory epithelium, which contains millions of sensing cells, selected from 100 – 200 different genotypes that interact with the odorous molecules in unique ways. The e-Nose has a variety of sensors that interact differently with the sample. The human receptors convert the chemical responses to electronic nerve impulses. The unique pattern of nerve impulses is propagated by neurons through a complex network before reaching the higher brain for interpretation. Similarly, the chemical sensors in the e-Nose react with the sample and produce electrical signals. A computer reads the unique pattern of signals, and interprets them with some forms of intelligent pattern classification algorithm. From these similarities we can easily understand the nomenclature (Stetter & Penrose, 2001).

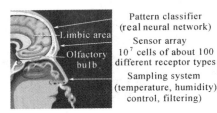

Fig. 4.6.1. The "Biological Nose"

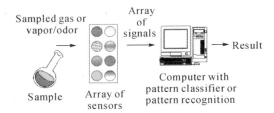

Fig. 4.6.2. The basic design of the "Electronic nose"

Although e-Noses are systems that, just like the human nose, try to characterize different gas mixtures, there are still fundamental differences in both the instrumentation and software. The Bio-nose can perform tasks still out of reach for the e-Nose, but the reverse is also true.

Table 4.6.1 Comparing e-Nose with the human nose

Human nose	Electronic nose (e-Nose)
10 million receptors, self generated	5 – 100 chemical sensors manually replaced
10 – 100 selectivity classes	5 – 100 selectivity patterns
Initial reduction of number of signals (– 1,000 to 1)	"Smart" sensor arrays can mimic this?
Adaptive	Perhaps possible
Saturates	Persistent
Signal treatment in real time	Pattern recognition hardware may do this
Identifies a large number of odors	Has to be trained for each application
Cannot detect some simple molecules	Can detect also simple molecules (H_2, H_2O, CO_2, …)
Detects some specific molecules	Not possible in general at very low concentrations
Associative with sound, vision, experience, etc.	Multisensor systems possible
Can get "infected"	Can get poisoned

Accordingly, an e-Nose is composed of two main components: the sensing system and the pattern recognition system, capable of recognizing simple or complex odors. And an individual sensor used for the detection of a particular substance, e.g., CO-sensor, is thus no e-Nose.

4.6.1.2 Sensing system

The sensing system, which consists of a sensor array, is the "reactive" part of the instrument. When in contact with volatile compounds, the sensors react, which means they experience a change of electrical properties. Each sensor is sensitive to all volatile molecules but each in their specific way. Most e-Noses use sensor arrays that react to volatile compounds on contact: the adsorption of volatile compounds on the sensor surface causes a physical change of the sensor. A specific response is recorded by the electronic interface transforming the signal into a digital value. Recorded data are then computed based on statistical models.

The most commonly used sensors include metal oxide semiconductors (MOS), conducting polymers (CP), quartz crystal microbalance (QCM), surface acoustic wave (SAW), and field effect transistors (MOSFET).

- *Gas sensor array*

Ping Wang et al., at Zhejiang University designed an electronic nose instrument CN e-Nose II used in lung cancer early stage diagnosis based on metal oxide gas sensor array. According to the research of Phillips et al., the exhaled gas of lung cancer patients contains some volatile organic compounds (VOCs) that can be taken as the biomarker of lung cancer, the corresponding diagnosis results can be obtained by detecting these VOCs.

In the CN e-Nose II, five TGS MOS gas sensors and three MQ MOS gas sensors were used in the gas sensor array. The cross sensitivity between these sensors can be seen from Table 4.6.2, and they have different sensitivities to the homogeneous substances. The original intention of the CN e-Nose II lies in examining the concentration of biomarkers in a human's breath to represent the health condition. Taking this into consideration, 30% of the breath gas comes from the alimentary tract, which makes a contribution to health representation, and taking the digestive tract disease of an ulcer as an example, it will have ammonia in micro-scale from breathing. Therefore the sensor array includes not only eight metal-oxide semiconductor gas sensors shown in Table 4.3, but also one high sensitivity NE-NH_3 electrochemical sensor.

Table 4.6.2 Characteristic parameter list of 8 MOS gas sensors according to the datasheets

Sensor model	Detectable gas	Detection range
TGS813	H_2, Isobutene, Ethanol, CH_4, CO	500 – 10,000 ppm
TGS822	Acetone, Ethanol, Benzene, n-Hexane, Isobutane, CO, CH_4	50 – 5,000 ppm
TGS2600	H_2, Ethanol, Isobutene, CO, CH_4	1 – 100 ppm
TGS2602	Toluene	1 – 30 ppm
	H_2S	0.1 – 3 ppm
	NH_3	1 – 30 ppm
	Ethanol	1 – 30 ppm
	H_2	1 – 30 ppm
TGS2620	Methane, CO, Isobutene, H_2, Ethanol	50 – 5,000 ppm
	Ethanol	100 – 200 ppm
MQ-2	H_2	300 – 5,000 ppm
	CH_4	5,000 – 20,000 ppm
	Butane	300 – 5,000 ppm
MQ-3	Ethanol, Benzene, n-Hexane, LPG, CH_4	0.1 – 10 mg/L
MQ-6	LPG, LNG, Butane, Propane	100 – 10,000 ppm

In the experiments of breath examination, low-concentration gas mixtures were prepared employing the possible biomarkers in the lung cancer patient's breath. Then the analysis was carried out after sample preconcentration.

Taking peak height, stable value and peak area as the characteristic values, the response curves from 8 MOS gas sensors are shown in Fig. 4.6.3.

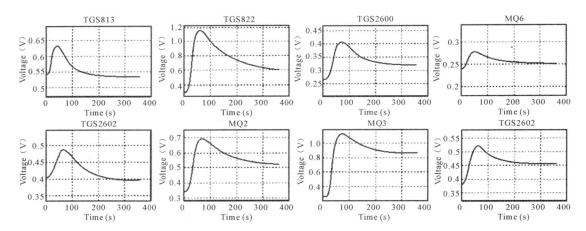

Fig. 4.6.3. Response curves of the mixed gas samples from 8 MOS gas sensors

4.6.1.3 *Virtual gas sensor array*

Ping Wang et al. (2009), at Zhejiang University designed an electronic nose instrument used in lung cancer early stage diagnosis based on a SAW gas sensor combined with a capillary separation technique.

The structure of the e-Nose is shown in Fig. 4.6.4. The respiratory gas is enriched by an adsorption tube, desorption happens in the inlet of the capillary at a high temperature, then the VOCs is carried into the capillary to be separated by the carry gas. When the VOCs come out from the capillary, there will be a frequency change because the VOCs can attach to the surface of the SAW sensor independently by reason of condensation, then the PCA and image analysis are used for pattern recognition after the signal is obtained and processed.

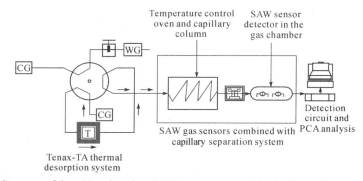

Fig. 4.6.4. Structure of the e-Nose based on SAW gas sensors combined with capillary separation system

One detect result of the mixed VOCs sample by the e-Nose system is shown in Fig. 4.6.5. As seen from the figure, the VOCs can be easily detected by the e-Nose system.

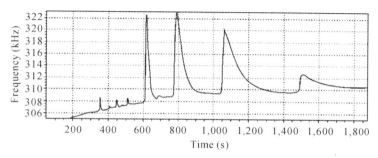

Fig. 4.6.5. Spectrogram of mixed VOCs sample

This e-Nose, which is based on the gas chromatography technology, causes its responses to have two components: the appearance time and the response intensity. We may know from the capillary separation technique that the peak time represents the time of each component used to pass through the capillary, as a result of the difference of physical and chemistry characteristics, different components need difference time to pass through the capillary, so the appearance time of the peak can be used for determining material. Because the capillary separating technique is applied, some disturbance factor in the environment is separated in the peak time; there will be no influence on the substance we need to detect, regardless of whether its density is high or low compared to the environmental disturbance. So the SAW gas sensor combined with the capillary separation technique can simulate a virtual sensor array containing hundreds of orthogonal (non-overlapping) sensors, which can detect and distinguish hundreds of different kinds of gases.

4.6.1.4 Pattern recognition system

The raw signal generated by an array of odor sensors is a typical collection of different electrical measurements vs. time curves (Fig. 4.6.6). These signals need to be processed in a more or less sophisticated manner in order to allow the recognition of a particular odor.

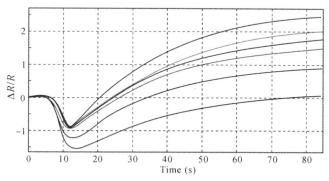

Fig. 4.6.6. Typical sensor response of a conducting polymer sensor array to a certain odorant

A basic method for observing a data set is simply to plot all variables, or a subset of variables, in a bar chart. Another form of output is a scaled polar plot (Fig. 4.6.7). Both forms can be obtained from the raw signal by integration of the curve over a distinct period of time. This way of visually displaying data is simple to interpret. Each vector on the polar plot represents the output from one sensor. As the relative response of each sensor changes when the sensor array is exposed to vapors from differing samples, the overall shape and appearance of the polar plot will vary.

Biomedical Sensors and Measurement

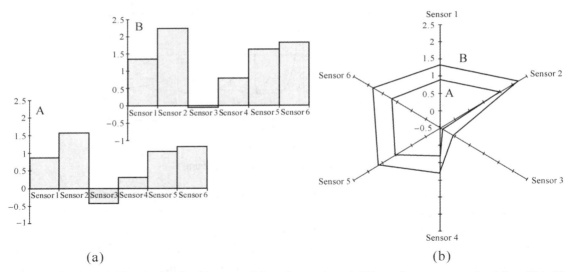

Fig. 4.6.7. Bar graph (a) and polar plot (b) generated from the raw data. A: Values taken as average signal from 15 to 75 s; B: Values taken as average signal from 55 to 75 s

In order to express the similarity or difference of two odors, it may be useful to calculate the distance of the two corresponding data sets.

As a chemical sensor system providing several variables, a multivariate distance measure is therefore more appropriate than a simple univariate distance measure. A multivariate distance is calculated in the original or a reduced variable space. There are two main methods to calculate multivariate distances. The euclidic distance (ED) is the length of the vector connecting two points in the variable space.

The ED can be calculated according to

$$ED = \sqrt{\sum_1^n (x_a - x_b)^2} \qquad (4.6.1)$$

where x_a is the response of sensor number n produced by sample A and x_b is the response of the same sensor of x_a contacting with sample B.

However, the euclidic distance does not take the variation within classes into account. A more appropriate distance measure between classes is the statistical distance (also called Mahalanobis distance). The statistical distance is calculated as the ratio of the euclidic distance and the class variance in the direction of vector among class centres. Directions of high variance within the classes will thus give a low statistical distance.

- **Classification and dimension reduction**

Classification is the task of making a model capable of assigning observations into different classes. A classification is often combined with a dimension reduction in variable space. A multi-sensor system produces data of high dimensionality, i.e. a large number of variables characterizing each observation. It is difficult to visualize more than three dimensions simultaneously. Hence, methods to reduce the dimensionality of multivariate data sets are important. The variable space is an essential concept in order to grasp the ideas behind many data processing techniques. In the variable space, variables are seen as orthogonal basis vectors. An observation corresponds to a point in the sensor space, and a whole data set can be seen as a point swarm in this space. A way to reduce the dimensionality is to find new directions in the variable space and use only the most influential directions as new variables. A basis change is made and a dimensionality reduction is performed. In a principal component analysis, a transformation (projection) in the variable space is made (Fig. 4.6.8). Directions are found explaining as much of the variance in a data set as possible. These new directions, called principal components, are then used as the new variables. Keeping only principal components with high variation, leads to a dimension reduction. There are other methods to reduce the dimensionality in a variable space. All these methods are performed by finding new directions optimizing a specific criterion, and only the most influential directions are kept for the following visualization and classification.

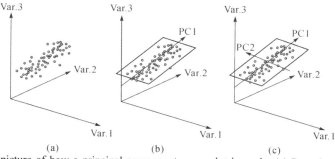

Fig. 4.6.8. Schematic picture of how a principal component score plot is made: (a) Raw data; (b) The first principal component is the direction with the most of the variance in the data set; (c) The low-dimensional projection of the data can be used as a simple but good approximation of the data set

- *Artificial neural network*

An artificial neural network (ANN) is an information processing paradigm that was inspired by the way biological nervous systems, such as the brain, process information. The key element of this paradigm is the novel structure of the information processing system. It is composed of a large number of highly interconnected processing elements (neurons) working in unison to solve specific problems (Fig. 4.6.9). ANNs, like people, learn by example. An ANN is configured for an application such as identifying chemical vapours through a learning process. Learning in biological systems involves adjustments to the synaptic connections that exist between the neurons. This is true of ANNs as well. For the electronic nose, the ANN learns to identify the various chemicals or odors by examples.

The basic unit of an artificial neural network is the neuron. Each neuron of the input layer receives a number of inputs, multiplies the inputs by individual weights, sums the weighted inputs, and passes the sum through a transfer function, which can be, e.g., linear or sigmoid (linear for values close to zero, flattening out for large positive or negative values). An ANN is an interconnected network of neurons. The input layer has one neuron for each of the sensor signals, while the output layer has one neuron for each of the different sample properties that should be predicted. Usually, one hidden layer with a variable number of neurons is placed between the input and output layer. During the ANN training phase, the weights and transfer function parameters in the ANN are adjusted such that the calculated output values for a set of input values are as close as possible to the known true values of the sample properties. The model estimation is more complex than that for a linear regression model due to the non-linearity of the model. The model adaptation is made using the specific algorithm like back-propagation algorithm involving gradient search methods, where each weight is changed in proportion to the error which is caused.

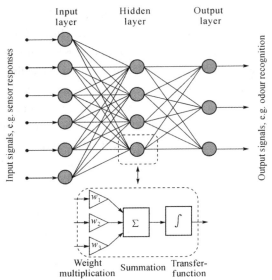

Fig. 4.6.9. Schematic of an artificial neural network. It consists of three interconnected layers of neurons. The computing neurons (hidden and output layers) have a non-linear transfer function. The parameters of the neurons are chosen with a minimization of the output error for a known training set

4.6.2 e-Tongue

Electronic tongue (e-Tongue) is a sort of analytical equipment using multi-sensor array to detect the characteristic response signal of the liquid sample and process it by pattern recognition and expert system for learning identification to obtain qualitative or quantitative information. The most obvious difference between e-Nose and e-Tongue is that the former is for the gases while the latter is for the liquids.

The research on e-Tongue began only a few decades ago, so it is still not very mature. The most successful company in marketing e-Tongue systems is Alpha-MOS whose production accounts for more than 99% of the world's market. The e-Tongue systems are very useful in food, medical, environmental and chemical industry.

4.6.2.1 Principle

The taste of organisms (Fig. 4.6.10) comes from taste buds on the surface of the tongue. The taste buds respond to different chemicals in the solution to generate signals which are transferred through the nerves to the brain. Then the brain does the analysis and processing to obtain the overall features of the signals and gives the distinction between different chemicals as well as the sensory information.

The initial design idea of e-Tongue originates from the biological mechanism of taste recognition (Fig. 4.6.11). Just like the tongue of organisms, the sensor array of e-Tongue responds to different chemical substances and collects a variety of signals to be transferred to the computer. Instead of the brain of the organism, the computer distinguishes the different signals, makes identification and finally gives sensory information of the various substances. Just as taste buds on the surface of the tongue, each individual sensor in the sensor array has cross sensitization. That is, a separate sensor not only responds to a chemical, but to a group of chemicals. In addition, while responding to specified chemicals, the sensor also responds to some other chemicals of a different nature.

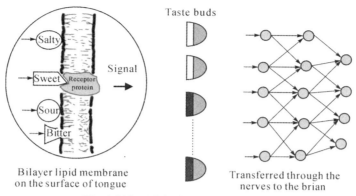

Fig. 4.6.10. Biological taste recognition

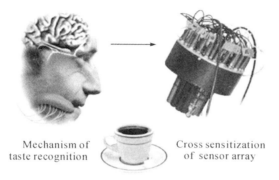

Fig. 4.6.11. Design idea of e-Tongue

The realization of the e-Tongue technology is based on multi-sensor multicomponent analysis in the traditional analytical chemistry. Supposing that a specific sensor system with an array consisting of M sensors are applied to the analysis of a solution containing N components whose concentrations are C_1, C_2, ..., C_N and all of the N components will be responded. P_i ($1<i<M$) represents the signal of sensor i. The M-sensor N-component analysis system can be written as the following mathematical expressions:

$$P_1 = A_{1,1}C_1 + A_{1,2}C_2 + \cdots + A_{1,N}C_N$$
$$P_2 = A_{2,1}C_1 + A_{2,2}C_2 + \cdots + A_{2,N}C_N$$
$$\cdots$$
$$P_M = A_{M,1}C_1 + A_{M,2}C_2 + \cdots + A_{M,N}C_N \tag{4.6.2}$$

All of the M sensors are specific which means that a sensor only responds to one component. The constants $A_{i,j}$ which are the ratio of the signal of sensor i to the concentration of component j are already known. As long as $M \geqslant N$, Eq. (4.6.2) can be solved by matrix operations to obtain the concentrations of all the N components.

The difference between the e-Tongue technology and the traditional multi-sensor multicomponent analysis is that the e-Tongue employs cross-sensitive sensors instead of specific sensors in the array. In this way, $A_{i,j}$ in Eq. (4.6.2) become non-linear functions related to the concentration of component j. So Eq. (4.6.2) must be solved by a non-linear pattern recognition method such as artificial neural networks.

The e-Tongue system should be trained by lots of sample solutions to establish self-learning expert system and then do the calculation.

4.6.2.2 Characteristics

The structure of e-Tongue can be divided into three main parts which are the cross-sensitive sensor array, the self-learning expert system and the smart pattern recognition system which are respectively equivalent to the tongue, memory and brain calculation of organisms. The main characteristics of e-Tongue can be summarized as follows:

(1) The detection object is liquid samples;

(2) The signal obtained is the overall response to a solution, rather than the response to a specific component in the solution;

(3) The attributes of different samples are able to be distinguished through processing the original signal collected from the sensor array;

(4) The sample attributes derived by e-Tongue are different from the concept of taste of organisms.

4.6.2.3 Functional membranes

- **PVC membrane**

The e-Tongue system based on PVC membrane sensor array invented by the research team of Toko (Toko, 1998). in Kyushu University was the first e-Tongue system in the world. The PVC membrane sensor shown in Fig. 4.6.12 measured the open circuit potential with the Ag/AgCl reference electrode. The intensity of the affinity between taste substances and the PVC membrane modified with a variety of active materials was transformed into a potential signal. Such an e-Tongue system is generally composed of several electrodes respectively, providing a response to different taste substances, had the advantage that the data was relatively limited. So the test results represented by radar charts were directly corresponding to the characteristics of the taste substances.

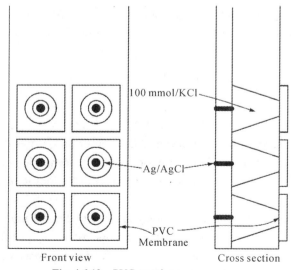

Fig. 4.6.12   PVC membrane sensor array

- *Chalcogenide glass membrane*

Chalcogenide glass membrane sensor is a kind of solid-state ion selective electrode which has been applied in the detection of heavy metal ions for more than 30 years. E-Tongue with chalcogenide glass sensor array (Fig. 4.6.13) was originally invented by the research team of Legin A. and Vlasov Y. G (Vlasov et al., 1994). They developed many non-specific sensors based on chalcogenide glass materials such as $GeS-GeS_2-Ag_2S$, $Ag_2S-As_2S_3$, Ge-Sb-Se-Ag and $AgI-Sb_2S_3$. According to the principle of e-Tongue, a variety of chalcogenide glass sensors of high sensitivity and low selectivity were used to fabricate the sensor array to detect heavy metal ions and H^+ in the solution. This e-Tongue system has wide application in environmental assessment of water pollution, food quality assessment, etc.

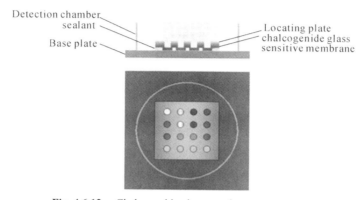

Fig. 4.6.13. Chalcogenide glass membrane sensor array

- *Langmuir-Blodgett membrane*

The research team of Riul A. Jr. in Brazil invented an e-Tongue system based on Langmuir-Blodgett membrane sensor array (Fig. 4.6.14). They modified the platinum electrode surface with 10 nm-thick Langmuir-Blodgett membrane that consists of stearic acid, polyaniline, polypyrrole, etc. It was easy to detect the signal of the interaction between the sensors and the taste substances such as sour, sweet, bitter, salty and so on by electrochemical impedance spectroscopy. The results showed that the e-Tongue system was very sensitive to the taste substances and able to distinguish mineral water, beverages, wine, coffee, etc.

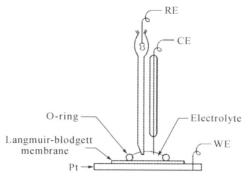

Fig. 4.6.14. Electrochemical detection device based on Langmuir-Blodgett membrane

- *Bionic taste chip*

The research team in Austin University utilized ion-sensitive polymer microspheres as bionic taste buds to detect the constituents of a solution based on photochemical principles. The bionic taste chip was able to do the parallel, real-time and quantifiable measurement of a variety of constituents in the same solution. Synthetic microspheres whose diameters in the range of 50 – 100 μm were fixed in micro grooves etched in the silicon wafer surface (Fig. 4.6.15). The chip and CCD were separately fixed on top of and under the platform. Modulated lights emitted by the blue light-emitting diodes went through the ball and the bottom of the platform and were absorbed by the CCD detector. The taste chip could preliminary determine the concentrations of H^+, Ca^{2+}, Ce^{3+} and sugar in the solution.

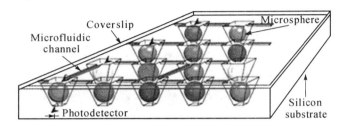

Fig. 4.6.15. The bionic taste chip

Based on the above work, the researchers developed algorithms of signal recognition to achieve the multi-ion identification automatically. Fig. 4.6.16a shows the detection system of the bionic taste chip. The analyte was pumped into the reaction chamber and reacted with the sensitive materials adsorbed on the surfaces of microspheres whose colors changed in the reaction. Images of the microspheres were recorded by CCD through the microscope. RGB values were extracted from specified areas on the microspheres as the output of the taste chip. Principal component analysis (PCA) was applied to process the data to achieve the qualitative and quantitative measurement of the constituents of the analyte. Fig. 4.6.16b shows the PCA results which successfully distinguished the six kinds of metal ions in the same solution.

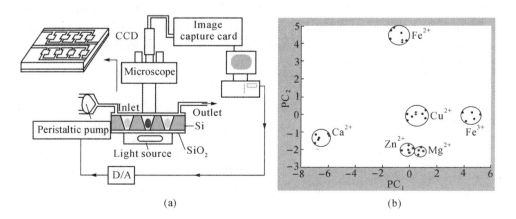

Fig. 4.6.16. Experiment of the bionic taste chip: (a) The detection system; (b) The PCA results of 6 kinds of metal ions

4.7 Micro Total Analysis System

Micro Total Analysis Systems (μTAS) have become the research hotspot in the world since its first appearance in 1990s, which is a distinct and novel field based on multidisciplinary fields such as analytical chemistry, micro-electro-mechanical systems (MEMS), computer science, electronics, materials sciences, and biology. It is also known as microfluidics or "lab-on-a-chip".

4.7.1 Design and Fabrication

As a unique and multidiscipline field, microfluidics is expanding into new areas of applications and the systems under development are becoming more complex. There is an increasing demand both for theoretical and experimental work on fundamental physical and chemical phenomena, and also for better modeling tools.

4.7.1.1 Basic principles of microfluidic chips

Fluid mechanics is one of the important disciplines to be further addressed, which may have great influences on the design of microfluidic devices and understanding the special effects related to fluid flow in micron-scale. Most of researchers working on microfluidics employ fluid mechanic modeling as a design tool, or as a way to correlate and explain experimental results. When dealing with flow in configurations of microns or less, some special effects and unexpected phenomena can be observed. Sir Eddington (1928) once said, "We used to think that if we know one, we know two, because one and one are two. We are finding that we must learn a great deal more about 'and'" (Eddington, 1928). Basically, the flows in macro and micro configurations are quite different. The unique features in micron-scale fluid flow are still far from being completely understood due to not much being known about the complex surface effects that play major roles in these events. This may excite researchers for years to search for the answers to these issues (Ho and Tai, 1998).

In many simple cases, the flow-pressure characteristics of a device are the fundamental quantities that can usually be dominated by one single restriction, which make it sufficient to use a simple analytical model well known from macroscopic fluid mechanics. This approach has been successfully applied to the modeling of some microfluidic components such as valves, and channels. In more complex structures or systems, numerical simulations are used. The most common method of numerical simulation is based on a subdivision of the complete structure into lumped elements, which can be described individually by simple analytical models, and for which simple relations between individual lumped elements can be formulated (Gravesen et al., 1993). These models and relations of interaction can then be fed into a dedicated or generally available computer program. In this way, micropumps, valves, flow sensors, and flow dispenser have been simulated using dedicated computer programs. In order to apply the approaches mentioned before to fluid mechanic modeling successfully by the direct utilization of analytical models or lumped element models, it is necessary to make correct assumptions as to types of flow. In micro flows, the Reynolds number is typically very small and shows the ratio between the viscous force and the inertial force. It is a common and practical method to determine whether a given flow pattern is laminar or tubular by evaluating the Reynolds number.

For the design of microfluidics, it is necessary to consider the following requirements carefully:

(1) The uniformity of the flow velocity in microfluidic chip, the reduction of dead volume;

(2) The evaporation and random flow of the fluid during reaction and storage, the uniformity of mixing and the avoiding of bubble formation;

(3) The calculation of reagent and production volumes, the compatibility, reliability, and reactivity of materials used in microfluidic chips;

(4) Interference of flow to the signal acquisition and improvements on signal to noise ratio;

(5) Proper handling of the waste and used microfluidic chips.

Computer modeling and simulation is an important approach for the design of microfluidics, which can also be used to interpret the experimental data. It can provide beneficial prediction on the liquid-phase process of flow mechanics as well as modeling of device thermal fields and chemical concentrations. By this approach, the time and cost of the microfluidic chip design can be significantly reduced and various parameters can be optimized. Currently, there are some commercial softwares, such as Flume (Coventor, INC) for microfluidic chip design which are available.

4.7.1.2 Microfluidic chip fabrication

It is very important to choose proper materials to fabricate microfluidic chips. Some of the main issues that should be addressed are as follows:

(1) Chemical and biological compatibility between microfluidic chip and working interface;

(2) Electric insulation and thermal properties of materials used;

(3) Optical properties related to the interference of signal detection;

(4) The modification properties of materials related to the generation of electroosmosis and solid immobilization biological molecules;

(5) The simplicity and low cost of microfluidic chip fabrication.

However, it is hard to have the kind of materials that can fully satisfy all the requirements mentioned above. The selection of materials for microfluidic chip fabrication is usually made according to the practical applications. At present, materials that are mainly used include silicon, glass, crystal, and organic polymer. Every type of material has its own advantages and disadvantages and some differences exist in the corresponding fabrication process. For example, silicon has excellent chemical inert and thermal stability. Also, the techniques for the production and micro fabrication are mature and have been widely used in semiconductor and integrated circuits. So, the initial microfluidic chips are usually fabricated on the basis of silicon.

The fabrication process of a microfluidic chip is very sensitive to the environment, and should be done in a clean room. The techniques used for microfluidic chip fabrication originated from the micro fabrication of semiconductors and integrate circuits, while they are different when compared to the silicon-based techniques for the two dimensional and depth fabrication of integrated circuit chips. The fabrication methods vary according to the different chip base materials. Some important methods are utilized such as lithography, etching techniques, soft lithography, molding, LIGA, ultraviolet laser, and deep reactive-ion etching (DRIE). For the details related to the fabrication process of microfluidic chips it is suggested that you refer to the related literatures.

Sealing is also a very important process for chip fabrication. Before sealing, chips with different structures and functional units must be very clean through strict washing and handling. For silicon and glass, some commonly used sealing techniques include heat sealing, anodic bonding, and cryogenic adhesive bonding. For polymer materials, various sealing methods, such as hot pressing method, heat or light catalysis adhesive technique, organic solvent adhesive method, automatic adhesive method, plasma oxidation sealing method, ultraviolet irradiation method, and cross-link agent regulation method, are available to be selected according to the different materials used.

After fabrication, the surface of micro channels usually needs to be modified according to their practical applications to improve some of the chemical or physical properties. The research on the techniques for chip surface modification is composed of a large percent of the research on the microfluidics. When the analysis system works, micro flow operation that depends on the inner surface properties of microchannels is usually worked in a passive mode. In the microfluidic chips that use electro osmotic flow driving, it is the most common used technique due to its simple operation, without requirements on extra devices, and without increase in the chip system volume. The basic principle of this technique is to determine the velocity and direction of the electro osmotic flow by changing the density and polarity of charges located in the inner surface of micro channels. As shown in Fig. 4.7.1, by the method of coating an electrolyte layer of polymer on the inner surface of micro channels, the direction of the micro flow can be changed and complex flow operations can be achieved (Barker et al., 2000). Furthermore, simultaneous flow in opposite directions can be achieved in a single micro channel. Another predominant application of this technique is to obtain the flow limitation. For example, hydrophobic glass micro channels can be achieved by the method of immobilizing a self-assemble monolayer of octadecyltrichlorosilane on the inner surface of micro channels. In the polymer micro channel, the hydrophobic and hydrophilic region can be achieved by depositing a layer of poly phenylene-2-methyl and silicon oxidize on the inner surface of micro channel.

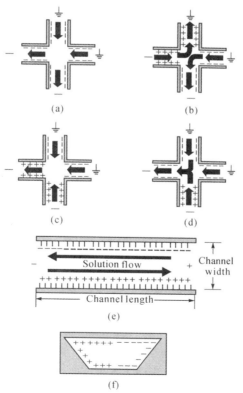

Fig. 4.7.1. Complex flow control can be realized by the surface modification of micro channel: (a), (b), (c), and (d) show various flow modes can be achieved by the control of charge polarity located on the inner surface of micro channel. (e) and (f) show the opposite direction flow can be realized in the single channel. (e) is the top view of the micro channel, while (f) is the cross Section view

4.7.1.3 Driving and control of micro flow

The basis of a microfluidic chip operation is the technique for micro flow driving and control. Since the invention of the microfluidic chip, it has always been an important topic in the basic research field of microfluidic chip and new techniques and methods are progressively appearing. Here, we will briefly introduce two techniques for micro flow driving, which are electro osmotic driving and micro pump driving. Also, two techniques for micro flow control will be introduced, which are electro osmotic control and micro valve control.

At present, electro osmotic driving is one of the most widely used methods in micro flow driving. Its basic principle is using the fixed charges on the inner surface of micro channels to drive the micro flow. Its advantages include lack of mechanical components, simple configuration, convenient operation, flat flow, and no pulsation. However, this method is sensitive to the influences of external electrical fields, channel surface, properties of micro flow, and the effect of heat transfer. So, it is not so stable and can only be applied to electrolyte solutions.

Electroosmosis cannot only be used to directly drive the charged flow, but also can be used as the energy source of the micro pump, which is called the electro osmotic pump. The method to achieve this kind of electro osmotic pump is as follows: electrodes with certain intervals are fabricated on the surface of a chip basis by using lithography. Then it is sealed with PDMS micro channels to form a hermetic electro osmotic driving system. When it works, the voltage is applied to the electrodes to generate an electro osmotic flow. As the electro osmotic flow only exists between the two electrodes, the flow outside the two electrodes can be driven by the electro osmotic flow. Consequently, the function of the electro osmotic pump can be realized as shown by Fig. 4.7.2.

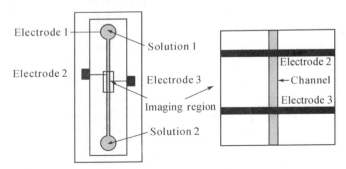

Fig. 4.7.2. Schematic diagram of electro osmotic pump structure

Pneumatic micro pump in the mechanical driving system is composed of multiple pneumatic micro valves. Its structure is shown in Fig. 4.7.3 (Unger et al., 2000). When the pressure is applied, PDMS thin film deformed under the effect of gas pressure, leading to the block of channel and the closing of the valve. When there is no pressure applied, the restitution of the PDMS thin film can be achieved by its own elastic force, thus the channel becomes unobstructed and the valve sits open. By the sequential control of the opening and closing of three valves, the driving of micro flow can be obtained.

Micro flow control is the central principle of microfluidic chip operations. It is related to almost all the processes such as sampling, mixing, reaction, and separation, which are necessary to be finished in the controllable flow. Valve is the central component for flow control both in macro and micro scale. Due to its importance, micro valve has been deeply studied before the invention of the microfluidic chip. In the primary stage of its development, microfluidic chips are normally a kind of capillary electrophoresis on a chip, which is dependent on the electro osmotic driving. So, until now, electro osmotic driving is still the most widely used technique for micro flow control. In addition, the structure of channel, surface modification of the chip, laminar flow, and the effect of diffuse also play an important role in micro flow control.

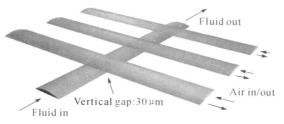

Fig. 4.7.3. Schematic diagram of pneumatic micro pump structure (reprinted from (Vnger et al., 2000), Copright 2000, with permission from AAAs)

There are various kinds of micro valves. Theoretically, all components that can control the opening and closing of the channel can be used as micro valves in microfluidic chips. An ideal micro valve can be characterized as follows: low leakage, low energy consuming, fast responses, linear operation, and wide range of adaptation (Gravesen et al., 1993). According to the necessity of the excitation source during micro valve operations, micro valves can be classified into passive valves and active valves. Fig. 4.7.4 shows the structure of a type of passive one-way valve and how it works.

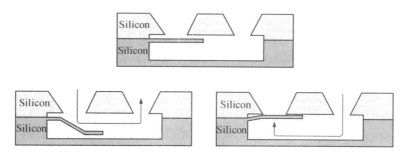

Fig. 4.7.4. Schematic diagram of passive one-way valve

Electroosmosis is a type of phenomenon that the solution in the micro channel can move in the desired direction along the inner surface of the channel under the effect of the electrical field. It has been widely used in micro flow control. Compared to other types of micro pumps, the most important feature of electro osmotic valves is their simple and flexible operation. The velocity and direction of flow can be controlled by adjusting the voltage applied to different nodes of the micro channel. Consequently, operations such as complex mixing, reaction, and separation can be realized. Besides the voltage, electro osmotic micro valves can be affected by such factors as the chemical composition of channel surface, ingredients of buffer solution, and temperature.

4.7.1.4 Sample introduction and pretreatment

Sample introduction is the first step of microfluidic chip analysis. It includes the process of sampling from the analysis object and the introduction of sample into the micro channel for sample handling. Before detection, a series of pretreatment and reaction steps are necessary to be done to the sample, such as pre-separation, pre-concentration, and dilution. Fig. 4.7.5 is the schematic diagram of microfluidc system sample operation mode.

Currently, in most microfluidic analysis systems, sample, reagent, and buffer solution are stored in a well type storage chamber located on the chip. Sample introduction methods usually add the sample to the well type storage chamber manually or automatically. Then the sample is imported into the channel for pretreatment or directly separated and analyzed. If various samples need to be measured, it is necessary to change the sample stored in the well type storage chamber manually or automatically in an intermittent way. Although the microfluidic analysis system has the ability of repeated measurement, the results of automatic continuous measurement are originated from the continuous measurement of the same sample. Very few of the results are originated from the different samples. There is no report on the applications of microfluidic analysis systems to the real-time process monitoring due to the lack of efficient sample changing methods.

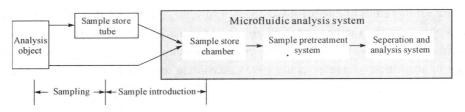

Fig. 4.7.5. Schematic diagram of microfluidc system sample operation mode

There are two methods being used to solve the changing sample problem. One is the once-off sample introduction, such as the utilization of disposable chips, multi sample chambers on a single chip, multi analysis units on a single chip, and multi sample introduction before measurement. Another method is to use recyclable chips to realize continuous changing of the samples either manually or automatically. Generally, the sample source tends to provide continuous sample flow. In order to get output from the sample zone, some auxiliary methods are necessary. A commonly used method is to set an auxiliary channel on the chip, which is perpendicular to the sample processing channel. The sample zone can be generated in the cross of the two channels. This method is called the single channel aid sample introduction. It is the most studied and most representative sample zone introduction method. It includes two steps, loading and sampling. Loading refers to the process of sample load to the auxiliary channel through a storage chamber and is filled within the cross of channels. Sampling is the process of introducing the sample that is located in the cross to the sample processing channel by electrical forces or pressure. Fig. 4.7.6 shows the principle of the simple sample introduction method by electrical forces.

Sample for microfluidic analysis systems are usually related to the biological sample containing complex composition. So, the techniques for sample pretreatment are very important, which often include sample pre-separation and pre-concentration. Pre-separation includes liquid-liquid extraction, solid phase extraction, filtration, chromatography, and membrane separation. Pre-concentration includes iso-electric focusing, isotachophoresis, and field amplified stacking. Furthermore, multi phase laminar flow techniques and various micro filters (Fig. 4.7.7) can be used to sample pretreatment.

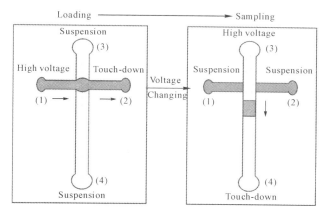

Fig. 4.7.6. Schematic diagram of simple sample introduction method by electrical forces: (1) Sample chamber; (2) Sample waste solution chamber; (3) Buffer solution chamber; (4) Buffer waste chamber

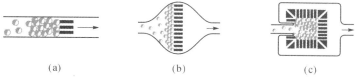

Fig. 4.7.7. Structure of various micro filters: (a) Micro channel-based filter; (b) Micro channel magnum-based filter; (c) Micro square cage-based filter

4.7.1.5 Micro mixer, reactor and separator

Reaction is the central process of chemical and biological experiments. Also mixing and separation are necessary for the process of reaction, especially in the micro scale. So, micro mixer, reactor, and separator are important components of microfluidic chips.

Micro mixer is very useful in the biological process that requires rapid responses such as the hybridization of DNA, cell activation, enzyme reaction, and protein folders. In microfluidic systems, the size of the channels is in the micro meter scale. The velocity of the solution is usually low and the solution mixing is mainly based on the mechanisms of laminar flow, which can be greatly influenced by the molecular diffusion. In order to improve the efficiency of laminar flow mixing, some principles should be followed: (1) extending the flow shear to increase the contact area of solution; (2) splitting and recombining the solution by the utilization of distributed mixing design, consequently reducing the solution thickness to realize more efficient mixing.

Micro reaction technique is the application of micro structure advantages to the process of chemical reaction. Micro reactor is a mini chemical reaction system with unit reaction interface in a micro meter scale. Its basic features include a small linear scale, high physical quality gradient, high surface to volume ratio, and low Reynolds number. Also, by its parallel units, a micro reactor can realize flexible and scale up production, and rapid and high throughput screening.

Recently, great progress has been achieved in the micro separation techniques. Now, various chromatography and electrophoresis separation modes can be realized on chips. Micro separator has become one of the fastest developing and maximum maturity technical units, which has greatly advanced the integration trends of microfluidic chips. Taking the integrated capillary electrophoresis chip as an example, microchannels and other functional units can be etched on the chip in a few centimeter square areas using micro processing technology. Consequently, a micro analysis device integrated with functions of sample introduction, separation, reaction, and detection can be realized, which is characterized with rapid, high efficiency, low sample consuming, low cost, and portability. Fig. 4.7.8 shows schematic diagram of the structure of an electrophoresis chip.

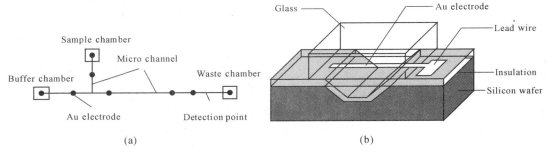

Fig. 4.7.8. Structure of (a) an electrophoresis chip and (b) its microelectrode and lead wire

4.7.1.6 Detection methods

The detector of microfluidic chips is used to measure the desired composition of the sample as well as its quantity. The overall performance of detectors will have a great influence on the sensitivity, detection limit, and detection speed of the whole system. So, it is the key component of microfluidic chips. Compared to the conventional analysis systems, microfluidic chips have some special requirements on its detector. Detection techniques, which are characterized with the higher sensitivity and signal-to-noise ratio, higher response speed, miniaturization, and low cost, are greatly preferred for usage in microfluidic chips. Currently, a large number of detection techniques have been used in microfluidic chips. However, the optical and electrochemical detection methods are two of the most widely used methods. Because microfluidic chips take over some characteristics of capillary electrophoresis, optical detection methods such as laser induced fluorescence (LIF), chemiluminescence, and UV absorption, are still the mainstream detection methods. Fig. 4.7.9 shows the basic optical structure of a confocal LIF detector. Various electrochemical detection methods have also been used in microfluidic chips due to their advantages of simply structure, low cost, and easy integration. In addition, mass spectra detector plays an irreplaceable role in the research of proteomics due to its powerful capability of distinguishing and identification.

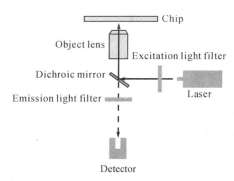

Fig. 4.7.9. Schematic diagram of basic optical structure of confocal LIF detector

4.7.2 Applications

Microfluidic chips have great potential in microminiaturization, integration, and portability of analytical devices, which provide wider prospects for microfluidic chips to be applied in many different fields such as biomedicine, high throughput screen for drug synthesis, optimization of priority crops sterile, environmental monitoring and protection, health quarantine, judicial expertise, and biological warfare agent detection. Currently, nucleic acid research is still one of the most widely used application fields for microfluidic chips. It has broadened its application fields from analysis of simple nucleic acid sequences to complex genetic analysis and diagnosis. In the clinical laboratory, microfluidic chips can perform the detection of multiple diseases of multiple patients on a single chip, which can provide very helpful diagnostic information for doctors.

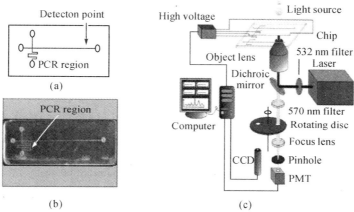

Fig. 4.7.10. Microfluidic chip for SARS virus detection: (a) Structure of the microchip; (b) PCR microchip; (c) SARS virus detection by the laser-induced fluorescence system.

In 2003, with the large-scale outbreak of SARS, it was necessary to establish a rapid, non-invasive method for SARS virus detection, which was very important to the diagnosis and control of SARS (Lin and Qin, 2006). For this undertaking, a microfluidic chip integrated with functions of PCR and electrophoresis separation was developed. The structure of the microfluidic chip is shown in Figs. 4.7.10a and b. By the utilization of detection system (Fig. 4.7.10c), it can realize the amplification, separation, and detection of virus genes. It can dramatically reduce the time for SARS virus detection compared to the conventional methods.

4.8 Sensor Networks

Benefiting from the progress of information and MEMS technology, sensor networks, which are novel style of sensors, begin to show promising advantages.

Sensor networks are special networks composed of a group of sensors, wired or wireless, utilizing given methods or rules, whose targets are feeling, collecting and processing specialized information of certain objects through a united approach. The breakthrough within sensor networks benefits from the progress of sensor techniques, built-in process techniques, distributing information process techniques and communication techniques (Wang and Akilydiz, 2002).

As to the communication mode, either a wired mode or wireless mode is utilized. The optical fiber sensor network is a freshman in the sensor network family. It has an optical fiber but not an electronic wire/cable that works as a communication media. Since the placement of optical fiber is similar to traditional cable, the optical fiber sensor network is usually classified a wired sensor network.

4.8.1 History of Sensor Networks

The 1st-generation sensor network is based on a traditional analog output sensor with point-to-point translation. Such sensor networks were utilized widely in 1980s, but dropped behind due to high cost, complex placement and poor EMC performance.

An obvious feature of the 2nd-generation sensor network is the application of smart sensors. A built-in processor, for example, MCU or DSP, acts as the center control unit of the sensor. It receives the analog signals from the sensitive unit, converts them to digital data, then stores or transmits accordingly. At the same time, more and more 2nd-generation sensor networks accept series digital bus as transport media, such as RS-232, RS-422 and RS-485 digital data bus.

The 3rd-generation sensor network is a smart sensor network, which has an advanced field bus with a transport network that is all-digital, double- direction and open. The requirement for wire/cable placement, communication bandwidth is much narrower than earlier sensor networks. MPS (Michigan parallel standard) bus and Inter-IC (I^2C) bus were two bus standards that were introduced successfully at an early stage. And now, controller area network bus (CAN bus) and Ethernet bus are two typical smart buses with universal applications.

The 4th-generation sensor network which is called a wireless sensor network (WSN) with features such as multi-function sensor units, self-organized network structures and wireless transport mode is in progress now. This is a new type sensor network, which is constructed of basic nodes. A sensitive device, built-in micro processor, wireless interface, power supply unit, application software and security strategy are integrated in each node. Each node can be a basic sensor unit, transmit relay unit, even a local information collecting unit. Due to the characteristic of WSN, specialized communication agreement and route arithmetic are the key points for developing WSN techniques. And we predict that advances in MEMS technology will produce WSN that is even more capable and versatile.

4.8.2 Essential Factors of Sensor Networks

4.8.2.1 Sensor

A sensor is composed of power, sensitive device, built-in processor, communication unit and respective software. Power provides energy to each part. Sensitive device detects and monitors information of targets, then changes the information to digital data. Built-in processor controls the status of each part, especially collects the information from the sensitive device, and then sends it to the communication unit after processing.

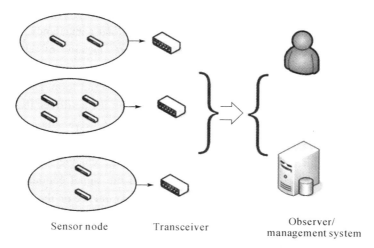

Fig. 4.8.1. A typical sensor network

4.8.2.2 *Observer*

The observer is the user of sensor networks and the accepter of information acquired. Observer is a man, a computer or other devices. For example, either a scientist or a computer station can be an observer of certain sensor networks. One sensor network can support multi observers at the same time. An observer can check the information provided by sensor networks, then judge, conclude the information accordingly, or take corresponding actions to the objects.

4.8.2.3 *Object*

An object is the monitor target of a sensor network. Physical parameter, chemical process and biomedical status can be included. One sensor network can monitor multi objects in certain area. At the same, an object can be monitored by different sensor networks.

4.8.3 Buses of Sensor Networks

In a sensor network, sensor nodes are generally connected to a controller or a computer which provides linearization, error correction, and access to the network. The interface between sensor node and controller becomes more and more important. Though maybe not introduced for sensor networks originally, some digital interface standards provide extensive applications in the field of sensor networks. Some brief introductions to typical digital interfaces are included in the following sections (Sichitiu, 2004).

4.8.3.1 *RS-232 bus*

RS-232 is a standard for serial binary data signal connections between data terminal equipment and data circuit-terminating equipment. It is commonly used in computer serial ports. The standard defines electrical signal characteristics such as voltage levels, signaling rate, timing and slew-rate of signals, voltage withstand level, short-circuit behavior, and maximum load capacitance, interface mechanical characteristics, pluggable connectors and pin identification. Details of character format and transmission bit rate are controlled by the serial port hardware, often a single integrated circuit called UART that converts data from parallel to asynchronous start-stop serial form. Details of voltage levels, slew rate, and short-circuit behavior are typically controlled by a line-driver that converts from the UART's logic levels to RS-232 compatible signal levels, and a receiver that converts from RS-232 compatible signal levels to the UART's logic levels.

For data transmission lines (TxD, RxD and their secondary channel equivalents) logic one is defined as a negative voltage, and logic zero is positive and the signal condition is termed spacing. Control signals are logically inverted with respect to what one would see on the data transmission lines. When one of these signals is negative, the voltage on the line will be between 3 V and 15 V. The active state for these signals would be the opposite voltage condition, between –3 V and –15 V. In order to convert TTL level to RS-232 level, a MAX232 chip or another chip with similar function is often be used.

4.8.3.2 I^2C bus

The I^2C bus was introduced by Philips as a standard for connecting integrated circuits (IC) which may or may not include sensors. I^2C is intended for application in systems which connect microcontrollers and other microcontroller-based devices or parts. It is a two-wire serial bus as shown in Fig. 4.8.2.

The serial data and serial clock carry information to every device connected to the bus, which has a unique address. The serial data wire is bi-directional but data may flow in only one direction at a certain time. Devices on the bus are defined as masters or slaves. A master, which is usually a microcontroller, initiates a data transfer on the bus and generates the clock, and generates the control signals which are placed on the data wire. The slave device is controlled by the master. A slave device can either receive or send data depending on the master. To save energy, some sensor nodes should be in sleep mode most of the time and woken up by a timer or sensing event.

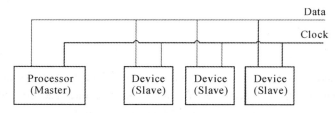

Fig. 4.8.2. I^2C bus structure

4.8.3.3 CAN bus

CAN bus is a vehicle bus standard designed to allow microcontrollers and devices to communicate with each other within a vehicle without a host computer, which is showed by Fig. 4.8.3. CAN bus standard was officially released in 1986 at the Society of Automotive Engineers (SAE) congress in Detroit, Michigan. The first CAN controller chip, which was produced by Intel and Philips, came on the market in the 1980s.

Chapter 4 Chemical Sensors and Measurement

A modern automobile may have as many as 70 electronic control units (ECU) for various subsystems, including the engine control unit, transmission, airbags, antilock braking, cruise control, audio systems, windows, doors, mirror adjustment, etc. Some of these form independent subsystems, but communications among others are essential. A subsystem may need to control actuators or receive feedback from the sensors. The CAN standard was devised to fill this requirement.

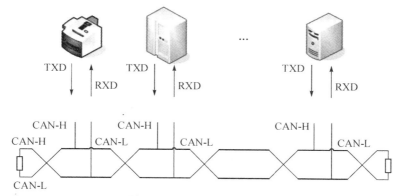

Fig. 4.8.3. CAN bus structure

CAN is a multi-master broadcast serial bus standard for connecting electronic control units. The devices that are connected by a CAN network are typically sensors, actuators and control devices. A CAN message never reaches these devices directly, but instead a host processor and a CAN controller are needed between these devices and the bus. If two or more nodes begin sending messages at the same time, the message with the more dominant ID will overwrite other nodes' less dominant IDs, so that eventually only the dominant message remains and is received by all nodes. Each node requires the support from the host processor, CAN controller and transceiver.

4.8.3.4 Serial peripheral interface

The serial peripheral interface bus (SPI bus) is a synchronous serial data link standard named by Motorola that operates in full duplex mode. Devices communicate in master/slave mode where the master device initiates the data frame. Multiple slave devices are allowed with individual slave select (chip select) lines. Sometimes SPI is called a "four-wire" serial bus, contrasting with three-, two-, and one-wire serial buses. The SPI bus specifies four logic signals, which are Serial Clock (SC), Master Output/Slave Input (MOSI/SIMO), Master Input/Slave Output (MISO/SOMI) and Slave Select (SS). The SPI bus can operate with a single master device and with one or more slave devices. Most slave devices have tri-state outputs so their MISO signal becomes high impedance when it is not selected. Devices without tri-state outputs cannot share SPI bus segments with other devices. At one time only one slave device could talk to the master with its chip select being activated.

To begin a communication, the master first configures the clock, using a frequency less than the maximum frequency of the slave device. The master then pulls the slave select low for the desired chip. Transmissions may involve any number of clock cycles. When there is no more data to be transmitted, the master stops toggling its clock. Normally, it then deselects the slave device. SPI bus is a full duplex communication standard, and has a higher speed than I^2C bus mentioned above. SPI bus requires only extra 4 pins in hardware design, so it is much easier for layout design. Furthermore, some chips combine MOSI and MISO into a single data line (SI/SO). Usually it is called three-wire signaling.

4.8.4 Wireless Sensor Network

At the beginning of this chapter, a typical WSN application example was given. With this example and moreover, the senses, technique challenges and wide applications of WSN will be introduced in detail.

4.8.4.1 Typical application

The early blue-green algae bloom in Taihu Lake (Fig. 4.8.4), which is the third largest fresh water lake of China, led to the tap water pollution and water supply crisis in May, 2007. It was a typical environmental hazard caused by chemical pollution and biological turbulence. To clarify the pollution status, a WSN monitor project has been carrying out since 2008.

Fig. 4.8.4. A photo of Taihu Lake

4.8.4.2 Typical structure

Sensor node and wireless network construction are two essential components of WSN.

- ***Sensor node***

Fig. 4.8.5 shows the structure of a WSN node, which will be deployed in the Taihu Lake project, focusing on chemical pollution monitoring. A WSN node is composed of sensitive device, processor unit, wireless communication module and power supply unit. The sensitive device detects the specialized information of a certain object. The processor unit receives the signal from the sensitive device, amplifies it if necessary, then converts it into digital data and sends it to the next process. The processor unit also controls the total sensor node as a commander. The wireless communication module answers the communications with the other nodes, sending/receiving data and exchanging information. The power supply unit provides energy to all the parts above. Usually a micro battery or solar energy solution is an advisable choice.

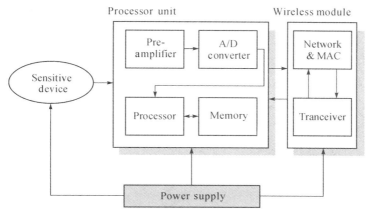

Fig. 4.8.5. Structure of sensor node

Furthermore, Fig. 4.8.6 shows the complex MEMS sensitive device of the WSN node in the Taihu Lake project. Up to 4 MEA (microelectrode array) and 4 LAPS (Light addressable potentiometer sensor), which all act as chemical sensors, are integrated on the complex chip. Based on different sensitive theories of MEA and LAPS, self-compensation and multi-parameter measurement are standout advantages of the complex device. Fig. 4.8.7 shows the pre-amplifier unit. Ultra low noise amplifier receives the weak signal from complex device and sends it to next stage after amplifying. The detect limit can be as low as 0.5 nA.

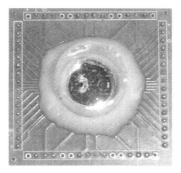

Fig. 4.8.6. MEMS sensitive device

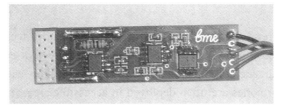

Fig. 4.8.7. Preamplifier unit

Biomedical Sensors and Measurement

- *Wireless network construction*

A typical wireless network contains a sensor node, and sink node and management node. Sensor nodes, which are placed randomly, can locate themselves and send information point by point. A sink node will collect the information from sensor nodes and then re-send it to the management node by internet, satellite, etc. Finally, the research team can get the quantity of information from the management node.

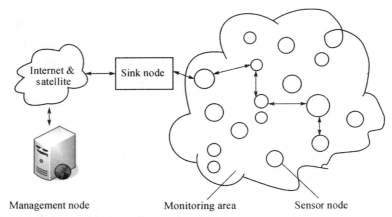

Fig. 4.8.8.   Wireless network construction

4.8.4.3   Key techniques

We now briefly describe three important techniques in WSN.

- *Power efficiency control*

Many WSNs must aggressively conserve energy in order to operate for extensive periods without wired power sources. Since wireless communication often dominates the energy dissipation in a WSN, several promising approaches have been proposed to achieve power-efficient multi-hop communication in ad hoc networks. Topology control aims to reduce the transmission power by adjusting nodal radio transmission ranges while preserving necessary network properties. Power-aware routing protocols choose appropriate transmission ranges and routes to conserve energy used for multi-hop packet transmission. Both topology control and power-aware routing focus on reducing the power consumption when the radio interface is actively transmitting and receiving packets (Sichitiu, 2004).

- *Network security*

Because sensor networks pose unique challenges, traditional security techniques used in traditional networks cannot be applied directly. Firstly, to make sensor networks economically viable, sensor devices are limited in their energy, computation, and communication capabilities. Secondly, unlike traditional networks, sensor nodes are often deployed in accessible areas, presenting the added risk of physical attack. And thirdly, sensor networks interact closely with their physical environments and with people, posing new security problems.

An adequate solution is in conjunction with secure group management, intrusion detection and secure data aggregation. First, interest in network data aggregation and analysis can be performed by groups of nodes, and the outcome of the group's computation is normally transmitted to a base station. Then, in order to look for anomalies, applications and typical threat models must be understood, and the use of secure groups may be a promising approach for decentralized intrusion detection. And last, depending on the architecture of the wireless sensor network, aggregation may take place in many places in the network. All aggregation locations must be secured (Perrig et al., 2004).

- ***Relative location estimation***

Self-configuration is a general class of estimation problems which we explore via the Cramer-Rao bound (CRB). Specifically, the sensor location estimation problem is explored for sensors that measure range via received signal strength (RSS) or time-of-arrival (TOA) between themselves and neighboring sensors. TOA ranging has been implemented using two-way or round-trip time-of-arrival measurements. Inquiry-response protocols and careful calibration procedures are presented to allow devices to measure the total delay between an original inquiry and the returned response. Ranging is also possible using RSS measurement, which can be measured from reception of any transmission in the network. In a frequency hopping radio, RSS measurements can be averaged over frequency to reduce frequency-selective fading error. RSS is attractive from the point of view of device complexity, but is traditionally seen as a coarse measure of ranges. Sensor location estimation with about 1 m RMS error has been tested using both TOA and RSS measurements (Meguerdichian et al., 2001).

Fig. 4.8.9 shows a multi-function sensor/sink WSN node, which is the basic unit of the Taihu Lake project, combined with key techniques mentioned above. With the special low power consumption MCU and related extern circuits. Average work current is lower than 20 mA, and idle mode current is as low as 100 μA. Moreover, the relative location estimation program has been planted into the MCU as part of the firmware. At the same time, the network security issue is checked from both the hardware and software view. Using CPLD and FPGA programmed devices, the hardware of the circuit is difficult to be copied. And security arithmetic is integrated in the application software. Further research is even in progress, focusing on simplifying the hardware circuit. The finial version node may include only 2 ICs (integrate chips): one is the MCU, including power management and interface module, the other is a complex analog chip, including all analog amplifiers, signal mixers, A/D and D/A converters.

Fig. 4.8.9. Multi-function sensor/sink WSN node

4.8.4.4 Senses and challenges

Recent advances in micro-electro-mechanical systems (MEMS) technology, wireless communications, and digital electronics have enabled the development of WSNs, which contain low-cost, low-power, multifunctional sensor nodes that are small in size and communicate by wireless media for short distances. These tiny sensor nodes, which consist of sensing, data processing, and communicating components, leverage the idea of sensor networks based on collaborative effort of a large number of nodes. WSN has the potential to revolutionize sensing (and/or actuating) technology in the future. Large numbers of cheap nodes can be placed in the area to be monitored. In contrast to traditional networks, WSN at least has the following advantages (Akyildiz et al., 2002).

The large number ensures that at least some of the sensors will be close to the phenomenon of interest and thus be able to have high quality measurements. In-network processing allows for the tracking of targets and the evolution of the studied phenomena. It also allows for substantial power savings and reduced bandwidth necessary to observe certain phenomena. The large number of sensors also increases the reliability of the system, as failure of a percentage of the sensor nodes will not result in system failure.

Sensors can be positioned far from the observers. In this sense, observers do not need to be near the actual position, which may be polluted, dangerous or hard to reach. So there are a wide range of applications envisioned for such sensor networks, including microclimate studies, groundwater contaminant monitoring, precision agriculture, condition-based maintenance of machinery in complex environments, urban disaster prevention and response, and military interests.

Several sensors that perform only sensing can be deployed. The positions of the sensors and communications topology are carefully engineered. They transmit time series of the sensed phenomenon to the central nodes where computations are performed and data are processed efficiently.

As a coinstantaneous result, due to its unique characters, WSN is facing some technical challenges, such as energy limitations, communication capacity/power confines, process performances and storage shortages (Chong and Kumar, 2003).

Micro sensor node is usually powered by small size batteries, whose capacity is limited. Since there are so many cheap sensor nodes in the target area, which sometimes human cannot reach, so it is very difficult, if not impossible, to change batteries and refresh the sensor notes by human operations. So the service time of sensor nodes is decided by capacity of the batteries and the power consumption of sensor nodes. A sensor node includes a sensitive unit, processor and wireless communication devices. The consumption of the processor and the sensitive device is reduced with the advance of integrated chipset progress. Most power consumption happens in wireless communication circuits. Such modules have four statuses, sending, receiving, idling and sleeping. It is the essential research and development to find a way to let wireless modules be more efficient and reduce unnecessary power consumption.

With the increase of communication distance, the power consumption arises accordingly. Considering that the communication ability of each node is limited and the object area is often large, it is necessary to adopting a multi-point route. The distance of each node should be no more than 100 – 150 m. The communication bandwidth and RF output power of each sensor node is also limited. The signal of each node may be degraded, or completely destroyed by complex surface circumstances or dreadful weather. In order to reduce the cost and power consumption of each sensor node, a low-end but power-saving built-in processor is used, with a small size memory device. The sensor node is expected to monitor the object, convert the analog signal to digital data, save/process the data, communicate with other nodes, etc. It is a challenge to complete the missions with limited processor ability and memory size.

Fortunately with the improvement of low consumption chips and IC system design, many ultra low consumption processors are now available. Besides reducing absolute work current, module power supply and dynamic voltage scaling is supported by the new generation processors. When the duty of the processor is light, some unnecessary units of the processor will be closed, and the power voltage and operation frequency may be controlled to a relative lower level. So the processor will not be jammed with heavy duties, and the operation current will be saved in idle time.

4.8.4.5 Forecasts

WSN has a lot of distributed sensor nodes. The concepts of micro-sensing and wireless connection of these nodes promise many new application areas, such as environmental monitors, military applications, etc.

Some environmental applications of sensor networks include tracking the movements of birds, small animals, and insects; monitoring environmental conditions that affect crops and livestock; irrigation; macro instruments for large-scale Earth monitoring and planetary exploration; chemical and biological detection; precision agriculture; biological and environmental monitoring in marine, soil, and atmospheric contexts; forest fire detection; meteorological or geophysical research; flood detection; bio-complexity mapping of the environment; and pollution study. A pollution monitor project based on WSN is a typical example for such an application.

Furthermore, the rapid deployment, self-organization and fault tolerance characteristics of sensor networks make them a very promising sensing technique for military applications. In chemical and biological warfare, being close to ground zero is important for timely and accurate detection of the agents. Sensor networks deployed in the friendly region and used as a chemical or biological warning system can provide the friendly forces with critical reaction time, which drops casualties drastically. For instance, we can make a nuclear reconnaissance without exposing a team to nuclear radiation.

References

Akyildiz I.F., Su W., Sankarasubramaniam Y. & Cayirci E., 2002. Wireless sensor networks: a survey. *Computer Networks*. 38, 393-422.
Anderson, J.L. & Bowden, E.F., 1996. Dynamic electrochemistry: methodology and application. *Analytical Chemistry.* 68: 379R-444R.
Barcelo D., 2006. *Comprehensive Analytical Chemistry*. Elsevier. 49, 87-100.
Barker S.L.R., Ross D., Tarlov M.J., Gaitan M. & Locascio L.E., 2000. Control of flow direction in microfluidic devices with polyelectrolyte multilayers. *Analytical Chemistry*. 72, 5925-5929.
Bergveld P., 2003. Thirty years of ISFETOLOGY-What happened in the past 30 years and what may happen in the next 30 years? *Sensors and Actuators B-Chemical*. 88, 1-20.
Bousse L., "The chemical Sensitivity of Electrolyte/Interface/Silicon structures," Ph.D. dissertation,

Twente University of Technology, Enschede, 1982.

Cammann K., Lemke U., Rohen J., Wilken H. & Winter B., 1991. Chemo-und biosensoren- grundlagen und anwendungen. *Angewandte Chemie.* 103: 519-541.

Chen Y. & Ge W., 2007. Principle and Application of Modern Sensors. *Science Press.* 239-250.

Chong C. & Kumar S.P., 2003. Sensor networks: evolution, opportunities, and challenges. *Proceedings of the IEEE.* 91, 1247-1256.

Cai W., Li Y., Gao X.M., & Wang P., 2011. Full automatic monitor for in-situ measurements of trace heavy metals in aqueous environment. *Sensor Letters.* 9, 137-142.

Dhanabalan A., Dabke R.B., Kumar N.P., Talwar S.S., Major S., Lal R. & Contractor A.Q., 1997. A study of Langmuir and Langmuir-Blodgett films of polyaniline. *Langmuir.* 13, 4395-4400.

Duffy D.C., McDonald J.C., Schueller O.J.A. & Whitesides G.M., 1998. Rapid prototyping of microfluidic systems in poly (dimethylsiloxane). *Analytical Chemistry.* 70, 4974-4984.

Dzyadevych S.V., Soldatikin A.P., El'skaya A.V., Martelet C. & Renault N.J., 2006. Enzyme biosensors based on ion-selective field-effect transistors. *Analytica Chimica Acta.* 568, 248-258.

Eddington A.S., 1928. *Nature of the Physical World. Cambridge/London/New York*: Cambridge University Press.

Feeney R., Herdan J., Nolan M.A., Tan S.H., Tarasov V.V. & Kounaves S.P., 1998. Analytical characterization of microlithographically fabricated iridium-based ultramicroelectrode arrays. *Electroanalysis.* 10, 89-93.

Huang FC., Chen YY., Wu TT., 2009. A room temperature surface acoustic wave hydrogen sensor with Pt coated ZnO nanorods. *Nanotechnology.* 20, 55-60.

Gravesen P., Branebjerg J. & Jensen O.S., 1993. Microfluidics-a review. *Journal of the Micromechanics and Microengineering.* 3, 168-182.

Gründler P., 2007. Chemical Sensors. Springer, Germany.

Ho C.M. & Tai Y.C., 1998. Micro-electro-mechanical-systems (MEMS) and fluid flows. *Annual Review of Fluid Mechanics.* 30, 579-612.

Heinze, J. 1993. Ultramicroelectrodes in electrochemistry. *Angewandte Chemie International Editiow.*, 32: 1268–1288.

Ismail A.B., Sugihara H., Yoshinobu T. & Iwasaki H., 2001. A novel low-noise measurement principle for LAPS and its application to faster measurement of pH. *Sensors and Actuators B-Chemical.* 74,112-116.

Janata J., Josowicz M., Vanysek P., & Devaney D.M., 1998. *Chemical sensors.* Anal. Chem., 70, 179R-20R.

Kaneyasu K., Otsuka K., Setoguchi Y., Sonoda S., Nakahara T., Aso I. & Nakagaichi N., 2000. A carbon dioxide gas sensor based on solid electrolyte for air quality control. *Sensors and Actuators B-Chemical.* 66, 56-58.

Koley G., Liu J., Nomani M.W., Yim M., Wen X. & Hsia T.Y., 2009. Miniaturized implantable pressure and oxygen sensors based on polydimethylsiloxane thin films. *Materials Science & Engineering C-Biomimetic and Supramolecular Systems.* 29, 685-690.

Koryta J., 1986. Ion-selective electrodes. *Annual Review of Materials Science.* 16, 13-27.

Lehmann, M., Baumann W., Brischwein M., Ehret R., Kraus M., Schwinde A., Bitzenhofer M., Freund I. & Wolf B., 2000. Non-invasive measurement of cell membrane associated proton gradients by ion-sensitive field effect transistor arrays for microphysiological and bioelectronical applications.

Biosensors and Bioelectronics. 15, 117-124.

Lin B.C. & Qin J.H., 2006. Microfluidic Chip Laboratory. *Science Press*, China.

Meguerdichian, S., Koushanfar F., Potkonjak M. & Srivastava M., 2001. Coverage problems in wireless ad-hoc sensor networks. *IEEE Conference on Computer Communications*. 3, 1380-1387.

Mourzina Y.G., Ermolenko Y.E., Yoshinobu T., Vlasov Y., Iwasaki H. & Schöning M.J., 2003. Anion-selective light-addressable potentiometric sensors (LAPS) for the determination of nitrate and sulphate ions. *Sensors and Actuators B-Chemical*. 91, 32-38.

Mourzina Y.G., Yoshinobu T., Schubert J., Lüth H., Iwasaki H. & Schöning M.J., 2001. Ion-selective light addressable potentiometric sensor (LAPS) with chalcogenide thin film prepared by pulsed laser deposition. *Sensors and Actuators B-Chemical*. 80, 136-140.

Perrig A., Stankovik J. & Wagner D., 2004. Security in wireless sensor networks. Communications of the ACM. 6, 47-58.

Rolf D., 2002. Electronic noses. *Eurocosmetics*. 10, 20-29.

Roveti D.K., 2001. Choosing a humidity sensor: a review of three technologies. Sensors. 18, 54-58.

Ryan M.D., Bowden, E.F. & Chambers, J.Q., 1994. Dynamic electrochemistry: methodology and application. Analytical Chemistry., 66: 360R-427R.

Sichitiu M.L., 2004. Cross-layer scheduling for power efficiency in wireless sensor networks. *IEEE Conference on Computer Communications*.

Stetter J.R. & Penrose W.R., 2001. Electrochemical nose. Electrochemistry Encyclopedia.

Stulik K., Amatore C., Holub K., *Mareceek*, & Kutner W., 2000. Microelectrodes. Definitions, characterization, and applications. *Pure and Applied. Chemistry*. 72: 1483-1492.

Swan P.N., 1980. Dissertation, University of Southampton.

Siu W.M., and R.S.C. Cobbold, "Basic properties of the Electrolyte-SiO_2 Si System: Physical and Theoretical Aspects," *IEEE. Transactions on Electron Devices*., Vol. 26, 1979, pp. 1805-1815.

Toko K., 1998. Electronic tongue. *Biosensors & Bioelectronics*. 13, 701-709.

Tallman, D.E. & Petersen, S.L., 1990. Composite electrodes for electroanalysis: principles and applications. *Electroanal*. 2: 499-510.

Unger M.A., Chou H.P., Thorsen T., Scherer A. & Quake S.R., 2000. Monolithic microfabricated valves and pumps by multilayer soft lithography. *Science*. 288, 113-116.

Vlasov Yu. G., Bychkov E.A. & Legin A.V., 1994. chalcogenide glass chemical sensor: Research and analytical applications. *Talanta*. 41, 1059-1063.

Wang P. & Ye X., 2005. Modern Biomedical Sensors, *2nd ed. Zhejiang University Press*, China.

Wang X. & Akilydiz I.F., 2002. A survey on sensor networks. IEEE Communication Magazine.

Whitesides G.M., 2006. The origins and the future of microfluidics. *Nature*. 442, 368-373.

Wightman R.M., 1981. Microvoltammetric electrodes. *Analytical Chemistry*., 53: 1125A-1134A.

Wightman R.M., 1988. Voltammetry with microscopic electrodes in new domains. *Science*. 240: 415-420.

Wollenberger U., Hintsche R. & Scheller F., 1995. Biosensors for analytical microsystems. *Microsystem. Technologies*. 1: 75.

Wang D., Wang L., Yu J., Wang P., Hu Y. & Ying K., 2009. A study on electronic nose for clinical breath diagnosis of Lung cancer. *AIP Conference Proceedings*. 1137: 314-317.

Wolfteis O.S., Chemical sensors-survey and trends, *Fresenius Journal of Analytical Chemistry*. 337, pp. 522-527, 1990.

Xie X., Stueben D. & Berner Z., 2005. The application of microelectrodes for the measurements of trace metals in water. *Analytical Letters*. 38, 2281-2300.

Zhang F., Niu W. & Sun Z., 1999. The research on the response mechanism of pH-LAPS. *Acta Scientiarum Naturalium University Nankaiensis*. 32, 13-16.

Zhou J. & Mason A., 2002. Communication buses and protocols for sensor networks. *Sensors*. 2, 244-257.

Zuo B. & Liu G., 2007. Principles and Applications of Chemical Sensors. Tsinghua University Press, China.

Chapter 5

Biosensors and Measurement

In this chapter, the enzyme biosensors and microorganism biosensors are introduced as catalytic biosensors. The antibody and antigen biosensors (immune sensors), nucleic acid biosensors (DNA sensors), as well as receptor and ion-channel sensors are introduced as affinity biosensors. Then, sensors for measurement of cellular metabolism, impedance, and electrophysiology are discussed in cell and tissue biosensors. Finally, some novel techniques such as biochips for microarrays and nanomaterials are also described, including typical sensing applications for biosensors.

5.1 Introduction

The basic concept, most important properties, and the primary bioreceptor components of biosensors are described as the introduction for biosensing and measurement.

5.1.1 History and Concept of Biosensors

5.1.1.1 History

In 1956, Professor Leland C. Clark, as the father of the biosensor concept, published his definitive paper on the oxygen electrode (Clark, 1956). Based on this experience and addressing his desire to expand the range of analytes that could be measured in the body, the concept was illustrated by glucose oxidase entrapped in a Clark oxygen electrode using a dialysis membrane (Historically, glucose sensing has dominated the biosensor literature and has delivered huge commercial successes to the field). So, the earliest biosensors were catalytic biosensor systems that integrated enzymes, cellular organelles, tissues or whole microorganisms with transducers that convert a biological response into a digital electronic signal. The principal transducers used were electrochemical, optical, and thermometric.

The next generation of biosensors, affinity biosensors, capitalized on a similar range of measurement principles but with the addition of piezoelectric transducers (that interconvert mechanical deformation and voltage to measure mass or viscoelastic effects) and magnetic transducers. Affinity biosensors delivered real-time information about the binding of antibodies to antigens, cell receptors to their ligands, and DNA and RNA to nucleic acid with a complementary sequence.

5.1.1.2 Concept

Generally, a biosensor is an analytical device which converts a biological response into an electrical signal. The term "biosensor" is often used to cover sensor devices to determine the concentration of substances and other parameters of biological interest even where they do not utilize a biological system directly. This very broad definition is even used by some scientific journals (e.g., Biosensors & Bioelectronics, Elsevier) but will not be applied to the coverage in this chapter.

Professor Anthony P. F. Turner (editor-in-chief of Biosensors & Bioelectronics, Cranfield University) once defined a biosensor as a compact analytical device incorporating a biological or biologically-derived sensing element either integrated within or intimately associated with a physicochemical transducer (Turner, 1996). The usual aim of a biosensor is to produce either discrete or continuous digital electronic signals which are proportional to a single analyte or a related group of analytes. Professor Turner's name is synonymous with the field of biosensors. Generally, his definition has been used as the concept of biosensors.

Biosensors represent a rapidly expanding field, at the present time, with an estimated growth rate of more than 60% annually. The major impetus comes from the health-care industry (e.g., 6% of the western world population are diabetic and would benefit from the availability of a rapid, accurate and simple biosensor for glucose) but with some pressure from other areas, such as food quality appraisal and environmental monitoring. Research and development in this field are wide and multidisciplinary, spanning from biochemistry, bioreactor science, physical chemistry, electrochemistry, and electronics to software engineering.

5.1.2 Components of a Biosensor

A biosensor generally consists of a biological sensing element, such as an enzyme, antibody or cell, in close contact with a physico-chemical transducer, such as an electrode or optical fiber. Measurement of the target analyte(s) is achieved by selective transduction of a parameter of the biomolecule-analyte reaction into a quantifiable electrical or optical signal (Fig. 5.1.1). The key part of a biosensor is the transducer which makes use of a physical change accompanying the reaction (Mohanty and Kougianos, 2006).

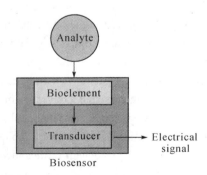

Fig. 5.1.1. A schematic representation of biosensors

A detailed list of differently applied bioelements and sensor-elements is shown in Fig. 5.1.2. Different combinations of bioelements and sensor-elements constitute several types of biosensors to suit a vast pool of applications (Mohanty and Kougianos, 2006).

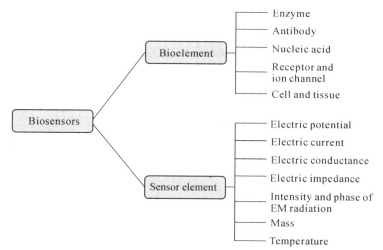

Fig. 5.1.2. Elements of biosensors

The high specificity of biomolecules and biological systems with respect to intermolecular interactions of interest can be successfully exploited in biosensor devices only if there is highly efficient coupling between the biological and transducer components. The bio and the sensor elements can be coupled together in one of the four possible ways as illustrated in Fig. 5.1.3: membrane entrapment, physical adsorption, matrix entrapment, and covalent bonding (Mohanty and Kougianos, 2006).

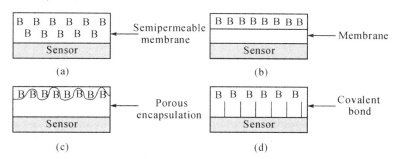

Fig. 5.1.3. In biomaterial-sensor coupling, the bio and sensor elements can be coupled together in one of four ways: (a) Membrane entrapment; (b) Physical adsorption; (c) Matrix entrapment, and (d) Covalent bonding (reprinted from (Mohanty and Kougianos, 2006), Copyright 2006, with permission from IEEE)

In the membrane entrapment scheme, a semi-permeable membrane separates the analyte and the bioelement; the sensor is attached to the bioelement. The physical adsorption scheme is dependent on a combination of van der Waals forces, hydrophobic forces, hydrogen bonds, and ionic forces to attach the biomaterial to the surface of the sensor. The porous entrapment scheme is based on forming a porous encapsulation matrix around the biological material that helps in binding it to the sensor. In the case of the covalent bonding, the sensor surface is treated as a reactive group to which the biological materials can bind.

5.1.3 Properties of Biosensors

The two most important properties of any proposed biosensor are: (a) its specificity and (b) its sensitivity towards the target analyte(s) (Byfield and Abuknesha, 1994). The specificity of a biosensor is entirely governed by the properties of the biological component because this is where the analyte interacts with the sensor. The sensitivity of the integrated device, however, is dependent on both the biological component and the transducer because there must be a significant biomolecule-analyte interaction and a high efficiency of subsequent detection of this reaction by the transducer.

In comparison with chemical sensors, an inherent advantage that can be exploited in biosensor technology is the significantly higher specificity that can generally be achieved as a direct result of biologically-optimized molecular recognition. This is best typified by an antibody-antigen interaction (the antigen is the substance against which the antibody has been generated *in vivo*) where an antibody can recognize and bind its antigen with extremely high specificity. Minor chemical modification of the molecular structure of the antigen can dramatically lower its affinity for the original antibody (which has low cross reactivity). Similarly, enzymes such as glucose oxidase will recognize their natural substrate (glucose in this case) with a far higher affinity than other components in the operating environment. Chemically similar molecules, in this case other small sugars, can elicit a small amount of cross-reactivity if present in sufficiently high concentrations. The level of specificity and, in many cases, sensitivity of detection and measurement that has been achieved to date in biosensors far exceeds that obtained for almost all chemical sensors.

In addition, arguably the most obvious disadvantage in exploiting the exquisite specificity and sensitivity of complex biological molecules is their inherent instability. Many strategies may be employed to restrain or modify the structure of biological receptors to enhance their longevity. So, stability is also a very important issue for biosensors.

5.1.4 Common Bioreceptor Components

The primary bioreceptor components can be classified into five groups.

- *Enzymes*

Enzymes are proteins that catalyze specific chemical reactions. These can be used in a purified form or be present in a microorganism or in a slice of intact tissue.

- *Antibodies and antigens*

An antigen is a molecule that triggers the immune response of an organism to produce an antibody, a glycoprotein produced by lymphocyte B cells, which will specifically recognize the antigen that stimulated its production.

- *Nucleic acids*

The recognition process is based on the complementary nature of the base pairs (adenine and thymine or cytosine and guanine) of adjacent strands in the double helix of DNA. These sensors are usually known as genosensors. Alternatively, interaction of small pollutants with DNA can generate the recognition signal.

- *Cellular structures or whole cells*

The whole microorganism or a specific cellular component (receptors), for example, a non-catalytic receptor protein, ion channels and lipid membrane, is used as the biorecognition element.

- *Other biomimetic receptors*

Recognition is achieved by use of receptors, for instance, genetically engineered molecules, artificial membranes, or molecularly imprinted polymers (MIP), that mimic a bioreceptor or ion channels. Those bioreceptors will not be discussed in this chapter.

Biosensors are typically classified by the above type of recognition element or transduction element employed (Turner, 2000). A sensor might be described as a catalytic biosensor if its recognition element comprised an enzyme or series of enzymes, a living tissue slice (vegetal or animal), or whole cells derived from microorganisms such as bacteria, fungi, or yeast. However, the sensor might be described as a bioaffinity sensor if the basis of its operation is a biospecific complex formation. Accordingly, the reaction of an antibody with an antigen or hapten, or the ligand with a receptor, could be employed.

5.2 Catalytic Biosensors

The most commonly used biological components in catalytic biosensors are enzymes. There is currently a wide selection range of enzymes available; especially, microorganisms which also take advantage of the complex enzyme systems, coenzyme, and all the physiological functions supplied by microorganism themselves for catalytic biosensors. Therefore, the enzyme biosensors and microorganism biosensors are described with some commercial applications.

5.2.1 Enzyme Biosensors

Enzyme biosensors utilize one or more enzyme types as the macromolecular binding agent and take advantage of the complementary shape of the selected enzyme and the targeted analyte. Enzymes are proteins that perform most of the catalytic work in biological systems and are known for highly specific catalysis. The shape and reactivity of a given enzyme limit its catalytic activity to a very small number of possible substrates. Enzymes are also known for rapidness, working at rates as high as 10,000 conversions per second per enzyme molecule. Enzyme biosensors rely on the specific chemical changes related to the enzyme-analyte interaction as the means for determining the presence of the targeted analyte. For example, upon interaction with an analyte, an enzyme may generate electrons, a colored chromophore or a change in pH (due to release of protons) as the result of the relevant catalytic enzymatic reaction. Alternatively, upon interaction with an analyte, an enzyme may cause a change in a fluorescent or chemiluminescent signal that can be recorded by an appropriate detection system.

5.2.1.1 Enzyme

Enzymes are proteins (polypeptide structures) which catalyse specific chemical reactions *in vivo*. They accelerate the rate of reaction of a particular chemical (the substrate) without being consumed in the process.

Fig. 5.2.1 shows the working principle of enzymes. An enzyme, upon reaction with a substrate, forms a complex molecule that, under appropriate conditions, forms the desirable product molecule releasing the enzyme at the end (Mohanty and Kougianos, 2006).

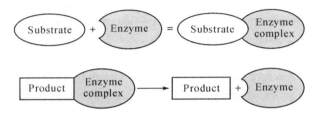

Fig. 5.2.1. Working principle of enzymes

Compared with chemical catalysts, enzymes demonstrate a significantly greater level of substrate specificity, primarily because of the constraints placed on the substrate molecule by the active site in the environment involving factors such as molecular size, stereochemistry, polarity, functional groups and relative bond energies.

The sensitivity of enzyme-based biosensors is directly dependent on the maximum limiting affinity underlying enzyme-substrate complication and the rate of subsequent transformation. And, after being isolated and incorporated into biosensor systems, the decrease in activity of enzymes can be minimized by suitable pH, temperature and other environment conditions. Enzyme immobilization techniques, such as adsorption, cross-linking, covalent bonding, physical entrapment, electrochemical polymerization and self-assembled monolayer, have well preserved the activity of lots of enzymes for many potential applications.

5.2.1.2 Enzyme sensors

The enzymes are extremely specific in their action: enzyme will change a specific substance to another specific substance, as illustrated in Fig. 5.2.2. This extremely specific action of the enzymes is the basis of biosensors.

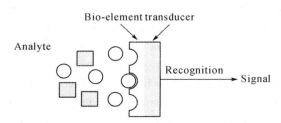

Fig. 5.2.2. Specificity of enzymes is the basis of enzyme sensors

Chapter 5 Biosensors and Measurement

The mechanisms of operation of enzyme-based biosensors can involve: (a) conversion of the analyte into a sensor-detectable product, (b) detection of an analyte that acts as enzyme inhibitor or activator, or (c) evaluation of the modification of enzyme properties upon interaction with the analyte.

The relationships between the parameters of relevance in enzyme catalysis are given by Michaelis-Menten equation as shown in the following enzyme-catalyzed reaction (Byfield and Abuknesha, 1994):

$$E + S \underset{k_{-1}}{\overset{k_1}{\rightleftarrows}} ES \overset{k_2}{\longrightarrow} E + P \tag{5.2.1}$$

where E is the enzyme, S is the substrate, P is the product(s), k_1 and k_{-1} are the forward and reverse rate constants for formation of the enzyme-substrate intermediate, and k_2 is the rate constant for formation of product(s). The second part of the reaction can be considered as irreversible at the beginning of the reaction, when the concentration of product is significantly lower than the substrate concentration. In many reactions the *in vivo* and *in vitro* concentration of the enzyme is much lower than the concentration of the substrate. In this case, all enzyme molecules are involved in the catalysis, or they will become saturated. The Michaelis-Menten equation relates the initial reaction rate v_0 to the substrate concentration $[S]$.

$$v_0 = \frac{v_{max}[S]}{K_M + [S]} \tag{5.2.2}$$

Fig. 5.2.3 is the corresponding graph of a hyperbolic function. The maximum velocity is described as v_{max}. K_M is Michaelis constant (Note that often the experimental parameter k_{cat} is used, but in simple cases this parameter equal to the kinetic parameter k_2 for the elementary reaction from $[ES]$ to $E + P$).

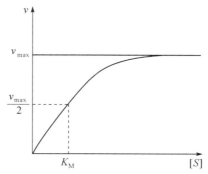

Fig. 5.2.3. Michaelis-Menten plot relating the reaction rate v_0 to the substrate concentration $[S]$

In the medical diagnostic field, several manufacturers have marketed biosensors for the measurement of common blood chemistry components including glucose, urea, lactate, and creatinine. In general, enzyme-based biosensors employ semi-permeable membranes through which target analytes diffuse toward a solid-phase immobilized enzyme compartment. The major drawback of this type of sensor is that many enzymes are inherently unstable, necessitating a packaging approach to limit degradation of the biosensor performance.

5.2.1.3 Clark oxygen electrode sensor

The most commercially successful biosensors are amperometric glucose biosensors. These biosensors have been made available in the market in various shapes and forms, such as glucose pens and glucose displays. The first historic experiment that served as the origin of glucose biosensors was carried out by Leland C. Clark. He used platinum (Pt) electrodes to detect oxygen (Fig. 5.2.4).

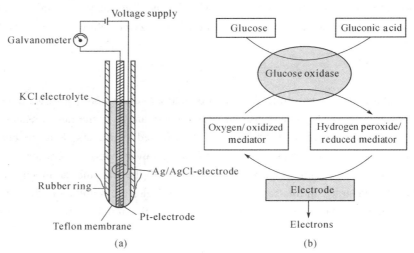

Fig. 5.2.4. Clark electrode for glucose detecting: (a) Clark-type electrode; (b) Glucose detecting

The Clark-type electrode is the most widely used oxygen sensor for measuring oxygen dissolved in a liquid (Mohanty and Kougianos, 2006). The basic principle is that there is a cathode and an anode submersed in an electrolyte. Oxygen enters the sensor through a permeable membrane by diffusion, and is reduced at the cathode, creating a measurable electrical current.

Electron flow to oxygen as a result of oxidative phosphorylation can be demonstrated using an oxygen electrode. The electrode compartment is isolated from the reaction chamber by a thin Teflon membrane. The membrane is permeable to molecular oxygen and allows this gas to reach the cathode, where it is electrolytically reduced. The reduction allows a current to flow. This creates a potential difference which is recorded on a flatbed chart recorder. The trace is thus a measurement of the oxygen activity of the reaction mixture. The current flowing is proportional to the activity of oxygen provided the solution is stirred constantly (stir bar) to minimize the formation of an unstirred layer next to the membrane.

The enzyme glucose oxidase (GOD) was placed very close to the surface of platinum by physically trapping it against the electrodes with a piece of dialysis membrane. The enzyme activity changes depend on the surrounding oxygen concentration. Fig. 5.2.4 also shows the reaction catalyzed by GOD. Glucose reacts with GOD to form gluconic acid while producing two electrons, and two protons, thus reducing GOD. The reduced GOD, surrounding oxygen, electrons, and protons (produced above) react to form hydrogen peroxide and oxidized GOD, which is in the original form. This GOD can again react with more glucose. The more the glucose content, the more oxygen is consumed. On the other hand, less glucose content results in less hydrogen peroxide. Hence, either the consumption of oxygen or the production of hydrogen peroxide can be detected with the help of platinum electrodes, and this can serve as a measurement for glucose concentration.

With oxygen electrodes or hydrogen peroxide electrodes and whether transfer mediators are used or not, glucose enzyme electrodes can be classified into oxygen electrode glucose sensors, hydrogen peroxide electrode glucose sensors, oxidized mediator glucose sensors, and direct electrochemical glucose sensors. Of course, Clark oxygen electrodes also have been well used for microorganism based biosensors and immune biosensors, which will be presented subsequently in this chapter.

5.2.2 Microorganism Biosensors

Many enzymes are extracted from the microorganism cells. In 1977, GA Rechnitz et al., proposed the direct use of microorganism cells as recognition elements for biosensors to avoid separation and purification of the enzymes. There is not only the problem of high cost, but also the reduction and even the loss of enzyme activities in the separating and purifying process. Compared to immobilized enzyme sensors, those novel microorganism sensors also take advantage of complex enzyme systems, coenzyme and all the physiological functions supplied by microorganism themselves.

5.2.2.1 *Microorganism*

A microorganism is an organism that is microscopic (usually too small to be seen by the naked eye). Microorganisms are very diverse. They include bacteria, fungi, archaea, and protists; microscopic plants (called green algae); and animals such as plankton and the planarian. Microorganisms live in all parts of the biosphere where there is liquid water. Microorganisms are critical to nutrient recycling in ecosystems as they act as decomposers. Microorganisms are also exploited by people in biotechnology, both in traditional food and beverage preparation, and in modern technologies based on genetic engineering. However, pathogenic microbes are harmful, since they invade and grow within other organisms, causing diseases that kill millions of people, other animals, and plants.

The structure of a microorganism sensor is immobilizing microbial cell membranes onto sensors, which is just like the enzyme membrane immobilized onto the enzyme sensors. The methods of immobilizing microorganisms are basically the same as those of immobilizing enzymes. Therefore, the methods of adsorption, embedding, cross-linking or covalent combining can also be used. Currently, the commonly used method is making microorganisms adhere to the membranes, such as a cellulose acetate membrane, is filtering paper or nylon membrane by centrifugation, filtration or combined culturing. Such a method of embedding microorganisms has high sensitivity.

5.2.2.2 *Microorganism sensors*

Microorganism pathways are activated by some analytes, such as pollutants. These pathways are involved in metabolism or nonspecific cell stress that results in the expression of one or more genes. For example, immobilized yeast is one of the most commonly used sensors. It has been used in the detection of formaldehyde and toxicity measurements of cholanic acids. The changes in metabolism of the analyte were detected via O_2 electrode measurements or extracellular acidification rates.

In the electrochemical detecting system, the cultured (or immobilized) microorganism was covered appropriately on the surface of the corresponding electrochemical sensor. It uses the selective catalytic reactions of enzyme in cells, such as hydrolysis, ammonolysis, or oxidation, as well as the selective detecting reactants of electrochemical sensor elements to measure the information of substrate.

Respiratory activity microorganism sensors consist of the aerobic microorganism immobilized membrane and oxygen electrodes, based on the activity of microbial respiration (Fig. 5.2.5). When the sensor is inserted in the test solution with dissolved oxygen maintained, the organic compounds in the test solution are assimilated by microorganisms. The microbial respiration enhancement leads to the reduction of diffusion oxygen on the electrode and the sharp drop of current value. Finally, the oxygen diffusion rate in the solution and the oxygen consumption rate of microorganism achieve a balance. When the oxygen spreading to the electrode tends to be constant, a constant current value can be obtained, which has correlation with the concentration of organic compounds in the tested solution.

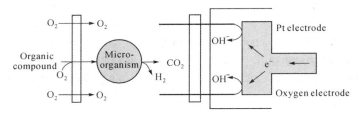

Fig. 5.2.5.   The respiratory activity microorganism sensors

The measurement of metabolic microorganism microbial sensors is based on microbial metabolic activities (Fig. 5.2.6). When microorganisms take organic compounds and generate a variety of metabolites containing electrode active materials, the ammeter can measure the hydrogen, formic acid, a variety of NADPH and other metabolites, while the electricity meter can measure CO_2, organic acids (H^+) and other metabolites. From this we can get the concentration information of organic compounds.

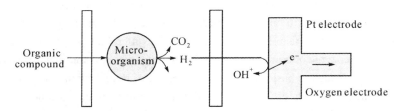

Fig. 5.2.6.   The measurement of metabolic activity microorganism sensors

5.2.2.3   *Microorganism sensors for BOD*

One of the areas where microbial biosensors are widely used is in environmental treatment processes. This is done by detecting the biochemical oxygen on demand (BOD). Most of the BOD sensors consist of a synthetic membrane with immobilized microorganisms as the biological recognition element. The bio-oxidation process is registered in most cases by means of a dissolved O_2 electrode. A wide variety of microorganisms have been screened during the construction of BOD sensors.

Recently, it has been shown that certain redox-active substances could be reduced by certain microorganisms (Yoshida et al., 2000). It can serve as electron shuttling between microorganisms and electrodes. These mediators have been applied to the fabrication of microbial fuel cells and to microbial detection. It has been suggested that reduction of the redox mediator, rather than oxygen, is due to metabolic reactions of microorganisms. Thereafter, instead of oxygen, potassium hexacyanoferrate(III) [HCF(III)] has been used as an electroactive compound for the development of amperometric biosensors using microorganisms. Fig. 5.2.7 shows the principle of the amperometric microbial sensor using HCF(III). Usually, organic substances are oxidized by microorganisms during aerobic respiration. However, when HCF(III) is present in the reaction medium, it acts as an electron acceptor and is preferentially reduced to HCF(II) during the metabolic oxidation of organic substances. The reduced HCF(III) is then reoxidized at a working electrode (anode) which is held at a sufficiently high electric potential. As a result, a current is generated and detected using the electrode system.

The microbial strains selected are chosen for their ability to assimilate a suitable spectrum of substrates. BOD sensors based on a pure culture have the advantages of relatively better stability and longer sensor lifetime, but are restricted by their limited detection capacity for a wide spectrum of substrates.

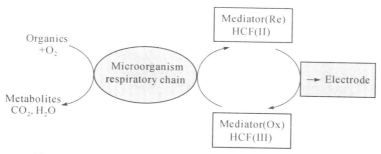

Fig. 5.2.7. Principle of the amperometric-mediated BOD biosensor

5.3 Affinity Biosensors

Affinity biosensors depend on an essentially irreversible binding of the target molecules (e.g., affinity sensors based on antibodies, nucleic acids, or receptors). The primary sensing event does not result from catalysis (e.g., enzymes or microorganisms).

5.3.1 Antibody and Antigen Biosensors

Immunosensors transduce antigen-antibody interactions directly into physical signals. The design and preparation of an optimum interface between the biocomponents and the detector material are the key parts of immuosensor development. Recently, lots of novel investigated sensing techniques have greatly improved the detection schemes.

5.3.3.1 Antibody and antigen

Antibodies are serum proteins which are produced by B lymphocytes and plasma cells, in response to a foreign (non-self) substance (Byfield and Abuknesha, 1994). The foreign substance is termed an immunogen, so-defined because it evokes an immune response. With very high affinity constant and low cross-reactivity, the antigen-antibody reaction may therefore be regarded as highly specific.

Antibodies consist of four polypeptide sub-units comprising two heavy chains (H chains) and two light chains (L chains) (Fig. 5.3.1). The carbohydrate residues in antibodies are covalently bonded to the C-terminal half (Fc). The key portions of the antibody molecules that contain the antigen binding sites are termed the Fab fragments; each Fab fragment comprises an entire light chain and a segment of the heavy chain. An antigen (Ag) will interact with an antibody (Ab) raised to one of its antigenic determinants (portion of structure to which antibodies are produced) with a high binding affinity. The strength of the interaction is dependent on the complementarities of the fit of the antigenic determinant to the binding site of the antibody. The binding forces present in the Ag-Ab complex are non-covalent forces, such as electrostatic attraction (major contribution) hydrogen bonding, hydrophobic bonding and Van der Waals forces.

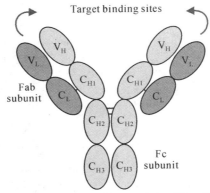

Fig. 5.3.1. Structure of an antibody showing two identical heavy chains (V_H - C_{H1} - C_{H2} - C_{H3}) and two identical light chains (V_L-C_L). The locations of the antigen binding sites are shown

A very important step in the preparation of immunoassay biosensor is to immobilize antibodies or antigens on the sensor surface, so as to detect the corresponding antigen or antibody. Besides common methods of immobilizing bio-recognition elements, there are biotin-avidin system (BAS), self-assembled monolayer (SAM), glutaraldehyde cross-linking method, protein A and other indirect fixation methods which can be used in the preparation of immune sensors with well immobilizing efficiency, fixed-layer adaptability and reacting sensitivity.

5.3.1.2 Immunosensor

The high specificity and high sensitivity which typifies an antigen-antibody reaction (Ag-Ab) has been used in a vast range of laboratory-based tests that incorporate an antibody component (immunoassay). The analytes detected and measured have included many medical diagnostic molecules such as hormones (for example, pregnancy related, steroids), clinical disease markers, drugs (therapeutic and abused), bacteria, and environmental pollutants such as pesticides. There is a crucial distinction to be made, between an immunoassay and an immunosensor, which is the technology relevant to this analysis.

The affinity of an antigen for a corresponding antibody, which is a measurement of the strength of the binding forces in the resultant Ag-Ab complex, is expressed as a molar association constant (K_a) detailed in Eq. (5.3.2) for a reaction scheme shown in Eq. (5.3.1).

$$Ag + Ab \underset{k_r}{\overset{k_f}{\rightleftarrows}} Ag - Ab \qquad (5.3.1)$$

$$K_a = \frac{[\text{Ag}-\text{Ab}]}{[\text{Ag}][\text{Ab}]} = \frac{k_f}{k_r} \quad (5.3.2)$$

where k_f is the forward rate constant (for complex formation) and k_r is the reverse rate constant for dissociation of the antigen away from the antibody. The vast majority of antigen-antibody interactions under conventional pH, temperature and buffer conditions, $k_f \gg k_r$, gives a very high value of k_a. Values of k_a up to 10^{15} mol/L have been measured. For comparison, the majority of enzymatic K_m values, or protein affinities, are in the range $10^3 - 10^6$ mol/L, while binding constants for typical chemical binding reactions are usually lower still. The much greater magnitude of k_f relative to k_r for Ag-Ab interactions is a consequence of the highly optimized, multi-site attractive forces between Ag and the binding pocket of the Ab.

An immunoassay is a laboratory-based multi step diagnostic test which is based on the recognition and binding of the analyte by an antibody. The two major types of immunoassay are shown in Fig. 5.3.2 and are known as (a) sandwich assay (non-competitive assay), and (b) competitive assay (Byfield and Abuknesha, 1994). Both are based on a solid-phase immobilization matrix. It can be seen in both cases that the use of a specific and high affinity Ab-Ag binding reaction to detect and measure the concentration of an analyte (the Ag) requires the presence of a signal-generating tracer in immunoassay formats. A tracer is therefore used to generate a signal (for example, optical, electrochemical, radioisotopic decay) which enables quantification of the amount of bound Ag relative to unbound Ag. The most common types of tracers are fluorescent molecules (fluoro-immunoassay (FIA)), enzymes (enzyme-linked immunosorbent assay (ELISA)) or radio-isotopes (radio-immunoassay (RIA)).

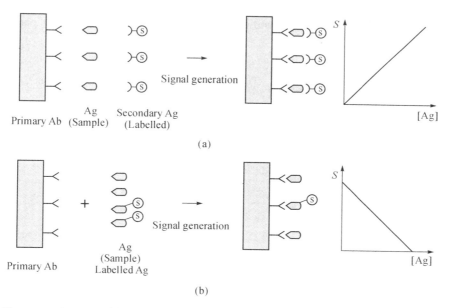

Fig. 5.3.2. The two main types of immunoassay: (a) Sandwich assay (non-competitive); (b) Competitive assay. The signal is proportional to analyte concentration in the former configuration, and inversely proportional in the latter

In contrast to immunoassays, the aim of immunosensor research is essentially to develop more simple antibody-based diagnostic tests, which combine the great specificity and sensitivity of laboratory-based immunoassays, with the much greater versatility with respect to operating environment offered by enzyme biosensors or chemical sensors. Direct (-acting) immunosensors therefore are those in which analyte quantification is carried out by direct detection and measurement of the Ag-Ab binding reaction, where generally either Ag or Ab is immobilized on a surface prior to sample addition. The binding of Ag to immobilized Ab, or vice versa, is accompanied by a small amount of conformational perturbation of the Ab polypeptide structure and alteration in localized electrostatic interactions.

In the study of electrochemical immune sensors, compared to non-labeled immunoassay sensors, the labeled immune sensors have more practical applications at present, and a number of enzyme immune sensors have been applied in clinical research with the biochemical amplification based on marked enzyme catalyzing to its substrate. These types of sensors require fewer samples, generally several microlitre to a few tens of microlitre. They have high sensitivity, good selectivity and can be used as a routine method. The disadvantages are the requirement of adding markers and the complicated operation process.

5.3.1.3 SPR-based immunosensor

Conventional chemical analytical techniques such as optical spectroscopy, nuclear magnetic resonance, or electrochemical measurements are largely incapable of resolving the small change in signals above background noise or non-specific binding of other components in the sample. Interest has tended to focus on alternative methods of transducing changes in physicochemical properties of the solution-surface interface, such as refractive index and dielectric constant. The surface plasmon resonance (SPR) based biosensor systems (i.e., BIAcore^{TM}, AutoLAB^{TM}), which have developed and are commercially available, represents a significant breakthrough in molecular sensor, particularly used in immunosensor techniques (Fig. 5.3.3).

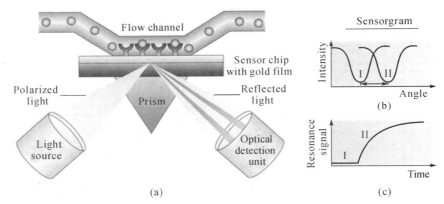

Fig. 5.3.3. SPR detects changes in the refractive index in the immediate vicinity of the surface layer of a sensor chip. SPR is observed as a sharp shadow in the reflected light from the surface at an angle that is dependent on the mass of material at the surface: (a). The SPR angle shifts (from I to II in the lower left-hand diagram) when biomolecules bind to the surface and change the mass of the surface layer (b). This change in resonant angle can be monitored non-invasively in real time as a plot of resonance signal (proportional to mass change) versus time (c) (reprinted from (Cooper, 2002), Copyright 2002, with permission from Nature Publishing Group)

SPR analysis has shown promise in providing direct measurement of Ab-Ag interactions occurring at the surface-solution interface (Cooper, 2002). SPR is a phenomenon which occurs when a beam of light is directed onto a glass-metal interface usually a glass prism on a gold or silver metal layer. At a specific angle (the resonance angle), a component of the electromagnetic lightwave propagates in the metal along the plane of the interface in the form of surface plasmons. The resonance angle is sensitive to changes in refractive index and dielectric constant at the interface up to a distance of ~1000 nm from the actual metal surface, with an exponential fall in sensitivity with distance from the surface. Immobilization of an Ab on the surface causes a measurable shift in the resonance angle; on binding of Ag to the immobilized Ab, a further change will occur. This binding-induced shift in resonance angle (expressed as resonance units, RU) is approximately linearly proportional to concentration of bound Ag (or Ab, if Ag is preimmobilized) for typical biological systems.

The AutoLAB™ system shows considerable promise in enabling determination of low concentrations of important clinical substances in whole serum samples (Dutra and Kubota, 2007). Determination of cTnT (a cardiospecific highly sensitive marker for myocardial damage and is immediately released to bloodstream during the acute myocardial infarction (AMI)) with a detection limit of 0.01 ng/mL in 5 min was achieved (Fig. 5.3.4). It was possible to measure the cTnT without dilution of the human serum with good specificity and reproducibility.

With the development of SPR, it has been increasingly used in many types of biosensors. It can be used to detect the dynamic relationship of the biomolecular combination, the structure and function of biomolecular. The BIA of BIAcore™ from the General Electric Company is the abbreviation of "Biomolecular Interaction Analysis", which just means that SPR is able to provide real-time observation of biomolecular interaction. Therefore, in the studies of DNA sensors, enzyme sensors, and even the cell biosensors, SPR sensors have played an important role.

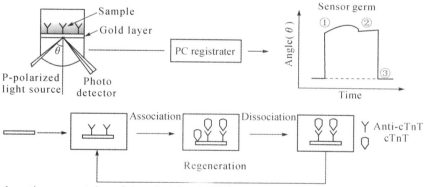

Fig. 5.3.4. Schematic representation of the principle of SPR immunosensor for cTnT determinations (reprinted from (Dutra and Kubota, 2007), Copyright 2007, with permission from Elsevier Science B.V.)

5.3.1.4 QCM-based immunosensor

Piezoelectric sensors are made from the characteristics of piezoelectric effect. The piezoelectric quartz crystal sensor has gained good business development (such as the commercialization of the Sweden Q-sense™ series) based on the wide application of biosensors. As the sensor is based on the principal theories of close relationship between the resonant frequency change of piezoelectric quartz crystal and the change of crystal surface mass loading, such piezoelectric sensors are often called quartz crystal microbalance (QCM) (Fig. 5.3.5).

As a sensitive surface mass sensor, QCM has been extensively applied as transducer in biosensing

(Carmon et al., 2005). A QCM (i.e., commercial Q-sense AB™) sensor utilizes an AT-cut piezoelectric quartz crystal film with gold electrodes deposited on both surfaces. Application of a radio frequency voltage at the resonant frequency of the crystal excites the crystal into oscillation. The principle of QCM detection is based on the frequency changes of the crystal that is proportional to the mass changes on the crystal surface. The details about QCM can be found in Chapter 3 sensors based on piezoelectric resonators. The quantitative relationship between the frequency shifts and the mass changes of the crystal is given by the well-known Sauerbrey equation:

$$\Delta f = -\frac{2f^2}{\sqrt{\rho_q \mu_q}} \Delta m = -C_f \Delta m \tag{5.3.3}$$

where Δf is the frequency shift resulting from the additional mass per area (Δm), f is the intrinsic crystal frequency, ρ_q is the density of the quartz, and μ_q is the shear modulus of the quartz film. For a 5 MHz crystal, C_f=56.5 Hz·cm^2/μg. When the added mass is not a rigid solid, as with many biological samples (i.e., antibody or nucleic acid), the frequency response is dampened, resulting in a frequency shift that is less than predicted using Sauerbrey equation. Recent papers have addressed this deviation from the Sauerbrey equation with efforts to account for the viscoelastic properties of the film on the surface. This expands the versatility of the QCM methodology and enables application of the technology to biological systems. Using QCM sensing, with an appropriate antibody immobilized on the sensor surface, it is possible to detect a specific antigen in solution.

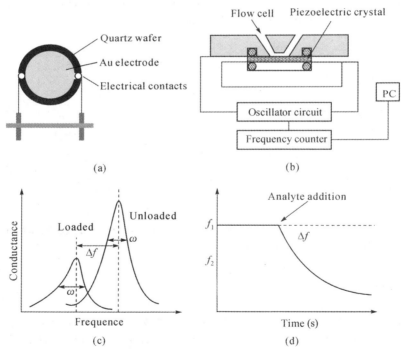

Fig. 5.3.5. QCM based sensor: (a) Typical QCM device; (b) Scheme of a flow cell for piezoelectric crystals in liquid media, including the oscillating system and frequency counter; (c) Impedance analysis is based on electrical conductance curve; (d) The central parameters of measurement are the resonance frequency and the bandwidth. The variation in quartz crystal frequency as a function of time. (f_1) initial frequency; (f_2) frequency after the addition of analyte. Schematic mass loading induced frequency change (Δf)

For example, the immune biosensor for *E. coli* O157:H7 detection can be constructed by the self-assembly monolayer (SAM) (Wang et al., 2008). In order to immobilize antibodies on the gold surface of QCM, a layer of dense protein- coupled interface is formed on gold with 16-mercapto-acid (MHDA) molecules using SAM firstly. MHDA is a long chain thiol acid with a carboxyl-terminal. One thiol end can form self-assembled film on gold surface through the strong Au-S bond, and the formed free end of SAM is carboxyl. Amino-terminal of antibody protein and self-assembled carboxyl-terminal should be cross-linked with an activator. Mostly, carboxyl group can be changed into a more active intermediate using EDC, making it more convenient for condensation reaction between amino group and carboxyl group. After the treatment of SAM and activation, its monoclonal antibodies are immobilized on the sensor surface. The preparation and testing process of QCM immunosensor are shown in Fig. 5.3.6.

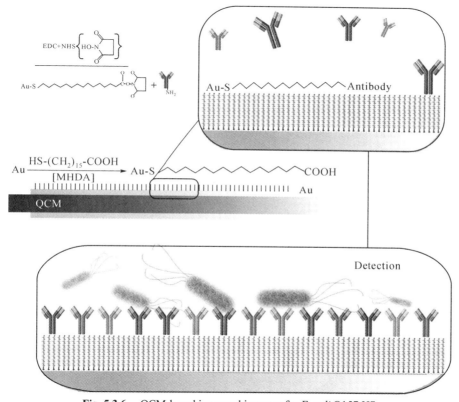

Fig. 5.3.6. QCM-based immune biosensor for *E. coli* O157:H7

E. coli O157:H7 is an enterohemorrhagic strain of the bacterium *Escherichia coli* and a cause of food-borne illnesses. Infection often leads to hemorrhagic diarrhea, and occasionally to kidney failure, especially in young children and the elderly. Its virulence is very strong, each intake of 10 viable cells can cause sickness (while 10^6 or more to normal *E. coli* pathogenic). QCM-based immune biosensors can be used to detect pathogenic O157: H7. Using such classic MHDA molecular self-assembly methods to fix antibodies and capture target bacteria have obtained good results with the detection scope between 10^2 CFU/mL and 10^8 CFU/mL.

Although the same as SPR sensors, the advantage of QCM biological sensors are real-time detection without labeling. However, the sensitivity of QCM sensors in the liquid environment is less than that in

Biomedical Sensors and Measurement

the gas environment. There are excessive and inconsistent interference factors in volatile liquid, such as pH, viscosity, action time and ion concentration, which may reduce the accuracy and precision for detecting the target. How to effectively eliminate or reduce the interference factors and optimize the test conditions, still needs continuous studies in the future.

5.3.2 Nucleic Acid Biosensors

Another biorecognition mechanism involves hybridization of deoxyribonucleic acid (DNA) or ribonucleic acid (RNA), which are the building blocks of genetics. In the last decade, nucleic acids have received increasing interest as bioreceptors for biosensor and biochip.

5.3.2.1 *Nucleic acid*

The complementarities of adenine:thymine (A:T) and cytosine:guanosine (C:G) pairing in DNA (Fig. 5.3.7) forms the basis for the specificity of biorecognition in DNA biosensors, often referred to as genosensors. If the sequence of bases composing a certain part of the DNA molecule is known, then the complementary sequence, often called a probe, can be synthesized and labeled with an optically detectable compound (e.g., a fluorescent label). By unwinding the double-stranded DNA into single strands, adding the probe, and then annealing the strands, the labeled probe will hybridize to its complementary sequence on the target molecule.

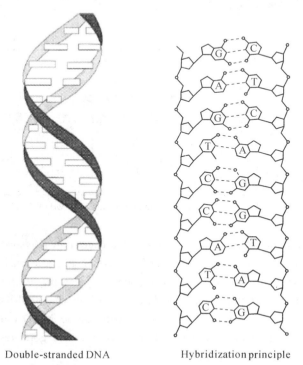

Double-stranded DNA   Hybridization principle

Fig. 5.3.7. DNA and the hybridization principle

Nucleic acid technology is based on the hybridization of known molecular DNA probes or sequences with complementary strands in a sample under test. In the case of nucleic acid biosensors, target DNA is captured at the recognition layer, and the resulting hybridization signal is transduced into a usable electronic signal for display and analysis (Fig. 5.3.8). In the case of electronic and electrochemical biosensors, signal transduction is greatly simplified, because the incoming signal is already electronic in origin (Vo-Dinh and Cullum, 2000).

Nucleic acid analysis in general requires extensive sample preparation, amplification, hybridization, and detection. In theory, nucleic acid analysis provides a higher degree of certainty than traditional antibody technologies, because antibodies occasionally exhibit cross reactivity with antigens other than the analyte of interest. The real time detection of hybridization events has been demonstrated in numerous optical or electrochemical systems.

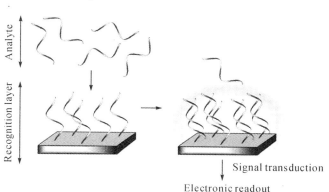

Fig. 5.3.8. General DNA biosensor design

In order to firmly immobilize DNA probes on the electrode surface, it often needs effective physical and chemical methods. In addition to the previously described methods for fixing biological elements, there are covalent bonding and electrical polymerization methods which can also be used to probe fixation. The fabrication of hundreds of thousands of polynucleotides at high spatial resolution in precise locations on a surface is a very important landmark of DNA chips developed by Affymetrix (Lipschutz et al., 1999). The combinatorial photolithographic process was originally destined for peptide syntheses. Highly structured lateral oligonucleotide libraries on glass supports are accessible by initially modifying the surface with photolabile protection groups (Fig. 5.3.9). Illumination through a microstructured photomask leads to the deprotection of selected areas, to which the first phosphoramidite building block is covalently attached. Since the coupled nucleotides also contain photolabile protection groups, the iterative repetition of the process generates new patterns, which lead to two dimensionally structured oligonucleotide arrays.

5.3.2.2 Nucleic acid sensors

Nucleic acid sensors allow an easy, fast and reliable DNA testing, giving the possibility to analyze simultaneously many samples without labeling. These sensor technologies exploit transducers to convert the hybridization event into electrical or optical signals.

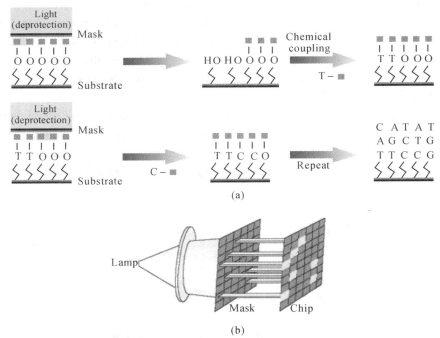

Fig. 5.3.9. Oligonucleotide array fabrication processes using polymeric photoresists: (a) Light directed oligonucleotide synthesis. A solid support is derivatized with a covalent linker molecule terminated with a photolabile protecting group. Light is directed through a mask to deprotect and activate selected sites, and protected nucleotides couple to the activated sites. The process is repeated, activating different sets of sites and coupling different bases allowing arbitrary DNA probes to be constructed at each site; (b) Schematic representation of the lamp, mask and array (reprinted from (Lipschutz et al., 1999), Copyright 1999, with permission from Nature America Inc.)

Electrochemical DNA sensors consist of the electrodes that support DNA probes and the detecting systems for electrochemical activities for target probes. Probe molecules are modified to the electrode surface to constitute a DNA electrode, and hybridized selectively with the target sequence to form double-stranded DNA, leading to electrochemical changes on the electrode surface. The differences in changes before and after hybrid can also be identified by hybrid indicators with electrical activities for target sequences or a specific gene. The electrochemical DNA sensors with enzyme-linked amplification have similarities with early enzyme marked immune sensors. Enzyme is marked in the DNA molecule as a recognition element. When it acts on the substrate, it leads to an electrochemical response. The target DNA can be detected indirectly by the electrochemical signals.

Recently, DNA sensors based on quartz crystal microbalance (QCM) and surface plasmon resonance (SPR) have been applied with success, for instance, for specific detection of DNA sequences. Fig. 5.3.10 was an example of a nucleic acid sensor based on QCM as a commercial example

Traditional QCM sensors have high sensitivity in mass changes, often respond to any loading and unloading on the electrode, however, without the effect of selectivity loading materials. DNA hybridization is based on the principle of base pairs. Only sequences with whole or partly complementarities can hybrid with high selectivity. In respect to the above advantages, piezoelectric DNA sensors can be constructed with high sensitivity and specificity. For example, ssDNA of staphylococcal enterotixn B (SEB) and *E. coli* O157:H7 can be modified on the QCM gold electrode by silane methods to construct a DNA sensor for rapidly detecting the genes of these pathogens. The amount of the electrode modified DNA probe can be determined by the amount of nucleic acid before and after solidification by the protein nucleic acid analysis system. The amount of ssDNA solidified on the crystal surface can be determined by the DNA dot blot color revealed by biotin-labeled DNA phosphoryl enzymes. Studies have found that, DNA fragments with relatively short length have better hybridization and well reusing capabilities.

And, the study on SPR-based DNA sensors also has become more and more developed. The relevant mechanism can be seen in the Section of immune sensors with SPR.

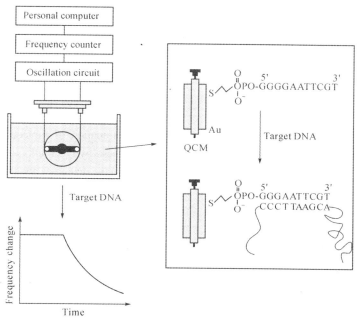

Fig. 5.3.10. DNA biosensor design based on QCM

5.3.2.3 *Nucleic acid sensor and DNA chip*

Compared to conventional nucleic acid detection, DNA biosensors have the following advantages (Kricka, 2001; Wang, 2000):

(1) *Hybridization in liquid-phase.* Conventional nucleic acid detecting is mainly hybridization in solid-phase. DNA sensors can detect DNA of target materials quantitatively by the changes of sound, light, electricity signals in the liquid phase reaction.

(2) *It can carry on real-time detection of DNA.* Combined with microfluidic chips, the dynamic reaction process of DNA can be monitored in real-time. And, DNA can be measured quantitatively and timely, achieving online and real-time detection of DNA.

(3) *It can carry on dynamic detection of nucleic acid in vivo.* Currently, there is no effective method studying nucleic acid *in vivo* directly. DNA sensors provide the possibility for studying of dynamic processes of nucleic acid metabolism transfer *in vivo*.

(4) *It can carry on a large number of intelligent DNA detections.* The combination of DNA sensors and artificial neural networks can filter out sensitive components with better selectivity and activity. The developing multi-functional or intelligent DNA sensors will detect a variety of DNA samples at the same time.

(5) *It has high sensitivity*. DNA sensors can directly detect the target materials. If combined with polymerase chain reaction (PCR) and other techniques, the sensitivity of sensors will be greatly improved for low copy nucleic acid.

(6) *It has high specificity*. A DNA sensor is based on the principle of complementary combination. Thus, it has high specificity.

(7) *It is clean*. It does not need isotopic labeling, and avoids harmful substances.

In the varieties of reported DNA sensors, most of the hybridization time is about 10 min to 1 h, which is a great progress when compared to typically traditional overnight hybridization (20 h), but still too long for sensors. Not only does it greatly reduce the advantages of dynamic monitoring, but also the deadly disadvantage of measuring large quantities of samples. The hybridization time and hybridization volume (sensitivity) is in itself a pair of contradictory elements. How to shorten hybridization time in the premise of ensuring adequate sensitivity is the main problem that needs to be solved. The development of a biochip (mainly DNA chips) based on micro-array hybridization has played a significant role for DNA sensors.

DNA sensors and DNA chips both developed with the combination of molecular biology, modern physics, modern chemistry, and micro-electronics technologies. These two technologies penetrate mutually and develop parallelly, which will bring major breakthroughs both in the diagnosis and therapy of diseases.

Recognizing molecules of DNA sensors and DNA chips are both based on the principle of base pairing complementarily. DNA sensors have dynamic monitoring functions with ultra-miniature, micro-system, multi-parameter, and finally for bedside monitoring, *in vivo* monitoring, non-invasive monitoring, and cell monitoring. However, DNA chips have the characteristics of high-density, which can detect tens and even thousands of genes simultaneously. The research goal of the DNA chip is to achieve overall and rapid diagnosis of a patient's entire genome change by a chip and to make it become conventional technology. Now, DNA sensor researches still focus on improving detection sensitivity and shortening response time as well as enhancing the stability and usability; while DNA chip researches have paid emphasis on specific DNA microarrays for disease-related gene fragments. The new generation of DNA chips should be based on microelectronics principles, with the dynamic monitoring characteristics of DNA sensors and multi-gene synchronous detection function of the DNA chip. A good hybrid of these two chips is also one of the main directions of the DNA biosensor development.

5.3.3 Receptor and Ion Channel Biosensors

Molecular receptors are cellular proteins (often membrane-bound) which bind specific chemicals in a manner which results in a conformational change in the protein structure. The conformational change triggers a cellular response, for example, opening an ion channel or secreting an enzyme.

Receptors and ion channels are interesting and provide important opportunities for the development of biosensors for three principal reasons (Subrahmanyam et al., 2002). First, receptors and ion channels possess high affinity and specificity refined by the evolutionary process. Second, receptors and ion channels are natural targets for toxins and mediators of physiological processes, and due to this, they can be used for monitoring these compounds in clinical and environmental analyses and in the development and screening of drugs. Third, the research of receptors and ion channels is an important area and novel sensors can be useful for real-time elucidation of receptor-ligand interactions.

5.3.3.1   Receptor and ion channel

The fact that many ion channels and receptors can be purified and reconstituted in black lipid membranes (BLMs) for studies of function and pharmacology has spurred initial interest in the development of ion channel/receptor-based biosensors. However, ion channels especially those pertaining to mammalian physiology, cannot be considered robust in BLMs or isolated membrane patches due to the well-known property of ion channel "rundown" or "washout". In the absence of integral intracellular machinery provided by cells needed to maintain function, ion channels typical of mammalian physiology presently do not constitute practical biosensors.

To preserve the integrity of the receptor with the immobilizing conditions to closely resemble the natural environment, biosensors are often achieved by reintegrating receptors into lipid membranes.

Fig. 5.3.11 shows a selection of membrane sensor platforms to support receptors and ion channels (Reimhult and Kumar, 2008). (a) Supported lipid bilayer on hydrophilic support. Typically a hydrophilic semiconductor or oxide substrate has to be used for direct assembly of a supported lipid bilayer (SLB). The support will act as a working electrode (WE) for electrochemical measurements; (b) Tethered supported lipid bilayer. Covalently attached hydrophobic molecules with a hydrophilic linker, often derived from lipids, are used to tether and support the lipid membrane to a gold WE; (c) Free-spanning membrane or black lipid membrane as it could look in a self-assembled membrane sensor array. Lipid membranes containing ion-channels are separated into functional spots, for example by non-fouling polymer barriers, and span apertures in a solid support. It has free liquid access on both sides to accommodate large proteins and to easily conduct electrochemical measurements. The schematics show how such a platform combined with microfluidics could be used for parallel voltage clamp measurements on different single transmembrane proteins.

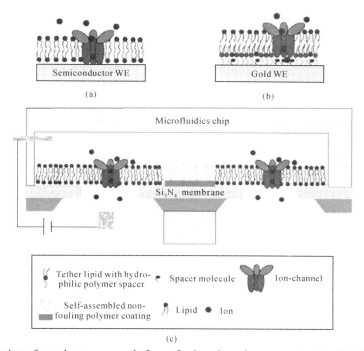

Fig. 5.3.11.   A selection of membrane sensor platforms for ion channels or receptors (reprinted from (Reimhult and Kumar, 2008), Copyright 2007, with permission from Elsevier Science B.V.)

Based on the above lipid supported biosensor system, "Gigaohm seal" enabling measurements of single ion-channels and transporters can be obtained and single ion channel sensitivity can be also achieved. Membrane stable >1 d for screening applications also has been achieved in several formats.

5.3.3.2 Receptor and ion channel sensor

The *in vitro* receptor molecule sensors have advantages in sensitivity and selectivity. However, there are many difficulties in the separation and purification of membrane receptors, fixation of *in vitro* receptor molecules and conversion and amplification of signals of receptor sensors, all of which affect the development of *in vitro* receptor biosensors. The glutamate receptor and dopamine receptor sensors using lipid membrane supporting system have created a certain foundation on receptor sensors.

The ion channel sensors are mainly designed with gramicidin supported by lipid bilayer, as shown in Fig. 5.3.12 (Liu et al., 2008).

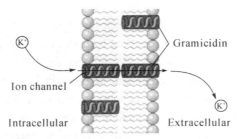

Fig. 5.3.12. The model of ion channels based on Gramicidin

Gramicidin A is a kind of polypeptide extracted from the *Brucella* consisting of 15 hydrophobic amino acid alternating array. One of its main biological functions is to form selective transmembrane ion channels, especially the unit price ion channels, enabling passive transportation of easily diffused ions such as Na^+, NH_4^+ and K^+. As the formation of lipid bilayer, gramicidin polymerization forms ion channels in the lipid bilayer in the form of dimer, which can dramatically increase membrane ion permeability. The dielectric properties of lipid bilayer, such as the membrane capacitance, change with the gramicidin embedded in. Thus, channel currents can be measured to obtain information about its permeability properties over a very wide range of permeate ion concentrations or applied potentials.

5.3.3.3 Electrifying cell receptor sensors

Modifying receptors on the surfaces of cells to enable them to interact with proteins that are not their natural partners is one way of controlling signaling processes in cell biology. And, when isolated from their natural environment, these re-engineered proteins could be used for various sensing and drug screening applications.

The challenge in developing such biosensors lies in functionally connecting a molecule detector to an electrical switch. In order to facilitate the study of ion channel-coupled receptors, recently the Western Reserve University and the University of North Carolina researchers have started G protein-coupled receptor clone and expression, based on the use of six amino acid (hexaglycine) such as C-terminal cross-linking the receptors with the inward rectifier potassium channel (Kir 6.2) of the N-terminal for coupling (Moreau et al., 2008). This method of using bio-engineering will be the coupling of receptor proteins and ion channel technology, which is easy to set up as a new type of bio-molecular sensing detection technology, allowing receptor-binding information to be detected by electrophysiological parameters, which are called electrical activity of cell receptors (electrifying cell receptor) (Fig. 5.3.13) (Abbas and Roth, 2008).

The studies were carried out on platforms of lipid bilayer, which facilitates by keeping the biological activity of receptors through the support of lipid bilayers. Also it can simulate the transduction of receptor signals based on ion channel coupling and transform stimulated signals into action potentials, which are bioelectrical signals available for biosensor detection.

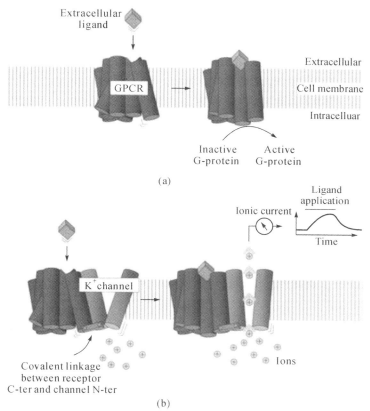

Fig. 5.3.13. Electrifying cell receptor of ion channel-coupled receptors: (a) Upon binding of its ligand at an extracellular site, a transmembrane G-protein-coupled receptor (GPCR) adopts a new conformation that triggers activation of intracellular G-proteins; (b) The GPCR is attached to an ion channel in such a way that both proteins are mechanically coupled. When the GPCR binds a ligand and changes conformation, this change is directly transmitted to the channel and results in a change in gating and in the ionic current through the channel (reprinted from (Moreau et al., 2008), Copyright 2008, with permission from Nature Publishing Group)

5.4 Cell and Tissue Biosensors

Cell and tissue biosensors on the other hand offer a broad spectrum detection capability. Moreover, using cells as the sensing elements provides the advantage of *in-situ* physiological monitoring along with analyte sensing and detection (Pancrazio et al., 1999; Wang and Liu, 2009). A cell by itself encapsulates an array of molecular sensors. Receptors, channels, and enzymes that may be sensitive to an analyte are maintained in a physiologically stable manner by native cellular machinery. In contrast with antibody approaches, cell and tissue sensors are expected to respond optimally to functional, biologically active analytes. Cell and tissue biosensors have been implemented using microorganisms, particularly for environmental monitoring of pollutants. Sensors incorporating mammalian cells have a distinct advantage of responding in a manner that can offer insight into the physiological effect of an analyte. Several approaches for transduction of cell sensor signals are available and these approaches include measurements of cell metabolism, impedance, and extracellular potentials.

In many cell and tissue biosensor applications, primary cells or tumor-derived cell lines are selected as the source of cells. Primary cells are extracted directly from animals, while a cell line refers to a genetically identical cellular population that actively divides *in vitro*. Primary cells have the advantages of cell type availability, and the likelihood to have *in vitro* functionality similar to that found *in vivo*. However, the process of harvesting is inefficient and presents ethical issues for large-scale operations. Tumor-derived cells, when the desired cell type is available, offer the advantage of convenience of preparation. Nevertheless, they are typically derived from abnormal cells and may suffer the disadvantage of not retaining the desired functionality of *in vivo* cells.

Since the last decade, there has been significant interest in the characterization of totipotent embryonic stem cells which are capable of maturing into any cell types. Stem cells provide an alternative between tumor-derived cell lines and terminally differentiated primary cells, because they can be dissociated from tissue, grown indefinitely in a culture, and induced to differentiate into mature cells. The technique of "embryonic stem cell test" has also been developed and used *in vitro* to screen new medicines and other chemicals. The test has shown merits such as no requirement of the sacrifice of animals, no side effects on humans, higher sensitivity, and even higher accuracy. Therefore, stem cells have provided a renewable novel source of cells for cell and tissue biosensor applications (Liu et al., 2007a; 2007c).

5.4.1 Cellular Metabolism Biosensors

A category of cellular biosensors relies on the measurement of energy metabolism, a common feature of all living cells. This is especially useful in testing drugs as well as in cancer research. The combined application of microfabrication technology to microfluidics has aided in the development of portable sensors. The changes in the cell metabolism due to the effect of a chemical reagent are transduced into electrical signals that are read out and analyzed.

5.4.1.1 Cellular metabolism by cytosensor

In the 1990s, Molecular Device Corporation presented a biological application of light addressable potentiometric sensors (LAPS) and invented the Cytosensor®, which is often called a microphysiometer. It is a system for performing bioassay, which integrates living cells with LAPS. Acidic products of energy metabolism acidify cellular environments and the microphysiometer measures the rate of proton excretion from cells, as in Fig. 5.4.1 (Hafner, 2000).

A general feature of living heterotrophic cells is the uptake of metabolites (carbon sources), the production of energy (ATP) and the excretion of acid waste products (e.g., lactic and carbonic acid).

Carbon sources are sugars, amino and fatty acids. In regular culture conditions, glucose and glutamine are present in high concentrations and are taken up by cells and broken down into energy and waste products. Under aerobic conditions, glucose is converted via pyruvate and acetyl CoA into CO_2 yielding energy (respiration). The corresponding pathways are glycolysis, citric acid cycle and oxidative phosphorylation. The extracellular acidification of cells sitting in a flowing chamber can be measured with the Cytosensor and thus, a functional response of cells upon receptor stimulation can be monitored under non-invasive conditions and in real time. Upon stimulation of a membrane-bound receptor, which can be either G protein-coupled, tyrosine kinase-coupled or an ion channel, a signal transduction cascade is initiated. Many steps in this cascade are either directly or indirectly energy-dependent. Under steady state conditions, one cell produces about 10^8 protons per second. After receptor stimulation, this will be raised to between 10% and 100%, depending on the cell-type, the receptor and the coupling pathway. The result is an acidification of the extracellular medium, which directly can be detected with the cytosensor system.

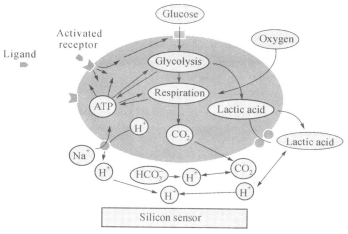

Fig. 5.4.1. Microphysiometer studies based on light addressable potentiometric sensor. Schematic representations of metabolic pathways in which extracellular pH changes occur (reprinted from (Hafner, 2000), Copyright 2000, with permission from Elsevier Science B.V.)

Large numbers of experiments on ligand-receptors have proved that the combination of receptor and agonist increases the rate of extracellular acidification (ECAR). A microphysiometer is able to estimate the effect of chemotherapy by ECAR. So it can be used to evaluate and screen drugs and even suborganelle, such as, G protein-coupled receptors, receptors with intrinsic catalytic activity, ligand gated ion channel, and receptors linked with cytoplasmatic tyrosine kinase.

5.4.1.2 *Microphysiometer based on LAPS*

Getting more information about the multi-functional cellular processing of input- and output-signals in different cellular plants is essential for basic research as well as for various fields of biomedical applications. The concentrations of the extracellular ions, such as Na^+, K^+, Ca^{2+}, may change along with the alteration of cell physiology. In order to analyze simultaneously the relations of the extracellular environmental H^+, Na^+, K^+, Ca^{2+} under the effects of drugs, our laboratory has developed a novel microphysiometer based on multi-LAPS (Wu et al., 2001; Zhang et al., 2001). The surface of the LAPS is deposited with different sensitive membranes by the silicon microfabrication technique and the PVC membrane technique. When we fabricated the K^+ sensitive membrane, we used 2 mg valinomycin as the electro active substance, 1.3 mL tetrahydrofuran (THF) and 1.3 mL cyclohexanone as solvent, 0.066 g PVC powder as bulk material, and 1 – 2 gutta di-butyl phthalate as plasticizing agent. When we fabricated the Ca^{2+} sensitive membrane, we used 0.24 mL formamide as an active substance, 4.72 mL tetrahydrofuran as solvent, 0.15 g PVC powder as bulk material, and paucity tetraphenylboron sodium as the plasticizing agent.

The different sensitive membranes are illuminated in parallel with light sources at different frequencies, and measured online by parallel processing algorithm Fig. 5.4.2.

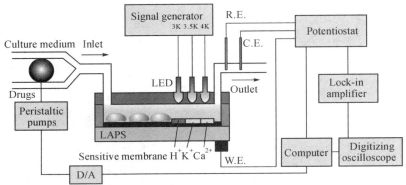

Fig. 5.4.2. The schematic drawing of the system of the multi-LAPS to different extracellular ions (H^+, K+, and Ca^{2+}) (reprinted from (Wu et al., 2001), Copyright 2001, with permission from Elsevier Science B.V.)

The amplitude of each frequency component can be measured on-line by the fast Fourier transform (FFT) analysis. On the sensor, a different sensitive (H^+, K^+, Ca^{2+}) membrane is illuminated simultaneously with three light sources at different frequencies (3 kHz for K^+, 3.5 kHz for Ca^{2+}, 4 kHz for H^+), the photocurrent comprises the three frequency components. The amplitude of each frequency component might be measured on-line by FFT (as shown in Fig. 5.4.3). Dilantin, i.e., phenytoin sodium, is a kind of anti-epilepsy drug, which has significant effects on transqulizing and hypnotic and anti-seizure. Moreover, dilantin is also one of the anti-arrhythmia drugs. It has been proved that dilantin has membrane stabilizing actions on neural cells because it can reduce pericellular membrane ions (Na^+, Ca^{2+}) permeability, inhibit Na^+ and Ca^{2+} influx, stave K^+ efflux, thus, prolonging the refractory period, stabilizing the pericellular membrane, and decreasing excitability. This mechanism can be proven by our experiments (Wu et al., 2001).

The extrusions or intrusions of protons and ions are very general parameters involved in the activation of nearly all kinds of membrane-bound receptors. These parameters can be measured with the novel microphysiometer. In addition, the microphysiometer works under regular cell culture conditions, so cells can be repeatedly stimulated with drugs within a few hours. Thus, a functional response of cells upon receptor stimulation can be monitored under non-invasive conditions and in real-time. The responses depend on the cell-type, the receptor and the coupling pathway. The microphysiometer makes it easy to complete a dose-response curve and EC_{50} (the concentration of agonist that provokes a response halfway between the baseline and maximum response) determination. Furthermore, by comparing different curves and EC_{50} values under different conditions (such as different cell types), drug effects may be evaluated. With the microphysiometer, no knowledge about the kind of receptor coupling is necessarily prior to the experiment. This makes it easier to work with orphan receptors and ligands from unknown receptors. However, by blocking certain signaling pathways inside the cell by means of signal transduction probes, specificity can be brought into the system and the exact coupling of a membrane bound receptor can be easily elucidated, and drugs like agonists can be screened.

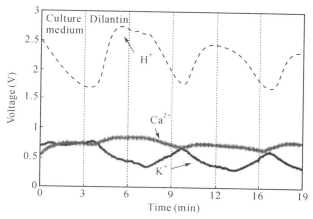

Fig. 5.4.3. Microphysiometer based on LAPS. Illuminate simultaneously at the three sensitive membranes with three light sources at the three modulation frequencies. H^+, K^+, Ca^{2+} analyze simultaneously by multi-LAPS (reprinted from (Wu et al., 2001), Copyright 2001, with permission from Elsevier Science B.V.)

5.4.2 Cellular Impedance Biosensors

Vertebrate cell behavior in tissue culture is normally studied by periodic microscopic examinations of cell density and morphology. If a continuous record of behavior is required, it is generally obtained by cinematographic arrangements. In recent years, lots of researchers have described novel methods in which cells can be monitored continuously using electric fields.

5.4.2.1 Electric cell-substrate impedance sensing

By culturing cells over one or more electrode contacts, changes in the effective electrode impedance permit a non-invasive assay of cultured cell adhesion, spreading, and motility (Giaever and Keese, 1993). Fig. 5.4.4 shows the schematic of an impedance sensor called an electric cell-substrate impedance sensing (ECIS™), which is used to monitor attachments and the spreading of mammalian cells quantitatively and in real time. This technology was invented by Drs. Ivar Giaever and Charles R. Keese while working at General Electric Corporate Research and Development, and then they formed Applied BioPhysics, Inc. as a private company to develop, commercialize and market this and other biophysical technologies.

A schematic view of a cell positioned over ECIS electrodes is shown in Fig. 5.4.5. The total measured impedance consists of the electrode impedance Z_e, the resistance between the electrode and the bulk electrolyte due to the thin layer of medium between the cell and the passivation layer R_{seal}, the membrane capacitance and ion channel resistance over the electrode C_{M1} and R_{M1}, the membrane capacitance and ion channel resistance of the top and sides of the cell C_{M2} and R_{M2}, the solution resistance and the counter electrode impedance (omit in the Figure). In reality, R_{seal} is distributed with the capacitance and conductance of the membrane in the region over the passivated layer and the result of AC impedance is mainly affected by R_{seal}. For this measurement, the counter electrode impedance, both the solution resistance and electrode impedance should be negligible (impedance of platinized electrodes is relatively small), so the impedance measurement is dominated by the seal resistance and the cell membrane properties. It is clear that R_{seal} must be on the same order (or larger than) the membrane impedance if changes in membrane properties are to be observed. When living cells are cultured on the surface of chips, some important factors including the speed of propagation, moving and adherence, etc., can be obtained from the impedance detecting of ECIS.

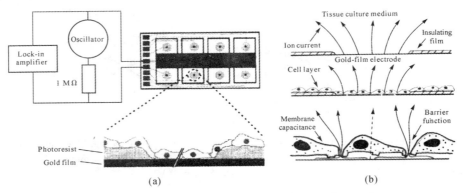

Fig. 5.4.4. The schematic (a) and principle (b) of an impedance cell sensor

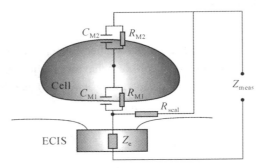

Fig. 5.4.5. Schematic of a cell positioned over an electrode (not to scale). The measured impedance consists of the electrode impedance, the resistance between the electrode and the bulk electrolyte due to the thin layer of medium between the cell and the passivation layer, the membrane capacitance and ion channel resistance over the electrode

This method is based on measuring changes in AC impedance of small gold-film electrodes deposited on a culture dish and used as growth substrates. The gold electrodes are immersed in the tissue culture medium. When cells attach and spread on the electrode, the measured electrical impedance changes because the cells constrain the current flow. This changing impedance is interpreted to reveal relevant information about cell behaviors, such as spreading, locomotion and motility. They involve the coordination of many biochemical events. They are extremely sensitive to most external parameters such as temperature, pH, and a myriad of chemical compounds. The broad response to changes in the environment allows this method to be used as a biosensor. Impedance techniques are theoretically capable of dynamic measurements of cellular movement at the nanometer level, with resolution above that of conventional microscopy. Impedance measurements have been used to assess the effect of nitric oxide on endothelin-induced migration of endothelial cells. From the standpoint of biosensors, changes in cell migration or morphology tend to be somewhat slow; and marked changes in impedance in the presence of cadmium emerged only after 2 – 3 h of exposure. Thus for real time tracking and monitoring the effects of analytes, a sensing technique based on impedance measurements would be slow and cumbersome.

5.4.2.2 Applications of cell impedance biosensors

Impedimetric analysis on adherently growing cells by micro-electrodes provides information related to cell number, cell adhesion and cellular morphology. In recent years, cancer is rapidly becoming the number one killer in many countries. And, chemotherapy (anti-cancer drugs) is still one of the most important treatment methods in clinics. In pre-clinical testing studies, there is a great demand to develop more rapid and simpler techniques for studying cancerous cells, especially for understanding their interactions with drugs and toxins. There is a need to develop minimally invasive, reliable, inexpensive, and easy to use instrumentation for studying real-time biological events *in vitro*.

In our study, cell-based biosensors with micro-electrode array (MEA) were used to monitor the culture behavior of mammalian cancer cells and evaluate the chemosensitivity of anti-cancer drugs using electrochemical impedance spectroscopy (Liu et al., 2009). The platinum electrode arrays were fabricated by semiconductor technology to a 10×10 pattern, with a diameter of 80 μm for each electrode (Fig. 5.4.6).

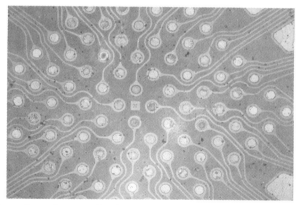

Fig. 5.4.6. Photo of the micro-electrode arrays (reprinted from (Liu et al., 2009), Copyright 2009, with permission from Elsevier Science B.V.)

The human oesophageal cancer cell lines (KYSE 30) were cultured on the surface of the electrode with the help of fibronectin, the connecting protein for tumor cells metastasis and adhesion in an extracellular matrix. Morphology changes of cell adhesion, spreading, and proliferation can be detected by impedimetric analysis in a real time and in a non-invasive way. The anti-cancer drug cisplatin was added to cells for potential drug screening applications (Fig. 5.4.7). The experimental results show that this well-known drug has characterized chemosensitivity effects detected by MEA. The cancer cell chip provides a useful analytical method for cancer research.

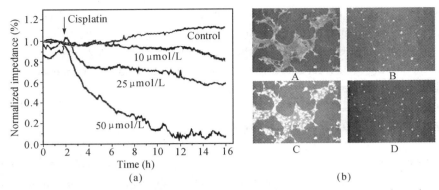

Fig.5.4.7. Analysis chemosensitive of cisplatin to KYSE 30 cells: (a) Cells were grown onto electrodes until confluent and treated with cisplatin (10 μmol/L, 25 μmol/L, and 50 μmol/L), and impedance was determined for up to 16 h; (b) Fluorescence imaging of KYSE 30 cells cultured on the surface of the micro-electrode. A: Phalloidin imaging specific for microfilament after cells cultured 48 h; B: Phalloidin imaging after cells treated with cisplatin 12 h; C: Propidium iodide specific for nucleic acid imaging after cells cultured 48 h; D: Propidium iodide specific for nucleic acid imaging after cells treated with cisplatin 12 h (reprinted from (Liu et al., 2009), Copyright 2009, with permission from Elsevier Science B.V.)

Cell adhesion in an extracellular matrix is the precondition for tumor metastasis, and then a tumor clone is formed with unbounded cell proliferation. The current adhesion assay is an *in vitro* method which is used to determine the rate or strength of adhesion for different cell types to extracellular matrix proteins by fluorescent labeling. At the same time, cellular morphology is recognized as one of the most important parameters in cancer biology. Especially, most of anti-cancer agents currently employed that target the cytoskeleton, do so through interactions with the microfilaments. The cytoskeleton consists of a complex network of filamentous proteins which are involved in regulation of cell morphology and adhesion. Moreover, the concept that when tumor cells are exposed to anti-cancer drugs, they usually die from apoptosis has become a widely held tenet of modern cancer treatment. Therefore, the cell chip with micro-electrode was used to monitor the culture behavior of mammalian cancer cells and evaluate the chemosensitivity of anti-cancer drugs using an electrochemical impedance spectroscopy.

5.4.3 Extracellular Potential Biosensors

Living cells grown *in vitro* are particularly attractive as potential detector elements. The electrophysiological signals of excitable cells depend on the cellular functional information. Excitable cells, such as neural cells, and myocytes, are evoked into action potential when subjected to stimulus. Patch clamps can be used to detect the cellular membrane potential directly, but the measurement is limited to a fixed recording site, so it cannot be used to measure the coupling of cells. Furthermore, the invasive nature of intracellular recording, as well as the voltage-sensitive and optical-sensitive dyes, limits the utility of standard electrophysiological measurements and optical approaches. As a result, planer microelectrode arrays (MEA) were developed and have become powerful tools, which offer a non-invasive and long-term approach to the measurement of excitable cells action potential, or communication between cardical or neural tissues.

5.4.3.1 *Extracellular potential detection by MEA*

MEA typically consisting of 16 – 64 recording sites presents a tremendous conduit for data acquisition from networks of electrically active cells. As a result, planar MEA have emerged as a powerful tool for long-term recording of network dynamics. Extracellular recordings have been achieved from dissociated cells as well; which is more useful in specific chemical agent sensing applications.

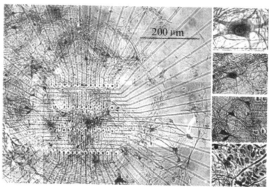

Fig. 5.4.8. Typical dissociated neuron (mouse spinal) culture atop a microelectrode array. Insets at right illustrate the variety of morphologies seen in the culture (reprinted from (Gross et al., 1995), Copyright 1995, with permission from Elsevier Science B.V.)

Gross and colleagues at the University of North Texas over the past 20 years have demonstrated the feasibility of neuronal networks for biosensor applications (Gross et al., 1995). Neurons cultured over microelectrode have shown regular electrophysiological behavior and stable pharmacological sensitivity for over 9 months. Fig. 5.4.8 shows neuronal cultures on a 64 microelectrode array of Multichannel Systems (Roboocyte®, Germany) (Kovacs, 2003). In fact, their precise methodological approach generates a co-culture of glial support cells and randomly seeded neurons, resulting in spontaneous bioelectrical activity ranging from stochastic neuronal spiking to organized bursting and long-term oscillatory activity.

As shown in Fig. 5.4.9, the coupling model of cell with MEA for extracellular potential detection. R_{seal} is the seal resistance between cell and electrodes, and it is necessary to make the seal resistance R_{seal} as large as possible. The current I_{total} through the electrode is:

$$I_{total} = C_M \frac{d(V_M - V_J)}{dt} + \sum_i I^i_M = C_J \frac{dV_J}{dt} + \frac{V_J}{R_{seal}} \qquad (5.4.1)$$

In Eq. (5.4.1), I_{total} is the total current, C_M is the membrane capacitance, V_M is the membrane voltage, V_J is the coupling voltage, I_M is the total current of extracellular ions, C_J is the capacitance of the coupling layer. So, when R_{seal} is lager, it means that the creepage current is less and the recoded current and it is mainly coming from the cell's electrophysiological signals.

When the electrode is coupled with cells, R_{seal} is the seal impedance, which is used to illustrate the leaking current between the interfaces of the cell and electrode and the width of the gap. When R_{seal} is large, the gap is small and more available for a larger voltage signal V_J with a high S/N ratio. However, when the R_{seal} value is small, it means there is a large part of the effective current leakage to the ground through point A as shown in Fig. 5.4.9, and the junction voltage V_J will be small. So in experiments, the whole impedance should be smaller or in the same magnitude with R_{seal} for a better detection of the electrical extracellular signals.

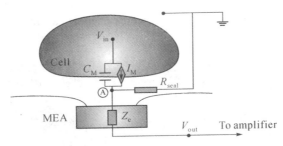

Fig. 5.4.9. Conceptional drawing of a cultured cell coupled to a microelectrode and the simplified circuit schematic of the cell-electrode junction

Microelectrode arrays coupled with "turnkey" systems for signal processing and data acquisition are now commercially available. In spite of the obvious advantages of the microelectrode array technology for determining the effect of chemical analytes at the single cell level, it becomes essential to pattern the dissociated cells accurately over microelectrodes. Single cell sensing forms the basis for determining cellular sensitivity with a wide range of chemical analytes and determining the cellular physiological changes. Also analyses of the extracellular electrical activity, which results in unique identification tags associated with cellular response to each specific chemical agent are also known as "signature patterns". Beside the merits of long-term and non-invasive applications, paralleled recording makes it possible to conveniently study the signal transmission at the synapse of neuronal network or at the junction gap between cardiac cells. With these abilities, MEA has been the fundamental tool in neuroscience for network study and widely used in the pharmacological field.

5.4.3.2 Applications of extracellular potential detection

Since olfactory and gustatory systems play important roles in detecting environmental conditions, a lot of olfactory and gustatory research has been carried out due to their potential commercial applications. Electronic nose and electronic tongue belong to these technologies, which mimic animals' olfactory and gustatory systems to detect smell and taste by exploiting sensitive materials. The detection ability of these devices mainly depends on the absorbability or catalysis of sensitive materials to special chemicals. Although great achievements have been made, this method still has limitations in sensitivity and specificity, compared with the biology binding of specific odorants or tastants to the olfactory receptor neurons or taste receptor cells.

It was considered to be beneficial to establish both artificial olfaction systems for odor detection and useful platforms for olfactory research. In recent years, some groups have used olfactory cells and receptors coupled with sensors, such as surface plasmon resonance (SPR), quartz crystal microbalance (QCM) and field effect transistor (FET), as a bioelectronic nose to detect odorant molecules. These methods have relatively high specificity and sensitivity with the interaction between the olfactory receptor and the odorant successfully recorded. However, subsequent intracellular changes, which result from changes in ionic currents and the membrane potential in cells, especially in olfactory receptor neurons, cannot be detected.

Based on microelectronic sensor chips, cell and tissue biosensors can collect the electrophysiological responses directly relating to cellular functions (Liu et al., 2006). In our studies, we have extracted olfactory cells and tissues from rats and cultured them on MEA, and recorded extracellular potentials of olfactory receptor neurons following odorant-receptor binding (Liu et al., 2010a; 2010b). We find that MEA might be suitable for recording the electrical signals produced by olfactory cells and thus could be used as olfactory cell based biosensors (Fig. 5.4.10). Especially, we have managed to combine the intact olfactory epithelium with MEA for a bioelectronic nose of olfactory receptor neurons. Compared to the cultured olfactory cells, the intact olfactory epithelium can be obtained conveniently with the primary cell structure well-preserved, which can mimic the *in vivo* process of gas sensing, and is a good candidate for the biological elements of a bioelectronic nose.

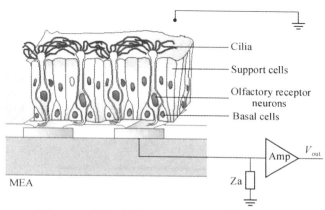

Fig. 5.4.10. Recording extracellular potentials of olfactory receptor neurons in intact epithelium by microelectrodes (reprinted from (Liu et al., 2010a), Copyright 2010, with permission from Elsevier Science B.V.)

MEA can record the multisite potentials simultaneously and has great potential in electrophysiological detection in a long-term and non-invasive way. The *in vitro* tissue was kept bioactive, where olfactory receptor neurons fired spontaneously, with amplitude and duration of peaks were about 50 – 100 μV and 10 – 20 ms. For example, Fig. 5.4.11 displays the 16-channel signals as an example. It can be seen that Channel 03 recorded the negative peaks, while Channels 05, 06 and 07 recorded spontaneous potentials in cluster with similar positive peaks, whose average amplitude was larger than that in the other channels. Channel 08 recorded small positive peaks with baseline drifting. Meanwhile, Channels 10 and 11 recorded both positive and negative peaks with smaller amplitudes. Other channels had no obvious peaks but were only recorded with steady baseline levels during this time. There were peaks appearing in these channels at other times. The correlation pattern analysis of neighboring channels provides more information about the neural activity during various experimental conditions.

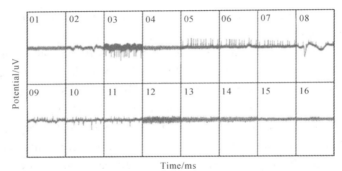

Fig. 5.4.11. Multi-channel recording electrophysiological signals of olfactory epithelium (reprinted from (Liu et al., 2010a), Copyright 2010, with permission from Elsevier Science B.V.)

On the basis of effectively recording spontaneous potentials, we recorded electrophysiological signals after odor stimulations. Fig. 5.4.12 shows the stimulated potentials compared with spontaneous potentials in one of the channels on the MEA. Spontaneous potentials were fired every second. The release of potentials after acetic acid stimulation was mainly about sustained signals with significant amplitudes in a short time, while the release of potentials after butanedione stimulation was mainly about long-term signals with low amplitudes. The frequency and amplitude of the olfactory epithelium potential changed with stimulation of different odors, showing some characteristics of the discharge modes.

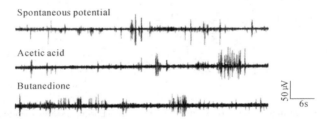

Fig. 5.4.12. Changes of electrophysiological signals after the stimulation of acetic acid and butanedione (reprinted from (Liu et a!., 2010a), Copyright 2010, with permission from Elsevier Science B.V.)

And, the recorded signals may be analyzed for individual spikes, the extracellular correlates of an action potential generated by a neuron, or the spikes of small populations of cells detected by one electrode. The overlapping potentials of larger populations of cells creates low-frequency components in the recording (local field potentials, LFP) that may reveal additional information not to be gained from spike data. So, the defined epithelial strata afford facile identification of extracellular electrophysiological recording sites with microelectrodes. In the study, we can distinguish the different discharge modes of spontaneous signals from those after the stimulation of butanedione and acetic acid both in time and frequency domain. The differences of spatio-temporal analysis may provide powerful support with pattern recognition for a practical bioelectronic nose system in the future.

5.5 Biochips

The completion of the human genome has revolutionized the traditional medical and biological research. It opens the door to tremendous analytical opportunities ranging from diagnostic tests for mutations to the assessment of medical treatment. Wide-scale DNA testing, the analysis of complex DNA samples and acquisition of sequence and expression information require the high throughput, parallel process analyzing, and sensitive techniques. The urgent demand for global analytical tools drives the rapid development of the biochip technology. The most attractive features of these devices are the miniaturization, speed and accuracy. DNA chips are the first kind of biochip. A number of terms, like DNA microarrays, gene chips or biochips, are often being intermixed to describe these devices. Such use of DNA microarrays is currently revolutionizing many aspects of genetic analysis. Soon, other types of biochips emerged, such as protein microarrays, tissue chips, cell chips, and lab-on-a-chip (Kricka, 2001). The microarray technology integrates molecular biology, advanced microfabrication/ micromachining technologies, surface chemistry, analytical chemistry, software, robotics and automation. All these chips share some common design characteristics, based on the specific hybridization, microarrays, optical readout, and software analysis. Therefore, they all belong to the "biochip" category. In the following text, the basic fabrication processes, the detecting systems and the related applications will be presented.

5.5.1 Chips of Microarray

Biosensors are small devices which utilize biological reactions for detecting target analytes. Such devices commonly couple a biological recognition element with a physical transducer that translates the recognition event into a useful detectable signal, such as an electrical and optical signal. And the biochips are characterized by the high-throughput parallel analyzing arrays (Wang, 2000). However, at the individual spot of microarray, it can be seen as a biosensor, where the hybridization information or other characteristic information is converted to a recognizable signal, such as fluorescence intensity. Therefore, generally speaking, the biochip is a biosensor array. Compared to biosensors, biochips can detect multiple analytes at the same time in a small area chip. That improves the analyzing efficiency greatly and decreases the volume of the analytes and reagents.

However, there are many differences between biosensors and microarray chips:

The range of the sensitive element: For the biosensor, DNA and enzymes are the most common used sensitive elements; in addition, the antibody, microorganism and cells can also be used to construct the biosensors. In the field of microarray chips, DNA chips have been commercialized. Protein arrays, tissue arrays and cell arrays are being developed as well.

Signal transforming methods: Many detection methods are employed by the biosensors, such as electrochemical analyzing, ion-selective electrodes, ion-selective FET, piezoelectric devices, optical methods; but the biochip is primarily based on the florescence detection and the following image process. Therefore, the related information can be readout.

Application fields: Biosensors have been applied broadly, such as agriculture, food, and environment and have not been restricted in the biomedical fields; the biochip is mainly used in the basic research for analyzing the DNA, protein, cell, and tissue.

5.5.2 Gene and Protein Chips

Microarray technologies of gene chips and protein chips have become crucial tools for large-scale and high-throughput biology. They allow fast, easy and parallel detection of thousands of addressable elements even in a single experiment.

5.5.2.1 *DNA microarray*

DNA is a very uniform and stable molecule which binds its complementary targets by means of a well-defined base-pairing principle. Based on this principle, DNA microarray is well developed. Currently a variety of DNA microarray and DNA chip devices and systems have been commercialized, which allow the DNA and/or RNA hybridization analysis to be carried out in microminiaturized highly parallel formats. Such array technology integrates molecular biology, advanced microfabrication/micromachining technologies, surface chemistry, analytical chemistry, software, robotics and automation (Heller, 2002).

The afterward microarray chips just adopt the basic process of DNA microarray. Fig. 5.5.1 shows a complete generic process diagram for microarray experiments. A complete microarray process includes a designing array probe, namely a capture molecule, selecting support material and manufacturing the chip; on the other hand, the analytes need pre-treating and labeling. In appropriate experimental conditions, capture molecules hybridize with target molecules. Then the chip is scanned and the data is post-processed and interpreted. Combining the bioinformatics resources, the experimental design can be further optimized.

The microarray system mainly consists of four integrated parts: a disposable DNA probe array, the test sample preparation and pre-label, the interaction between the probe and the target molecules, and a scanner to read the data, along with corresponding software to control the instrument and process the data.

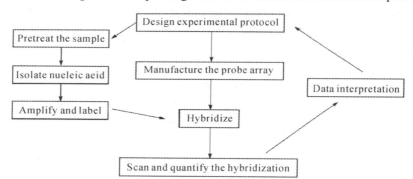

Fig. 5.5.1. Fabrication procedures and systems of DNA microarray

The fabrication of the DNA microarray is based on the supporting materials, such as glass, silicon, plastic, nylon membrane, and a nitrocellulose membrane. It is essential to activate the surface for a covalent attachment of the oligonucleotide probes. Then the probes are arranged in the solid support, and each site has its own physical ID. According to the density of the sites, the DNA microarray is classified into low-density and high-density DNA chips. The capture molecules can be immobilized by either a situ synthesis or spotting method. A situ synthesis method includes the use of photolithography for the *in situ* synthesis of high-density DNA microarrays, developed by Affymetrix, as well as the electronic-based addressing of microarrays developed by Nanogen. Many microarray spotting technologies and techniques now exist. Two of the most important spotting techniques used are the pin-based fluid transfer systems and the piezo-based inkjet dispenser systems. Once the probes are immobilized onto the solid support, a disposable DNA probe array is completed and can be used to detect the target molecules. The well-developed methods improve the density of the microarray and promote the manufacture of high-density DNA chips. Fig. 5.5.2 illustrates three kinds of microarray.

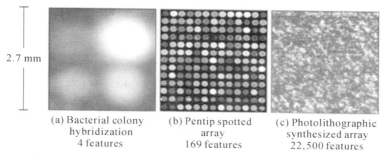

(a) Bacterial colony hybridization
4 features

(b) Pentip spotted array
169 features

(c) Photolithographic synthesized array
22,500 features

Fig. 5.5.2. Feature density of representative microarray. Each image shows a 2.7 mm subregion: (a) Bacterial colony spots on nylon from the 1980s; (b) Ink-jet in situ synthesized oligo spots on glass; (c) Photolithographic synthesized array (reprinted from (Stoughton, 2005), Copyright 2005, with permission from *Annual Reviews*)

Sample preparation includes isolating and amplifying the desired form of nucleic acid molecules. At the same time, label molecules can be incorporated during synthesis of amplification products. Nucleic acid amplification can be accomplished through reverse transcription of RNA or via polymerase chain reactions (PCR) or a combination of these. The most common dyes are florescence molecules, such as Cy3, Cy5, FITC, F12.

Hybridization is the most important step in the detection. The fundamental parameters are time, stringency, concentration, and complexity of the sample, as well as density of available binding sites. Secondary parameters include the distribution of fragment lengths, steric effects of dye molecules, and surface chemistry. Under optimal conditions, hybridization can be well conducted. Then washing off the unbound sample after hybridization is also a crucial step.

The detection of the DNA hybridization relies on the signal generated by the binding events, therefore scanning or imaging the chip surface is essential for obtaining the complete hybridization pattern. Florescence imaging and mass spectroscopy are commonly used for such "reading" of the chips. Bioinformatic tools are used to translate the complexity of the data into useful information. Electrochemical detection of DNA hybridization is another important method; it meets the needs for point-of-care diagnosis and is ideal for shrinking the hardware. Scanning of a fluorescent hybridization signal can be done with CCD imaging, but now it is more commonly done with laser confocal scanners. The laser confocal approach has fundamental geometric advantages that tend to provide better signal-to-background ratios and less photo bleaching of the labels. Most devices have lasers and filter sets compatible with common fluorescent label pairs such as Cy3 and Cy5. New options for brighter individual labeling units, such as quantum dots and plasmon resonance particles, may finally allow single-molecule detection efficiency, further easing requirements on amplification and on the amount of input required for the biological sample amounts.

In general, DNA microarray hybridization applications are usually directed at gene expression analysis or for point mutation/SNP analysis. In addition to these important molecular biological and genomic research applications, microarray systems are also used for pharmacogenomic research, for infectious and genetic diseases and cancer diagnostics, and for forensic and genetic identification purposes. Table 5.5.1 shows some typical uses of the DNA microarray (Stoughton, 2005).

Table 5.5.1 Modes of use for DNA microarray

Purpose	Target sample	Multiplexed reactions	Demultiplexing probes on array
Expressing profiling	mRNA or totRNA from relevant cell cultures or tissues	Amplification of all mRNAs via some combination of RT/PCR/IVT	Single-or double-stranded DNA complementary to target transcripts
Pathogen detection and characterization	Genomic DNA from microbes	Random-primed PCR, or PCR with selected primer pairs for certain target regions	Sequences complementary to preselected identification sites
Genotyping	Genomic DNA from humans or animals	Ligation/extension for particular SNP regions and amplification	Sequences complementary to expected products
Resequencing	Genomic DNA	Amplification of selected regions	Sequences complementary to each sliding N-mer window along a baseline sequence and also to the three possible mutations at the central position
Find protein-DNA interactions	Genomic DNA	Enrichment based on transcription factor binding	Sequences complementary to intergenic regions

5.5.2.2 Protein microarray

The cellular functions of a living cell are mediated by the proteins and the accurate description of biological processes which requires in-depth knowledge of protein expression and the functional state of proteins. Proteomics development demands a powerful high-throughput tool to identify and quantify the target proteins. DNA microarray technology revolutionized gene analysis successfully, and therefore a protein microarray can be fabricated in the same fashion. Compared to the DNA microarray, the protein microarray offers more diversity. It can be used for the analysis of interactions between proteins with other proteins, peptides, low molecular weight compounds, oligosaccharides or DNA (Fig. 5.5.3). Technologies established for DNA chip applications have been adopted to cater for the needs of protein microarray-based research. The basic process of the protein microarray is similar to the DNA chip, such as selecting the solid supports, modifying the support, immobilizing the capture molecule, detection methods and data interpret methods (Templin et al., 2003).

As shown in Fig. 5.5.3, there are many types of capture molecules in protein microarray, such as monoclonal antibodies, polyclonal sera, scFV/Fab, diabodies, affinity binding agents, scaffolds, affibodies, aptamers, and DNA/RNA/peptide. Monoclonal antibodies represent a virtually unlimited source of uniform, pure and highly specific binding molecules. However, antibodies, be they polyclonal or monoclonal, have some disadvantages in terms of generation, cost and overall application. The most promising alternative technologies in this field involve phage display techniques combined with highly diverse synthetic libraries. This enables the fast and efficient production of ultra diverse protein molecules and leads to the selection of binder molecules which can be directed against nearly any target within a few weeks. Aptamers are also a kind of good candidate for the capture molecules of the protein microarrays.

The lack of specific capture molecules is the main factor which still limits the broader use of protein microarray technology. There is no one-by-one interaction as observed in DNA base pairing. Proteins exhibit very diverse and individual tertiary molecular structures. Their binding interaction takes place by different means such as electrostatic forces, hydrogen bonds and/or weak hydrophobic Van der

Waals interactions. In addition, individual proteins can even interact with different binding partners at the same time and in a synergistic way. At present, there is no way which would allow the prediction of high affinity protein capture molecules only on the basis of their primary amino acid sequence. Steady or dynamic post-translational modifications like glycosylation, phosphorylation, and acetylation must also be taken into consideration. Table 5.5.2 illustrates some distinct properties in microarray technology between DNA and proteins (Templin et al., 2003).

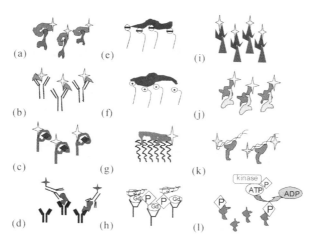

Fig. 5.5.3. Types of protein interaction and protein capture microarrays. Specific protein capture on microarrays by affibodies (a), antibodies (b), aptamers (c) or antibody sandwich formation (d). Unspecific capture is based on electrostatic (e, f), van der Waals-hydrophobic (g) or metal-chelate (h) interactions. Specific interaction microarrays have been described for receptor-ligand (i), protein-protein (j), protein-DNA (k) and enzyme-substrate interactions (l)

Table 5.5.2 Properties of DNA and proteins with respect to their applications in microarray technology

Properties	DNA	Protein
Structure	Uniform	Individual types
	Hydrophilic acidic backbone	Hydrophobic and/or hydrophilic domains
	Stable	Fragile
Functional state	Denatured, no loss of activity, can be stored dry	3 D structure important for activity, Avoid denaturing
Interaction sites	1 by 1 interaction	Multiple active interaction sites
Interaction specificity	High	Dependent on individual protein
Activity prediction	Well defined, based on primary nucleotides sequence	Not possible yet. Efforts are undertaken to predict models that are based on sequence homologies, structure, etc.
Amplification	Established (PCR)	Not available yet

Capture molecules are immobilized in a microarray format in the same way as in DNA microarrays. The principles of solid support selection and probe immobilization are similar to that of a DNA microarray, and for the individual capture molecule, some special details should be considered. To detect the target proteins efficiently, the sample should be labeled by the markers, such as fluorescent molecules, enzymes, chemiluminescent substances, etc., and then be incubated on the array. Subsequently, bound proteins can be detected with the instruments that are used for conventional DNA microarray technology, such as laser confocal scanners. In addition, the protein microarray can be detected directly by the mass spectrum instrument.

Protein microarrays have broad applications. For proteomics, protein-protein interactions, enzyme-substrate assays, protein-DNA interactions, carbohydrate- protein interactions, and protein-small molecule interactions can all be conducted by the protein microarray. Fig. 5.5.4 shows a proteomic array. The arrays are made by immobilizing antibodies or ligands in distinct spots. The array is incubated with a population of sample proteins, which are either directly or indirectly tagged with an enzyme or dye detected by fluorescence, light generation or colorimetric readout. The level of signal on each capture spot is proportional to the concentration of the target protein in the original sample (Liotta and Petricoin, 2000).

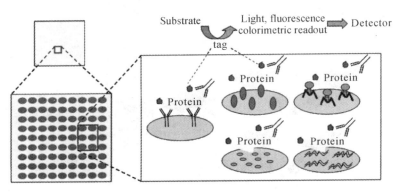

Fig. 5.5.4. Protein microarray

5.5.3 Tissue and Cell Chips

Tissue microarrays (also tissue chips) consist of paraffin blocks in which up to 1,000 separate tissue cores are assembled in array fashion to allow multiplex histological analysis. While, cell chips highlighted the multiparameters can be detected simultaneously in a range of cellular functions.

5.5.3.1 Tissue chips

Tissue chips follow the same modes of the DNA and protein microarray and a large number of specimens are arranged on one chip and treated simultaneously in an identical manner. In the past, immunohistochemistry or *in situ* hybridization, has been applied "one slide at a time" to a patient's tissue sections. Consequently, to screen hundreds of specimens from patients, it was necessary to stain hundreds of microscopic glass slides, each containing a tissue slice. That is a rather slow and labor-intensive process. Parallel processing of a large number of histological samples will dramatically increase the throughput. Therefore, tissue microarrays are powerful research tools when it comes to screening a large number of samples for well-defined parameters, but should be considered with caution when using them as a diagnostic tool for individual cases.

Taking Fig. 5.5.5 as an example, we will explain the fabrication process of tissue chips briefly (Liotta and Petricoin, 2000). The array consists of 1,000 cylindrical tissue samples, each from a different patient, all distributed on a single glass slide. Each tumor is represented by a minute disc-shaped tissue section, 0.6 mm in diameter and 4 – 8 μm in thickness. Fixed tumor tissue embedded in paraffin in blocks is sampled with a punch to generate cylindrical cores. The cores are packed together into a new block that contains cores from hundreds of patients. The composite array block is sliced into sections that are placed on a glass slide. The slide now contains hundreds of tumor samples. Immunohistochemical staining or *in situ* hybridization can be used to query the array for specific molecules such as insulin-like growth factor binding protein (IGFBP1), apoptosis related proteins (BCL2), heat-shock proteins (HSPs) and a transcription factor (GATA2).

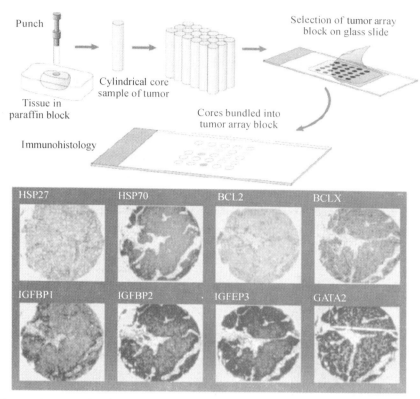

Fig. 5.5.5. Tissue microarray (reprinted from (Liotta and Petricoin, 2000), Copyright 2000, with permission from Macmillan Magazines, Ltd.)

The potential of tissue array technology has been tested by assembling two replicas of breast cancer tumors, each with 645 samples, and screening the arrays using known breast cancer markers. The data confirmed many of the clinic pathological correlations of gene amplifications, or immunostaining reactions, reported with conventional techniques on the basis of whole-tumor analysis. Tissue arrays cannot be applied to study normal epithelium or pre-malignant lesions. This is because the punch biopsy diameters are so small that they will often miss the branching ducts or glands dispersed in the tissue. Conversely, tumor arrays are ideal for comparing large numbers of solid-tumor samples. Full automation of tissue array creation and screening is envisioned as a means to expeditiously correlate marker levels over large panels of tumors. Many continuing studies are now using tumor arrays to amplify leads from DNA arrays.

5.5.3.2 *Cell chips*

Generally speaking, there are two kinds of cell chips, namely cell microarray and microfluidic cell chips. The concept of "cell chips" is a general description. Sometimes microfluidic chips and lab-on-a-chips are included in the same category when they are handling the cells as the target elements. Investigators took inspiration from printing strategies used to create DNA and protein microarrays and applied them to living-cell assays, this is called cell microarray. This platform arranges the cells in the microarray in order to improve the miniaturization and parallelization and is especially useful for the high-throughput analysis of gene expressing profiling and drug screening. The cell chip based on the microfluidic technology, explores the MEMS technology and sensing techniques, to detect the parameters related to the living cells, such as cellular electrophysical signals, cellular metabolism and proliferation, intracellular ions and proteins. The greatest advantage of the microfluidic cell chip is that the multiparameters can be detected simultaneously in one chip (Wu et al., 2002).

One typical manufacturing process of the cell microarray is illustrated in Fig. 5.5.6. DNA is mixed with gelatin and spotted as a high-density array on a glass slide, and it is dried and stored until needed for an experiment; To perform an experiment, the array is rehydrated, and cells are seeded uniformly across the array; Cells attach and spread across the array, and those cells residing above DNA spots take up the underlying nucleotides and become transfected in a spatially localized fashion. Importantly, results from cell microarray experiments, in contrast to microtiter plate assays, must always be quantified using automated microscopy and image-based analysis because the experiments are all performed on the same substrate in the same culture solution (Yarmush and King, 2009).

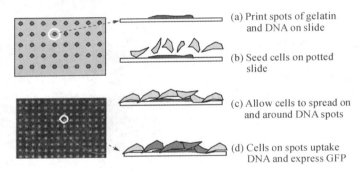

Fig. 5.5.6. Cell microarray

Cell microarrays have been applied to a broad range of applications in basic and applied biology, for example, overexpressing the defined cDNA to study the corresponding functions, to study intracellular localization of a library of genetic fusion proteins, or the Loss-of-Studies using plasmid-based siRNAs. Cell microarrays have advantages over traditional methods of expression cloning because the identity of each cDNA is known from its array coordinates. This eliminates the need to isolate the cDNAs responsible for phenotypes of interest, a process that often requires substantial work involving fluorescence-activated cell sorting or repeated rounds of sib selection. In addition, cell microarrays can be used to identify gene products that regulate cell cycle and cell growth. The high-throughput and real-time monitoring are the most distinguished advantages.

5.5.4 Lab-on-a-Chip

Lab-on-a-chip and microfluidic cell chips are all based on the microfluidic technology. Microfluidics is aimed at creating microscale closed-volume networks of channels that have emerged as a promising technology for miniaturizing and parallelizing fluid-addressable cultures and creating continuous-flow living-cell microarrays, and realizes the real-time spatial and temporal control of soluble cellular microenvironments (Yi et al., 2006). The detailed introduction of microfluidic chips has been presented in the Section of Chapter 4.

The cell chip in Fig. 5.5.7 integrated many functional units into one small chip (El-Ali et al., 2006). Advanced tissue organization and culture can be performed in this chip by integrating homogeneous and heterogeneous cell ensembles, 3D scaffolds to guide cell growth, and microfluidic systems for transport of nutrients and other soluble factors. Soluble factors—for example, cytokines for cell stimulation—can be presented to the cells in precisely defined spatial and temporal patterns using integrated microfluidic systems. Microsystems technology can also fractionate heterogeneous cell populations into homogeneous populations, including single-cell selection, so different cell types can be analyzed separately. Microsystems can incorporate numerous techniques for the analysis of the biochemical reactions in cells, including image-based analysis and techniques for gene and protein analysis of cell lysates. This makes microtechnology an excellent tool in cell-based applications and in the fundamental study of cell biology. As indicated by the yellow arrows, the different microfluidic components can be connected with each other to form an integrated system, realizing multiple functionalities on a single chip. However, this integration is challenging with respect to fluidic and sample matching between the different components, not least because of the difficulty in simultaneously packaging fluidic, optical, electronic and biological components into a single system. All these sequent events can be realized in one chip, and even much more complex functions can also be conducted. Therefore it is a kind of lab-on-chip or "micro-total-analysis-system" (μTAS) (Andersson and Berg, 2003).

Cell chips have great potentials in many aspects.

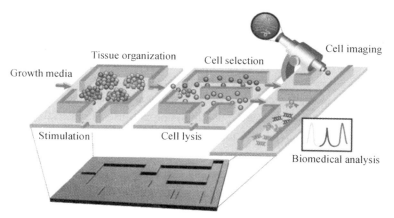

Fig. 5.5.7. Tissue organization, culture and analysis in microfluidic cell chip (reprinted from (El-Ali et al., 2006), Copyright 2006, with permission from Nature Publishing Group)

5.5.4.1 *Microfabricated cell cultures*

Culturing cells *in vitro* is one of the cornerstones of modern biology. Nevertheless, even for intensively studied tissues, many of the factors that induce or stabilize differentiated phenotypes are poorly understood and difficult to mimic *in vitro*. One approach to increase control over cell-cell and soluble cues typical of *in vivo* cell environments is to combine microfabrication of 3D ECM structures and microfluidic networks that transport soluble factors such as nutrients and oxygen. Microfluidics has the additional advantage of being capable of creating mechanical strain, through shear, in the physiological range.

5.5.4.2 Cell stimulation and selection

The control of cellular microenvironments via microfluidic systems potentially represents a valuable tool for fundamental studies of cell biology. Biological insight into the pathways that control cell phenotype and behavior can be gained by monitoring cellular responses to controlled perturbations in the extracellular environment. A wide range of microsystems are therefore emerging with the express aim of facilitating the basic study of biochemical pathways, cell-fate decisions and tissue morphogenesis. In the next two sections, we provide examples of some techniques being applied to cell-based assays.

5.5.4.3 Biochemical analysis of cell lysates

Currently, almost every analytical tool available in a conventional biology lab has an equivalent microfabricated counterpart. A significant research effort has been devoted to the development of integrated tools for microscale biochemical analysis. Quantitative analysis of complex biochemical mixtures, such as cell lysates, remains challenging, and with many devices success has only been achieved with low-complexity samples. The problems of low abundance and high complexity are generally handled in one of two ways: by linking sample preparation steps such as physiochemical separation and concentration before analysis, or by using high selectivity in the analytical system, typically through affinity methods based on antibodies. In this aspect, the microfluidic cell chip has profound application prospects.

5.6 Nano-Biosensors

In the past decade, nanotechnologies have greatly changed the state of science and technology. Nanotechnology involves the study, creation, manipulation and use of materials, devices and systems typically with dimensions ranging from 1 nm to 100 nm. The most commonly used nanomaterials include nanowire, nanotube, nanocapsule, nanopartical, nanochannel array, nanoporous membrane, etc. Now, nanotechnology is also playing an increasingly important role in the development of biosensors. Sensitivity and performance of biosensors can be improved by using nanomaterials, which display unique physical and chemical features due to effects such as the quantum size effect, mini size effect, surface effect and macro- quantum tunnel effect. Based on their submicron dimensions, nanobiosensors have allowed simple and rapid analyses *in vivo*. Even portable instruments which are capable of analyzing multiple components are becoming available.

A lot of works have reviewed the status of the various nanotechnology-based biosensors, especially at the molecular level (Chen et al., 2004; Helmke and Minerick, 2006; Kriparamanan et al., 2006). However, the application of nano- technology to biosensor design and fabrication is also promising to revolutionize diagnostics and therapy at the cellular level. The convergence of nanotechnology, biology, and photonics makes it possible to detect and manipulate atoms and molecules by using a new class of nanoprobes and nanosensors for a wide variety of medical application at the cellular level. Nanosensors have the potential for monitoring *in vivo* biological processes within/without a single living cell, e.g., the capacity of sense individual chemical species in specific locations of a cell, which will improve our understanding of cellular functions greatly, thereby revolutionizing cell biology. In this section, we will give some examples, such as nanomaterials, nanoparticles, nanopores, nanotubulars and nanowires for nano-biosensor studies.

5.6.1 Nanomaterials for Biosensors

Nano-biosensors combined the chemistry and materials science of nanomaterials and biomolecules with their detection strategies, sensor physics and device engineering. The important types of nanomaterials for sensory applications will be introduced in this section.

5.6.1.1 *Nanomaterials*

A long time ago, Nobel Prize winner Richard Feyman proposed that, when the dimension of the material is reduced to the similar values of its fundamental physical state, it would have a significant impact on the nature of the material. The emergence and development of nanoscience verified his prediction. In the early stages of nanomaterials development, nanomaterial is a nanoparticles, nanofilms or solid state, composed by nanoparticles. As shown in Fig. 5.6.1, nowadays generalized nanomaterial is composed of basic units of material; at least one dimension within the three-dimensional space is at least in the nanoscale range (Chopra, 2007).

For a dot with the diameter of 1 mm, the surface atomic proportion of the total volume is only 1%, and when the diameter decreases to 10 nm, the proportion is 25%, while the diameter is about 1 nm, it is 100%, that is, all of the atoms are distributed on the surface in this case. Forces between atoms as well as changes in the proportion of non-equivalent atoms (under-coordinated atoms) determine that nanomaterials are different from their bulk materials. Therefore, new physical and chemical properties will generate in such a system.

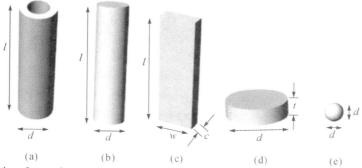

Fig. 5.6.1. Shows various forms of nanostructures with typical dimensions: (a) Nanotube, l: length (greater than 1,000 nm), d: diameter (less than 100 nm); (b) Nanowire, l: length (greater than 1,000 nm), d: diameter (less than 100 nm); (c) Nanobelt, l: length (greater than 1,000 nm), w: width (less than 500 nm), c: depth (less than 100 nm); (d) Nanodiskette, t: thickness (less than 100 nm), d: diameter (generally between 500 – 1,000 nm); (e) Nanoparticles, d: diameter (order of few nanometers) (reprinted from (Chopra, 2007), Copyright 2007, with permission from Taylor & Francis Group, LLC)

5.6.1.2 Nanomaterials in the biosensor application

A combination of biology and materials science in the nanoscale will have a revolutionary impact on many fields of science and technology (McNeil, 2005). The nanoscale in biology field is significantly relevant, for the reason that biomacromolecules such as proteins, DNA, as well as the scale of many important subcellular structures fall within the range of 1 – 1,000 nm (Fig. 5.6.2).

Recent research on the nanostructured materials, found that these nanomaterials have a huge potential to develop the unique capabilities of new devices and sensors. These nanomaterials have good biocompatibility, chemical stability, and changes of sensitive electrical characteristics when chemical composition changes. They are close to biomolecules in size, which become the sensitive materials in chemical-biological sensors. In the next few years, the development of material physics and chemistry will significantly change the development of biological molecules and tissue optical, magnetic, and electrical sensing. Controlling the state of material in the nanoscale will allow for the construction of new biosensors. This new system will allow for single molecule tests in living cells, which can also be integrated to test multiple signals in parallel, and handle a number of different reactions simultaneously. Some of the sensors based on the nano-technology platform will allow the carrying out of the electrical tests of biological and chemical substances directly, without labels. This platform uses functionalized nanomaterials such as nanoparticles, nanotubes or nanowires, which bind sensitively and specifically with the tested objects including DNA, RNA, proteins, ions, small molecules, cells and pH, and many other components. Therefore, the nanostructures are widely applied in biosensors with different characteristics and special uses, and the different nanostructure biosensors will be described respectively in the following sections.

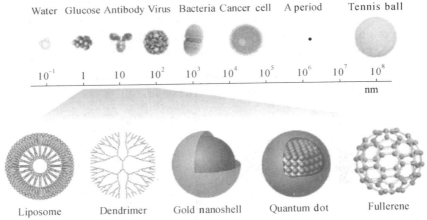

Fig. 5.6.2. The size distribution of different biological molecules and nanostructured materials (reprinted from (McNeil, 2005), Copyright 1997, with permission from the Society for Leukocyte Biology)

5.6.2 Nanoparticles and Nanopores Biosensors

Nanoparticles and nanoporous materials possess large specific surface areas, and high sensitivity to slight changes in environments. They are the most widely used nanomaterials in the biomedical field.

5.6.2.1 Nanoparticle biosensors

- **Characteristics of nanoparticles**

Nanoparticles are usually larger than 1 nm, and are one of the most interesting nanomaterials. The control of nano-particle size, particle size distribution, shape, surface modification as well as their photoelectric chemistry application is the key to nanoparticle research. This new material level between the micro- and macro-, has many unique characteristics.

Small size effect: changes of macrophysical properties caused by smaller particle size are known as the small size effect. For the nanoparticle, the size decreases, while its specific surface area increases significantly, the electronic energy levels of surface atoms are discrete, the energy gap widens, the lattice changes, and the surface atom density decreases, resulting in a series of new properties. For example, for special optical properties, all the metals show black in their ultrafine particles state. The size is smaller, the color is darker, and the light reflection rate is less than 1%. Another example is the special thermal properties. The melting point of ultrafine particles is much lower, and becomes even more obvious with a size of less than 10 nm.

Quantum size effect: nanoparticles between atoms, molecules and bulk solid, split a continuous energy band of the bulk material into discrete energy levels, and the energy gap increases as the particle size decreases. If the thermal energy, electric field or magnetic energy is lower than the average energy gap, nanoparticles will present a series of abnormal characteristics compared with macroscopic material.

Surface effect: the ratio of surface atom numbers and total atom numbers increases sharply as the particle size decreases, resulting in a change of characteristics. The surface area of spherical particles is proportional to the square of the diameter, while the volume is proportional to the cube of the diameter, so the surface area/volume ratio (namely, specific surface area) is inversely proportional to the diameter. As the diameter of spherical particles decreases, their surface area will increase significantly, therefore, they have higher surface chemical activity. Surface effect is manifested mainly in lower melting points, specific heat increases and so on. Surface effect manifests itself in a lower melting point and higher specific heat.

Macroscopic quantum tunnel effect: the tunnel effect is one of the fundamental quantum phenomena, that is, when the total energy of microparticles is less than the barrier height, the particles can still pass through this barrier. In recent years, scientists discovered some macroscopic physical quantities including microparticle magnetization, the magnetic flux and charge in quantum coherent devices, also have the tunnel effect. They can cross the barrier of macrosystems to cause changes, so it is called the macroscopic quantum tunnel effect.

Volume effect: the nanoparticles contain few atoms due to their minimal size, therefore, many physical and chemical properties related to the state of interfaces, such as adsorption, catalysis, diffusion, sintering are significantly different from that of traditional large particle materials. Consequently, the characteristics cannot be illustrated by the characteristics of the bulk material which has an unlimited number of atoms.

- **Preparation and modification of nanoparticles**

In addition to the characteristics of nanoparticles, their composition is also important for their applicability, for example, composition determines the compatibility and matching of nanoprobe and analyte, and also determines detection accuracy. The most common raw materials used for preparation of nanoparticles are gold, silicon and semiconductors (e.g., CdSe, ZnS, CdS) (Wang et al., 2008).

Gold nanoparticles are tiny gold particles, and usually form colloidal gold in aqueous solutions (Fig. 5.6.3). The sodium citrate reduction method is the most classic preparation method of colloidal gold. Different sizes of colloidal gold can be prepared by a reducing agent with different types and concentrations in the laboratory. Moreover, the method is simple and the raw materials are low cost. Colloidal gold has an absorption peak within the visible spectrum wavelength of 510 – 550 nm, and the absorption wavelength increases when the diameter of the gold particles increases. When the particle size changes from small to large, the apparent color shows a pale orange yellow, wine red, dark red, and blue-purple in turn. The properties of colloidal gold depend on the diameter and surface properties of gold particles. The diameters of the most important biomolecules (such as proteins, nucleic acids, etc.) are between 1 – 100 nm, so gold nanoparticle can be used as a probe into the biological tissue to detect biomolecule physiological functions, revealing life processes at the molecular level. Meanwhile the unique color change of gold nanoparticle particles is also an important basis for biochemistry applications.

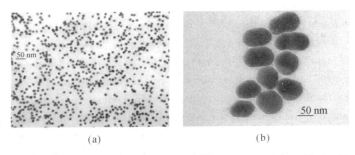

Fig. 5.6.3. Electron micrograph of gold nanoparticles: (a) and (b) are nanoparticles with the diameter of 18 nm and 70 nm, respectively (reprinted from (Wang et al., 2008), Copyright 2008, with permission from Springer)

Silicon has been widely used in bioanalysis, such as biosensors, biochips, etc. It can be synthesized by a variety of processing techniques to prepare nano- particles, transparent film and solid flat material. Preparation of silica nano- particles has two classic methods. One is the reverse microemulsion method, which is mainly used for synthesis of dye-doped silica particles and ultra-small magnetic silica particles; the other is the Stöber method for preparation of pure silicon particles and organic dye-doped silica particles. Synthetic silica particles are characterized by properties of dimension, optics or magnetics. The diameter of pure silicon particles can be confirmed by Transmission electron microscope (TEM), or scanning electron microscope (SEM), generally between 60 – 100 nm. Dye molecules in dye-doped silicon particles can be ruthenium bipyridine (RuBpy2), rhodamine, tetramethyl dextran, and fluorescein dextran, etc. The size and optical properties of this silicon particle are the most significant factors to determine the applications. Magnetic silica particles include Fe_3O_4/SiO_2 and Fe_2O_3/SiO_2 with diameter about 2 – 3 nm, which are close to super-paramagnetic material, consequently, the size and magnetic properties will determine the best synthesis conditions of magnetic silica particles.

Quantum dots (QDs) are semiconductor crystal materials in nanoparticles, within 10 nm in diameter, and the volume of normal cells is thousands of times larger than that of QDs. It has a wide range of absorption wavelength and a narrow range of emission wavelength. Using different materials will result in different fluorescents. Cadmium selenide (CdSe), zinc sulphide (ZnS), indium arsenide (InAs) are used for QDs, and the investigations have focused on CdSe in recent years. Synthesis of QDs offers a great deal of variety, such as the traditional fluorescent quantum dots and elongate nanobar for measuring anisotropy. The bottom-up one-step reaction is one of the classic methods, which allow for the inorganic chemical transformation and nanocrystallization processes in the same container.

- *Biological sensing applications of nanoparticles*

Gold nanoparticles are widely applied in the sensor field, and gold nanoparticles have good catalytic activity, a strong surface-enhanced resonance, surface tension, and nonlinear optical properties, which can be used as structural and functional units to fabricate sensors. The formation of multilayer film extends the properties and application of nanoparticles, while their biological effects can enhance the sensitivity of biosensors. Gold nanoparticles can serve as the surface modification material of DNA sensors to enhance their sensitivity. For example, when gold nanoparticles are introduced into the sensitive membrane preparation, the performance of chemical and biological sensors will be greatly improved.

Another example is identification of the target gene by gold nanoparticle-DNA probes (Liu et al., 2004). These studies have shown that nanotechnology plays an important role in the DNA sensor sensitivity, stability and specificity. Scientists found that gold nanoparticles can be used to enhance the fluorescence of fluorophore indirectly in the immune optical biosensors. Fluorescence effects can be enhanced after gold nanoparticles were fixed at the right distance from the fluorophore. They also found that nanoparticles combined with the biocompatible solvent can expand by tens of times the fiber optic biosensor signal, as well as cardiac markers which can be accurately quantified to the level of 0.1 pmol. Experiments of gold nanoparticles to enhance DNA sensor sensitivity have been carried out. Gold nanoparticles can be used as an amplifier. Besides, the sensitivity of DNA detection can exceed 10^{-16} mol/L in a quartz crystal microbalance (QCM) system modified by gold nanoparticles (Fig. 5.6.4), and is much higher than a QCM sensor without gold nanoparticles modification. The enhancement of antibody fragments adsorption is applied in immunosensor platforms, which is based on gold nanoparticles, and QCM was directionally used as the transducer to adsorb antibody fragments. This enhancement of immobilization technology in nano-gold particles is expected as the immunosensor platform in solid-phase measurement, affinity chromatography and so on. Experiments showed that the method has good performance with high sensitivity, fast response rate, and operational stability.

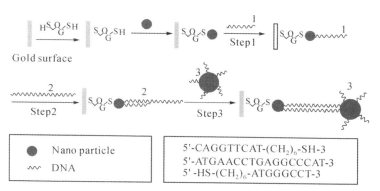

Fig. 5.6.4. DNA detection mechanism by QCM with gold nanoparticles (reprinted from (Liu et al., 2004), Copyright 2004, with permission from Elsevier Science B.V.)

At the same time, the study of enzyme immobilization by gold particles with diameter less than 10 nm also showed that gold nanoparticles can significantly improve the glucose oxidase (GOD) electrode sensitivity and lifetime. Hydrophilic and hydrophobic gold nanoparticles have good conductivity, which can be used as the electron transfer medium between the electrode surface and the GOD intermolecular to improve the electron transfer process in electrodes. Compared to using three kinds of nanoparticles, compound nanoparticles composed of SiO_2 and gold or platinum can significantly improve the glucose sensor's current response.

In the biological sensors for bacterial detection, nano-particles improve and enhance the bacterial concentration and separation, signal amplification and other processes. Immunoassay based on a fluorescence signal of biomodification nanoparticles can even detect individual bacteria. Many bacterial surface antigens can recognize antibodies and combine with them, and nanoparticles can amplify the signals. The surface of every bacterial will combine thousands of nanoparticles, providing an enhanced fluorescence signal (Fig. 5.6.5). And a single *E. coli* can be detected within 15 min by this approach (Tan et al., 2004).

In the detection process of *E. coli* O157:H7 DNA by QCM, frequency changes of QCM caused by DNA which captures bacterial is too weak (<1 Hz). Therefore, after the electrode surface is fixed with a probe and hybrid with a target DNA, it is necessary to introduce Fe_3O_4 nanoparticles, binding it to the target DNA chain to amplify the signal by increasing the mass of nanoparticles. The QCM with amplification nanoparticles DNA sensor can detect *E. coli* of 2.67×10^2 CFU/mL, and also found that the frequency change was linear in the concentration from 2.67×10^2 CFU/mL to 2.67×10^6 CFU/mL. From this result, scientists designed a resistance biosensor based on interdigitated microelectrode array for *E. coli* O157:H7 rapid detection. Biotin-labeled specific antibodies combined with streptavidin-coated magnetic nanoparticles, are used to separate enriched bacteria from the minced beef juice sample. The microelectrode array can detect bacteria concentration of 8.0×10^5 CFU/mL, and the time from preparing sample to detecting result is about 35 min.

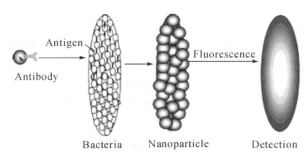

Fig. 5.6.5. The individual bacteria detection mechanism by nanoparticles fluorescence enhancement (reprinted from (Tan et al., 2004), Copyright 2004, with permission from Wiley Periodicals, Inc., a Wiley Company)

5.6.2.2 *Nanopores biosensors*

- ***Characteristics of nanopores***

Nonapore material can be divided into nanopore and nanopore membranes, and carbon, silicon, silicates, metal oxides, polymer materials are common materials that can be used. Alumina nanopore membrane prepared by anodizing is used as a template material to assemble and design nanostructured material and function devices, which is a hot topic in the nanomaterials research field. Such nanopore membranes have good heat resistance and insulativity, even and orderly holes distribution, controllable size, etc. In addition to the native characteristics of the material, nanopores with special structures also have other characteristics.

(1) Large specific surface area, controllable pore size, morphology and distribution, expand the chemical properties of the material surface. Under specific temperature and pressure conditions, nanopore materials with good penetrability and selectivity, can be used for selective separation of gases. Large specific surface areas of nanopore material can contain or absorb more biochemical substances with a small size. Therefore, it can be used as the catalyst carrier and the response platform in the catalytic reactions.

(2) The nanopore materials with large interior space may also be used for gas or liquid loading and storage, however, it is still in the preliminary study stage.

(3) Nanopore diameter of nanomaterials membrane match with the single biomolecule, so nano membrane material can also be used as the shell membrane for drug delivery and biological encapsulation to transport macromolecular drugs into the living body, and play a significant role in DNA sequencing, and single-molecule analysis as well.

- *Preparation of nanopores*

Using the nanopore membrane as a template to synthesize other orderly nanostructures, the pore size and length of the template determine the size of the nanomaterials to a certain extent. Furthermore, it will determine the properties and functions of the nanomaterials. Consequently, preparation of different diameter and thickness of the nanoporous alumina templates is the key step in the preparation of nanomaterials.

Anodic oxidation of aluminum began in the 1920s, which was mainly for the manufacturing of electrolytic capacitors. For decades, many scientists have carried out deep and extensive research on properties, micro-structure and the growth mechanisms of anodized aluminum. Anodic oxidation of aluminum is now mainly applied in the sulfuric acid, phosphoric acid, oxalic acid, or other acidic electrolytes. Sulfuric acid has been widely used in particular due to its low cost, high transparency oxide film, good corrosion wear resistance, easy dye and electrolytic coloring.

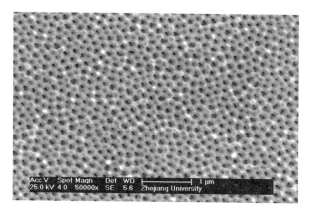

Fig. 5.6.6. SEM micrograph of porous anodic alumina membrane

As shown in Fig. 5.6.6, porous alumina film has a unique and highly ordered nanopore array structure, forming a number of hexagonal micropores, which are perpendicular to aluminum substrate with a small diameter. The diameter of micropores is between 10 – 500 nm depending on different oxidation conditions. The density of micropores is very large, between $10^{13} - 10^{14}$ per square meter. And the diameter of porous alumina film and other parameters changes under different conditions. However, the porous alumina film maintained its fixed shape and structure features, so porous alumina film has the benefit of good heat resistance, stability and insulation, even pore distribution and higher pore density compared to the polymers.

- **Biological sensing applications of nanopores**

In recent years, due to humidity sensors and ammonia sensors in food quality tests and the importance of meteorological research, development of these sensors have aroused people's attention. Among them, Pennsylvania State University, developed a humidity sensor and an ammonia body sensor, and did a print out of an Au electrode with different nano apertures on one side of an alumina film by an evaporation mask, and using an anodic porous alumina membrane as the sensitive medium. Studies have shown that the device's response to the behavior of ammonia and humidity strongly depends on membrane pore size and operating frequency. In the 5 kHz frequency, with an average pore diameter of 13.6 nm sensor ammonia and argon at work, the magnitude of its impedance increased two orders; while at the same frequency, the same sensor at 20% – 90% relative humidity environment, the magnitude of impedance can increase three orders, showing good sensitivity properties.

Nanopores with dimensions comparable to the sizes of biological polymers, such as short DNA and peptides, have been successfully. Researchers from Mexico State University have used single nanopores to sensor DNA hybridization via ionic conductance (Vlassiouk et al., 2005). Their study showed that nanoporous alumina modified with covalently linked DNA can be used to detect target DNA by monitoring the increase in impedance of the electrode upon DNA hybridization, which resulted from blocking the pores in the ionic flow (Fig. 5.6.7). Using cyclic voltammetry, direct current conductance, and impedance spectroscopy, they confirmed the importance of pore size: the effect is observed with 20-nm-diameter pores and is absent for 200 nm pores. So, nanoporous alumina with covalently linked ss-DNA on its surface can be used for electrical detection of complementary target DNA sequences without modification.

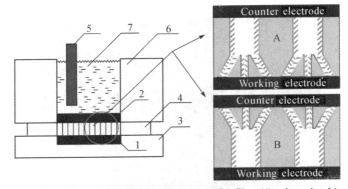

Fig. 5.6.7. DNA sensor based on nanoporous alumina. Two options for filter (4) orientation [A, working electrode (1) at the 20 nm side of the membrane; B, at the 200 nm side] in the homemade electrochemical cell with a stainless steel screen counter electrode (2) and reference minielectrode (5), immersed in solution (7). The working electrode is made of Pt. (reprinted from (Vlassiouk et al., 2005), Copyright 2005, with permission from American Chemical Society)

At the same time, solid-state nanopores have emerged as possible candidates for next-generation DNA sequencing devices. Researchers from Purdue University reported that the DNA sequence would be determined by measuring how the forces on the DNA molecules and also the ion currents through the nanopore, and ion current change as the molecules pass through the nanopore (Iqbal et al., 2007). Functionalized with a "probe" of hair-pin loop DNA, nanopores can selectively transport short lengths of "target" ssDNA that are complementary to the probe under an applied electrical field. As shown in Fig. 5.6.8, even a single base mismatch between the probe and the target can result in longer translocation pulses and a significantly reduced number of translocation events. The results can be explained in the conceptual framework of diffusive molecular transport with particle-channel interactions.

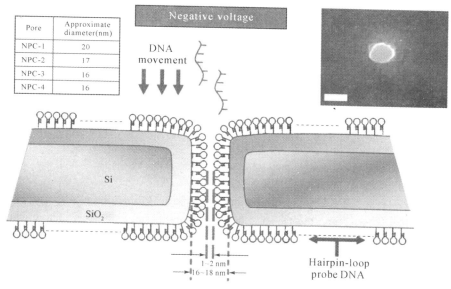

Fig. 5.6.8. Cross-section of the solid-state nanopore channels (NPC) functionalized with HPL-DNA molecules (not drawn to scale). The inset table shows dimensions of the various NPCs used in this study. The inset TEM image shows the NPC-2 before functionalization (scale bar, 20 nm) (reprinted from (Iqbal et al., 2007), Copyright 2007, with permission from Nature Publishing Group)

Putting the binding properties of receptors together with a transducer to form biosensor systems for a variety of applications has been achieved by reintegrating receptors or ion channels into lipid membranes, because of their immobilization conditions closely resemble the natural cellular environment. Recently, advances in the development of nanoporous substrates for electrochemical characterization of membrane protein-containing lipid bilayers have greatly improved techniques for lipid membrane self-assembly and membrane protein incorporation on these substrates. Anodic aluminum oxide is one of the particular interesting nanoporous membranes due to its excellent biocompatibility as well as its established fabrication process. With precise pore diameter and length achieved, anodic aluminum oxide can be used as solid supported membranes to lipid membranes, and appear to be well suited for the development of membrane biosensors with fully functional transmembrane ion channels. Instead of single nanopores, porous arrays consisting of densely packed pores will avoid current leakages and be capable of working with lower protein concentrations or alternatively with proteins that have a low charge translocation rate, such as ion-transporters. Several channel peptides and proteins have been studied with single channel recordings on the platforms. Ion-channel and receptor protein functionality was measured for up to 10 h and the blocking capacity of the membrane was observed for up to 6 d.

And, because of its excellent biocompatibility as well as the established fabrication process, nanoporous anodic aluminum oxide is of particular interest in cell culture and monitoring cell response. Alumina surfaces incorporating porous features on the nanoscale show significant biointegration and cell ingrowth; and, in addition, the cell response can be improved with nanoscale architecture. It has already been extensively used as a substrate for tissue construction.

5.6.3 Nanotubes and Nanowires Biosensors

Since carbon nanotube was discovered in 1991, one-dimensional nanomaterials have attracted the interest of researchers. One-dimensional nanomaterials have a new structure, which can be defined as material, whose mean free path for its nanomaterials charge carriers is greater than a two-dimensional scale. Based on morphology characteristics shown by the electron micrograph, one-dimensional nanomaterials are divided into nanowires, nanorods, nanobelts, nanotubes, etc. A lot of work on these materials has been carried out in synthesis, characterization, and applications.

5.6.3.1 Characteristics of nanotubes and nanowires

One-dimensional materials have a high apparent ratio (scale limitation in both directions), so the specific surface area is large, and the Debye length (the role distance of either charge electric field in the plasma) is close to the material size. As a result, the surface chemical processes have high sensitivity due to these properties. Because of the scale limitation effect, these materials have an adjustable band gap, high optical gain and rapid response, along with other characteristics.

One-dimensional nanomaterials have attracted extensive attention among scientists, for the reason that they provide many methods for mesoscopic physics research, and also provide material for the preparation of nanodevices. One-dimensional materials can be used to study dependence on the dimension and scale reduction caused by electronic heat transfers and mechanical properties, which is expected for connection parts and functional units to play an important role in the preparation of electronics, optoelectronics, electrochemical and electro-mechanical devices, etc., in nanoscales.

Some metal nanowires will act as the semiconductor, after the diameter reaches a certain scale. Some consider that the conduction band and valence band move backwards extending the band-gap due to the quantum constraints. Some metal generates the ballistic effects of electronic conduction in the nanowires pattern. Semiconductor materials in nanowire patterns maintain the original electrical properties (e.g., 17.6 nm GaN nanowires), others were insulators (e.g., 15 nm Si nanowires). In addition, nanowires have many important photoelectric properties, such as field emission, surface plasmon resonance, photoconductivity and optical switching characteristics, and so on. One-dimensional nanomaterials have many different characteristics from those of bulk materials, and will not be repeated here.

One-dimensional materials have a lot of unique characteristics, so that they are extensively studied as the sensitive material of sensors. The properties include the following aspects:

(1) One-dimensional nanomaterials have large specific surface areas, which mean that a large

proportion of atoms or molecules are on the material surface, and are involved in surface reactions.

(2) The Debye length of most semiconductor oxide nanowires is equivalent to the radius in a wide temperature range and the doping level, which means that their electrical properties will be strongly influenced by the surface processes. It can be seen that the conductance of nanowires can change with the surface processes in the state of complete insulation and high-conductivity, which can improve test sensitivity significantly.

(3) For oxide semiconductor nanowires, they have a more certain chemical ratio, more complete crystal form, and are more stable than the multi- particle oxide sensors.

(4) Nanowires can build field-effect transistor (FET) devices easily, which make it possible to combine with existing preparation device technology. The construction of three-terminal FET devices can control the energy band position in the Fermi level. Furthermore, it can influence and control surface processes by electrical means.

5.6.3.2 Preparation of nanotubes and nanowires

The growth mechanism and crystallization process of one-dimensional mainly includes nucleation and growth processes. When the concentration of solid construction components (atoms, ions or molecules) is high enough, they are gathered into small clusters (or cores) by homogeneous nucleation. Larger structures are produced by a continuous component supply. The growth process should be reversible, and the rate is controllable. According to the synthetic environment, preparation methods of one-dimensional materials can be generally divided into vapor deposition and liquid deposition. And there are also the corresponding physical and chemical reactions that need to be considered, during these processes. One-dimensional structures for many materials have been synthesized by different methods, including simple substance and compound materials, such as carbon nanotubes, silicon nanowires, tin nanowires, GaN nanowires, and various one-dimensional nanostructures of metal oxides. At present, carbon nanotubes, and silicon nanowires are two hot topics in science and technology studies, shown in Fig. 5.6.9.

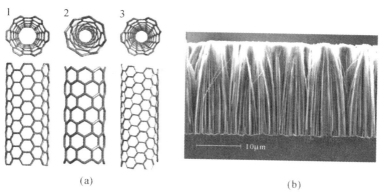

Fig. 5.6.9. Carbon nanotubes and silicon nanowires: (a) 1, 2, 3 are carbon nanotubes of the armchair, zigzag and chiral structures, respectively; (b) Electron micrograph of silicon nanowires

Arc discharge is the main method to produce carbon nanotubes. As early as 1991, the Japanese expert in electron microscopy, Sumio Iijima, discovered by accident, carbon molecules (namely carbon nanotubes) which consisted of tubular coaxial nanotubes, when he observed spherical carbon molecules produced by graphite are equipment under high-resolution TEM. A carbon nanotube is several layers to tens of layers of coaxial pipe, also consisting of hexagonal carbon atoms, with a fixed distance about 0.34 nm between layers and a diameter of 2 – 20 nm. Because of their unique structures, carbon nanotubes research has important theoretical significance and potential application value. For example, its unique structure is an ideal one-dimensional model material; with a large aspect ratio, that is expected to be used as a tough carbon fiber, and its strength is 100 times stronger than steel with only 1/6 the weight of steel; besides, it can be used as molecular wires, nanosemiconductor materials, catalyst carrier, molecular absorbent, near field emission materials and so on.

Semiconductor properties or metallic properties of carbon nanotubes are determined by the curl direction, which is difficult to control effectively during the preparation of carbon nanotubes, making it difficult to perfectly prepare semiconductor properties or metal properties of carbon nanotubes. Therefore, it limits the applications of carbon nanotubes in electronic devices. While one-dimensional silicon nanomaterials have stable semiconductor properties and are compatible with modern semiconductor technology, consequently, silicon nanomaterials have better practical value in microelectronic fields. Silicon nanowires were prepared by lithography technology as well as the method of using a scanning tunneling microscope, but the output was very low, which restricted the practical application. Until 1998, a large number of silicon nanowires were prepared successfully by a light burning technique, and this allowed for the faster development of silicon nanowires. Since then, synthesized silicon nanowires had also been prepared successfully by chemical vapor deposition (CVD), hot gas-phase deposition, organic solvent growth and other methods, respectively.

5.6.3.3 Applications

- **Carbon nanotube biosensors**

Carbon nanotubes are formed from carbon atom layers of graphene sheet rolled into a seamless, hollow tank; its radial diameter is between 1.4 nm to 60 nm, and with an axial length range from a few microns to more than one centimeter. According to the wall configuration, the number of layers of carbon atoms can be divided into single-walled carbon nanotubes (single-walled nanotube, SWNT) and multiwalled carbon nanotubes (multiwalled nanotube, MWNT). As a one-dimensional nanomaterial, carbon nanotubes have good mechanical properties, have better toughness and strong compression capabilities, and will help build biochemical sensors; wall carbon nanotubes, however, have a large number of topological defects, have greater reactivity, and offer both metal and semiconductor conductivity. These unique properties can make carbon nanotubes develop into different types of nano-scale bio-electrodes and sensors.

In enzyme-based biosensors, the electrical contacting of redox enzymes with electrodes has been a subject of extensive research in the last decade, with important implications for developing biosensing enzyme electrodes. Tethering of redox-relay units to enzymes associated with electrodes, the immobilization of enzymes in redox-polymers, and the reconstitution of enzymes on relay-cofactor units associated with electrodes were reported as means to establish electrical communications between redox proteins and electrodes. Single-walled carbon nanotubes (SWCNTs) were reported to have been used as a nanoconnector that electrically contacted the active site of the enzyme and the electrode (Patolsky et al., 2004). The electrons are transported along distances greater than 150 nm and the rate of electron transport is controlled by the length of the SWCNTs. The compatibility of SWCNTs with the preparation of novel biomaterial hybrid systems may have fascinating new properties for enzyme biosensors (Fig. 5.6.10).

In addition, the immune sensors, combined with the electrochemical analysis method, based on carbon nanotubes immune sensors provide the advantages of being low-cost, and offering fast and convenient methods for testing. The anti-lg (anti-immunoglobulinG) and lgG fixed with carbon nanotube arrays, using electrochemical impedance spectroscopy detection of unlabeled antigen-antibody binding conditions, can be tested by changing the parameters to improve the detection sensitivity. In addition, the carbon nanotubes-DNA and carbon nanotubes-RNA complex, are also increasingly applied to gene diagnosis *in vivo* experiments.

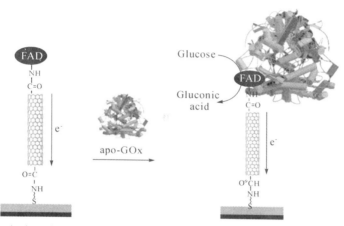

Fig. 5.6.10. The reconstitution of glucose oxidase on a flavin-adenine-dinucleotide (FAD)-functionalized single-walled carbon nanotube (SWCNT) associated with an Au electrode yields an electrically contacted biocatalyst. The efficiency of the electrical contact is controlled by the length of the SWCNT (reprinted from (Patolsky et al., 2004), Copyright 2004, with permission from Wiley-VCH Verlag GmbH & Co. KGaA, Weinheim)

After enzymes, antibodies, DNA, and other biological molecules have been fixed on carbon nanotubes, the promising nano-sensors can be designed. Produced through a variety of models with specific biological features, devices can be used for drug delivery and cell pathology research. The use of the spiral structure of carbon nanotubes is also an easy method for splitting chiral molecules. The disadvantage is that the current preparation processes, due to the complexity of the carbon nanotubesands, make it more difficult to obtain pure carbon nanotubes, for which scientific research has brought certain difficulties, and is not conducive to the carbon nanotubes in industrial applications. It is also believed that with improvements in the preparation and purification processes, carbon nanotubes will open up a whole new broader area for biosensor development.

- ***Silicon nanowire biosensors***

Silicon nanowire is a new one-dimensional nanomaterial, and their line diameter is about 10 nm, which is a single crystal silicon nuclei, and the outer coating is covered with SiO_2. Doped silicon nanowires, over carbon nanotubes and other nanomaterials, have better field emission properties, in the flat panel display technology, and also have better value than the carbon nanotubes, with excellent transmission performance and a better stability of the electronic properties of semiconductors (Patolsky and Lieber,

2005). Transistors are the basis for preparation of nano-electronic device components; doped silicon nanowires can be used to fabricate with excellent performance FET.

The ability of nanowire field-effect devices to detect species in liquid solutions was demonstrated for the case of hydrogen ion concentration or pH sensing (Cui et al., 2001). A basic *p*-type Si nanowire device had the silicon oxide surface modified with 3-aminopropyltriethoxysilane, which yields amino groups at the surface along with the naturally occuring silanol (Si-OH) groups of the oxide (Fig. 5.6.11). The amino and silanol moieties function as receptors for hydrogen ions, which undergo protonation/deprotonation reactions, thereby changing the net nanowire surface charge. Significantly, the nanowire devices modified in this way increase in conductance as the pH is increased gradually from 2 to 9. The nearly linear increase in conductance with pH is attractive from the standpoint of a sensor, and results from the presence of two distinct receptor groups that undergo protonation/deprotonation over different pH ranges.

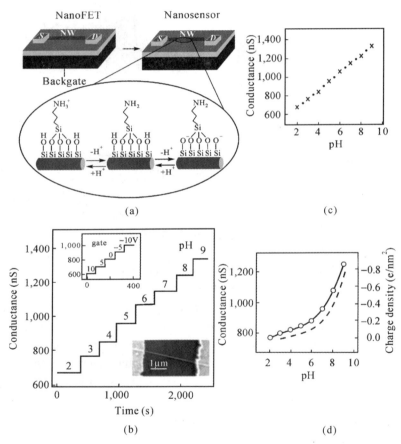

Fig. 5.6.11. Nanowire pH sensors: (a) Schematic of an amino-functionalized nanowire device and the protonation/deprotonation equilibria that change the surface charge state with pH; (b) Changes in nanowire conductance as the pH delivered to the sensor varied from 2 to 9; inset is a plot of conductance data versus pH; (c) Schematic of an unmodified nanowire sensor containing silanol groups and the protonation/deprotonation equilibria that change the surface charge state with pH; (d) Conductance of an unmodified Si nanowire device (red) versus pH. The dashed green curve is a plot of the surface charge density for silanol groups on silica as a function of pH (reprinted from (Cui et al., 2001), Copyright 2001, with permission from the American Association for the Advancement of Science)

The key role that the surface receptor plays in defining the response of the nanowire sensors was further tested by probing the pH response without modifying the silicon oxide surface. When only the silanol group function serves as a receptor for hydrogen ions, measurements of the conductance as a function of pH exhibit two different response regimes, unlike nanowire surfaces containing both amino and silanol receptors, where the conductance change is small at low pH (2 to 6) but larger compared to the high pH (6 to 9). This comparison in these early experiments clearly demonstrated that the sensing mechanism was indeed the result of a field effect analogous to applying voltage by a physical gate electrode.

- *Nanorods biosensors*

With the development of the nanotechnology, devices of nanoscale dimensions became capable of probing the inner space of single living cells, leading to new information on the inner workings of the entire cell. As a novel approach for system biology research, it can greatly improve our understanding of cellular functions. These nano-sensors could be fabricated to have extremely small sizes, which make them suitable for sensing intracellular/intercellular physiological and biological parameters in microenvironments.

Zinc oxide (ZnO) has received considerable attention because of its unique optical, semiconducting, piezoelectric, and magnetic properties. ZnO nanostructures exhibit interesting properties including high catalytic efficiency and strong adsorption ability. Recently, researchers have been focusing toward the application of ZnO in biosensors because of its high isoelectric point, biocompatibility, and fast electron transfer kinetics (Kumar and Chen, 2008). Such properties indicate that ZnO is one of the more promising materials for biosensor applications.

The focus of the current biosensor study is the fabrication of nanostructure ZnO nanorods suitable for intracellular pH sensing. Some authors have reported ZnO nanorods as an intracellular sensor for pH measurements (Al-Hilli et al., 2007). Their main effort has been directed toward the construction of tips capable of penetrating the cell membrane as well as optimization of the electrochemical properties. In this study, ZnO nanorods with a diameter of 80 nm and length of 700 nm, grown on the glass capillary (0.7 μm in diameter), were used to create a highly sensitive pH sensors for monitoring H^+ within single cells. The ZnO nanorods, functionalized by proton H_3O^+ and hydroxyl OH^- groups, exhibit a pH-dependent electrochemical potential difference versus an Ag/AgCl microelectrode (Fig. 5.6.12). The potential difference was linear over a large range, which could be understood in terms of the change in surface charge during protonation and deprotonation. Therefore the nanoelectrode devices have the ability to enable analytical measurements in single living cells and the capability to sense individual chemical species in specific locations within a cell.

Besides intracellular detecting, some groups have also reported nanosensors for extracellular studies. For example, field-effect transistor arrays of silicon nanowire have been used to record neuronal signals. In their studies, hybrid structures of nanowire arrays were integrated with the individual axons and dendrites of live mammalian neurons. And, they think those nanoscale junctions can be well used as biosensors for highly sensitive detection, stimulation, and/or inhibition of neuronal signal propagation, with simultaneous measurement of the rate, amplitude, and shape of signals propagating along axons and dendrites. It was also a very important and successful application of nanotechnology for biosensor research.

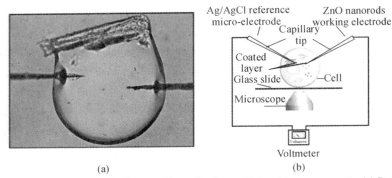

Fig. 5.6.12. Optical image and schematic diagram illustrating intracellular pH measurements: (a) Performed in a single human fat cell using ZnO nanorods as a working electrode with Ag/AgCl reference microelectrode; (b) Schematic diagram (reprinted from (Al-Hilli et al., 2007), Copyright 2007, with permission from American Institute of Physics)

As a summary and conclusion, biosensors have the potential to provide low cost detection and measurement technology for accurate and highly-specific quantification of low levels of important analytes in multi-component samples. The major commercial application areas for biosensor technology are medical diagnosis, food and hygiene analysis, environmental monitoring, security, and industrial process monitoring. The relatively slow rate of progress of biosensor technology from inception to fully functional commercial devices for these applications is due to both technology-related and market factors. Biosensor research has advanced to its present state in which a range of biological molecules and systems have been successfully coupled, in a relatively stable form, to transducers (generally electrochemical or optical) to provide specific analytical devices, often with high sensitivity of detection. Future work is likely to focus on the development of strategies for enhancing biomolecule stability (that is, extending sensor lifetime) and for permitting reversibility of response and/or regenerable sensors. Most importantly, the next generation of biosensors must be able to maintain acceptable performance in complex and often potentially interference-inducing sample matrices such as whole blood.

References

Abbas A. & Roth B.L., 2008. Protein engineering: electrifying cell receptors. *Nature Nanotechnology.* 3, 587-588.

Al-Hilli S. M. Willander M., Öst A. & Peter S., 2007. ZnO nanorods as an intracellular sensor for pH measurements. *Journal of Applied Physics.* 102, 084304.

Andersson H.A. & Berg V.D., 2003. Microfluidic devices for cellomics: a review. *Sensors and Actuators B-Chemical.* 92, 315-325.

Byfield M.P. & Abuknesha R.A., 1994. Biochemical aspects of biosensors. *Biosensors and Bioelectronics.* 9, 373-400.

Carmon, K.S., Baltus R.E. & Luck L.A., 2005. A biosensor for estrogenic substances using the quartz crystal microbalance. *Analytical Biochemistry.* 345, 277-283.

Chen J., Miao Y., He N., Wu X. & Li S., 2004. Nanotechnology and biosensors. *Biotechnology*

Advances. 22, 505-518.

Chopra N., 2007. Functional one-dimensional nanomaterials: applications in nanoscale biosensors. *Analytical Letters.* 40(11), 2067-2096.

Clark LC Jr. Monitor and control of blood and tissue O_2 teusions. *Trans Am Soc Artif Intern Organs.* 1956; 2, 41-48.

Cooper M.A., 2002. Optical biosensors in drug discovery. Nature Reviews Drug Discovery. 1, 515-528.

Cui Y., Wei Q., Park H. & Lieber C.M., 2001. Nanowire nanosensors for highly sensitive and selective detection of biological and chemical species. *Science.* 293, 1289-1292.

Dutra R.F. & Kubota L.T., 2007. An SPR immunosensor for human cardiac troponin T using specific binding avidin to biotin at carboxymethyldextran- modified gold chip. *Clinica Chimica Acta.* 376, 114-120.

El-Ali J., Sorger P.K. & Jensen K.F., 2006. Cells on chips. *Nature.* 442, 403-411.

Giaever I. & Keese C.R., 1993. A morphological biosensor for mammalian cells. Nature. 366, 591-592.

Gross G.W., Rhoaes B.K., Azzazy H.M.E. & Wu M.C., 1995. The use of neuronal networks on multielectrode arrays as biosensors. *Biosensors & Bioelectronics.* 10, 553-567.

Hafner F., 2000. Cytosensor® microphysiometer: technology and recent applications. Biosensors and *Bioelectronics.* 15, 149-158.

Heller M.J., 2002. DNA microarray technology: Devices, systems, and applications. *Annual Review of Biomedical Engineering.* 4, 129-153.

Helmke B.P. & Minerick A.R., 2006. Designing a nano-interface in a microfluidic chip to probe living cells: challenges and perspectives. *Proceedings of the National Academy of Sciences USA.* 103, 6419- 6424.

Iqbal S.M., Akin D. & Bashir R., 2007. Solid-state nanopore channels with DNA selectivity. *Nature Nanotechnology.* 4 (2), 243-248.

Kovacs G.T.A., 2003. Electronic sensors with living cellular components. Proceedings IEEE. 91, 915-929.

Kricka L.J., 2001. Microchips, microarrays, biochips and nanochips: personal laboratories for the 21st century. *Clinica Chimica Acta.* 307(1-2), 219-223.

Kriparamanan R., Aswath P., Zhou A., Tang L. & Nguyen K.T., 2006. Nanotopography: cellular responses to nanostructured materials. *Journal for Nanoscience and Nanotechnology.* 6, 1905-1919.

Kumar A.S. & Chen S., 2008. Nanostructured zinc oxide particles in chemically modified electrodes for biosensor applications. *Analytical Letters.* 41, 141-158.

Liotta L. & Petricoin E., 2000. Molecular profiling of human cancer. *Nature Reviews Genetics.* 1, 48-56.

Lipschutz R.J., Fodor S.P.A., Gingeras T.R. & Lockhart D.J., 1999. High density synthetic oligonucleotide arrays. *Nature Genetics.* 21, 20-24.

Liu Q., Cai H., Xu Y., Li Y., Li R. & Wang P., 2006. Olfactory cell-based biosensor: a first step towards a neurochip of bioelectronic nose. *Biosensors and Bioelectronics.* 22, 318-322.

Liu Q., Huang H., Cai H., Xu Y., Li Y., Li R. & Wang P., 2007a. Embryonic stem cells as a novel cell source of cell-based biosensor. *Biosensors & Bioelectronics.* 22, 810-815.

Liu Q., Cai H., Xu Y., Xiao L., Yang M. & Wang P., 2007b. Detection of heavy metal toxicity using cardiac cell-based biosensor. *Biosensors & Bioelectronics.* 22, 3224-3229.

Liu Q., Cai H., Xiao L., Li R., Yang M. & Wang P., 2007c. Embryonic stem cells biosensor and its application in drug analysis and toxin detection. *IEEE Sensors Journal*. 7, 1625-1631.

Liu Q., Yu J., Liu Z., Zhang W., Wang P. & Yang M., 2008. A novel taste sensor based on ion channels incorporated in nano-lipid bilayer membranes, *The Seventh Asian-Pacific Conference on Medical and Biological Engineering*, Apr. 22-25, Beijing, China.

Liu Q., Yu J., Xiao L., Tang J.C.O., Wang P. & Yang M., 2009. Impedance studies of bio-behavior and chemosensitivity of cancer cells by micro- electrode arrays. *Biosensors & Bioelectronics*. 24, 1305-1310.

Liu Q., Ye W., Xiao L., Du L., Hu N. & Wang P., 2010a. Extracellular potentials recording in intact olfactory epithelium by microelectrode array for a bioelectronic nose. *Biosensors & Bioelectronics*. 25, 2212-2217.

Liu Q., Ye W., Yu H., Hu N., Du L. & Wang P., 2010b. Olfactory mucosa tissue based biosensor: a bioelectronic nose with receptor cells in intact olfactory epithelium. *Sensors and Actuators B-Chemical*. 146, 527-533.

Liu T., Tang J. & Jiang L., 2004. The enhancement effect of gold nano- particles as a surface modifier on DNA sensor sensitivity. *Biochemical and Biophysical Research Communications*. 313, 3-7.

McNeil S.E., 2005. Nanotechnology for the biologist. *Journal of Leukocyte Biology*. 78(3), 585-594.

Mohanty S.P. & Kougianos E., 2006. Biosensors: a tutorial review. *IEEE Potentials*. MARCH/APRIL, 35-40.

Moreau C.J., Dupuis J.P., Revilloud J., Arumugam K. & Vivaudou M. 2008. Coupling ion channels to receptors for biomolecule sensing. *Nature Nanotechnology*. 3, 620-625.

Patolsky F. & Lieber C., 2005. Nanowire nanosensors. *Materials Today*. 8, 20-28.

Pancrazio J.J., Whelan J.P., Borkholder D.A., Ma W. & Stenger D.A., 1999. Development and application of cell-based biosensors. *Annals of Biomedical Engineering*. 27, 697-711.

Patolsky F., Timko B.P., Yu G., Fang Y., Greytak A.B., Zheng G. & Lieber C.M., 2006. Detection, stimulation, and inhibition of neuronal signals with high-density nanowire transistor arrays. *Science*. 313, 1100-1104.

Patolsky F., Weizmann Y. & Willner I., 2004. Long-range electrical contacting of redox enzymes by SWCNT connectors. *Angewandte Chemie International Edition*. 43, 2113-2117.

Reimhult E. & Kumar K., 2008. Membrane biosensor platforms using nano- and microporous supports. *Trends in Biotechnology*. 26, 82-89.

Stoughton R.B., 2005. Applications of DNA microarrays in biology. *Annual Review of Biochemistry*. 74, 53-82.

Subrahmanyam S., Piletsky S.A. & Turner A.F.P., 2002. Application of natural receptors in sensors and assays. *Analytical Chemistry*. 74, 3942-3951

Tan W.H., Wang K.M., He X.X., Zhao X.J., Drake T., Wang L. & Bagwe RP., 2004. Bionanotechnology based on silica nanoparticles. *Medicinal Research Reviews*. 24, 621-638.

Templin M.F., Stoll D., Schwenk J.M. Potz O., Kramer, S. & Joos T.O., 2003. Protein microarrays: promising tools for proteomic research. *Proteomics*. 3, 2155-2166.

Turner A.P.F., 1996. Biosensors: past, present and future. http://www.cranfield.ac. uk/biotech/ chinap.htm.

Turner A.P.F., 2000. Biosensors-sense and sensitivity. Science. 290, 1315- 1317.

Vlassiouk I., Takmakov P. & Smirnov S., 2005. Sensing DNA hybridization via ionic conductance through a nanoporous electrode. *Langmuir*. 21, 4776- 4778.

Vo-Dinh T. & Cullum B., 2000. Biosensors and biochips: advances in biological and medical diagnostics. *Analytical Chemistry*. 366, 540-551.

Wang J., 2000. From DNA biosensors to gene chips. *Nucleic Acids Research*. 28, 3011-3016.

Wang L., Liu Q., Hu Z., Zhang Y., Wu C., Yang M. & Wang P., 2009. A novel electrochemical biosensor based on dynamic polymerase-extending hybridization for E. Coli O157:H7 DNA detection. *Talanta*. 78, 647-652.

Wang L., Wei Q., Wu C., Hu Z., Ji J. & Wang P., 2008. The Escherichia coli O157:H7 DNA detection on a gold nanoparticle-enhanced piezoelectric biosensor. *Chinese Science Bulletin*. 53, 1175-1184.

Wang L., Wu C., Hu Z., Zhang Y., Li R. & Wang P., 2008. Sensing Escherichia Coli O157: H7 via frequency shift through a self-assembled monolayer based QCM immunosensor. *Journal of Zhejiang University-Science B*. 9, 121-131.

Wang P. & Liu Q., 2009. *Cell-Based Biosensors: Principles and Applications*. Artech House Publishers, USA.

Wu R.Z., Bailey S.N. & Sabatini D.M., 2002. Cell-biological applications of transfected-cell microarrays. *Trends in Cell Biology*. 12, 485-488.

Wu Y., Wang P., Ye X., Zhang Q., Li R., Yan W. & Zheng X., 2001. A novel microphysiometer based MLAPS for drug screening. *Biosensors and Bioelectronics*. 16, 277-286.

Yarmush M.L. & King K.R., 2009. Living-cell microarrays. *Annual Review of Biomedical Engineering*. 11, 235-257.

Yi C., Li C. & Yang M., 2006. Microfluidics technology for manipulation and analysis of biological cells. *Analytica Chimica Acta*. 560, 1-23.

Yoshida N., Yano K., Morita T., McNiven S.J., Nakamura H. & Karube I., 2000. A mediator-type biosensor as a new approach to biochemical oxygen demand estimation. *Analyst*. 125, 2280-2284.

Zhang Q., Wang P., Wolfgang J.P., George M. & Zhang G., 2001. A novel design of multi-light LAPS based on digital ompensation of frequency domain. *Sensors and Actuators B-Chemical*. 73, 152-156.

Index

A
Absolute humidity (AH) 183, 186
Accuracy 15
Action potential 154, 247, 258
Active measurement system 20
Ad hoc network 216
Addition and subtraction circuit 22
Additional error 51
Affinity biosensor 223, 233
Alternating current bridge 52, 55, 66
Amplitude modulation 24, 27
Amplitude-frequency characteristic 18, 22
Analyte 34, 36, 139, 172, 200, 223, 248, 272
Anode 141, 147, 169, 230
Antibody 5, 36, 223, 226, 233, 273
Antigen 3, 33, 223, 226, 233, 281
Anti-interference capability 23
Artificial neural network (ANN) 195
Auxiliary electrode 147, 149
Averaging technique 30
B
Bias potential 162
Biochip 7, 11, 41, 223, 240, 244, 258, 272
Bioelectrical signal 3, 247
Bioelectronic nose 38, 256
Biomagnetic field 3, 4
Biomedical material 32, 36
Biomimetic sensors 11
Biosensor 2, 7
Black lipid membrane (BLM) 245

Boltzmann rule 162
Buffered solution 153
C
Calibration curve 14, 151, 188
Calomel electrode 148
Capacitive sensor 47, 81, 85 95
Carrier 23-28, 110, 129, 159, 171, 275
Catalytic biosensor 223, 227
Cathode 141, 147, 169, 230
Chalcogenide glass 162, 199, 221
Charging current 164-166
Chemiluminescence 208
Chemotherapy 249, 253
Clark oxygen electrode 223, 229, 231
Clean room 202
Cold junction compensation 130
Compact layer 144
Computing circuit 13, 22
Conductance 2, 141, 238, 276, 282
Controller area network bus (CAN bus) 210
Counter electrode 149, 166, 168, 252, 276
Cramer-Rao bound (CRB) 217
Cross-linking 5, 228, 231, 234, 246
Cross-sensitive 197, 198
Crystal lattice 48, 144
Current-voltage characteristics 163
Cytosensor 248, 249
D
Dac and adc interface circuit 23
Deep reactive-ion etching (DRIE) 202

Detection limit 14, 139, 151, 165, 208, 237
Dew point 183, 186
Differential equations 16, 17
Differential rectification circuit 77
Differential transformer sensor 73
Diffusion coefficient 164, 166
Diffusion layer 144, 165
Digital signal processing 19, 23
Diode 87, 119, 130
Diode temperature sensor 130, 131
Direct measurement 20, 236
Direct current bridge 52, 55
DNA chip 241, 243, 244, 260-262
DNA microarray 244, 259-263
Drain current 155, 156
Drift 16, 23, 55, 60, 110, 152
Dynamic characteristic 13, 14, 16, 17

E

Eddy current sensor 69
Electric cell-substrate impedance sensing (ECIS™) 251
Electric double layer 161, 164
Electrochemical 8, 60, 137, 140, 156, 167, 199, 208, 242
Electrochemical gas sensors (EGS) 167
Electrochemical reaction 140, 168
Electrode potential 144, 145, 148, 164
Electrode spacing 165
Electrolyte-insulator-semiconductor (EIS) structure 159
Electrolytic cell 140, 141, 152, 164, 165
Electromotive force (EMF) 150, 178
Electronic nose (e-Nose) 139, 190
Electronic tongue (e-Tongue) 139, 196
Electroosmosis 202, 204, 205
Electrophysiology 223
Envelope detection 27, 28
Equilibrium electrode potential 144
Ethernet bus 210
Euclidic distance (ED) 194
Extracellular acidification rate 157, 158, 231
Extracellular potential 37, 38, 159, 248, 254-257

F

Faraday current density 165
Field effect transistor (FET) 256
First-order sensors 17
Fm: frequency modulation 9, 24, 27, 68, 87
Frequency response 62, 69, 94, 238

G

Gate voltage 156
Gene chip 259
Glucose oxidase (GOD) 230, 274

H

Hall effect 110-112, 114
Hall sensor 106, 110, 114
Histocompatibility 6, 33
Hybridization 207, 240-244, 259-261, 276
Hysteresis 15, 75, 183, 186

I

Immune sensor 11, 223, 234, 236, 242, 243
Immunoassay 121, 161, 234-236, 274
Indicator electrode 147, 148
Indirect measurement 19, 20
Inductive sensor 47, 62, 80
Integrated temperature sensor 126, 130, 131
Interdigital transducers (IDTs) 180
Interface reaction 144
Interfacial potential 162
Inter-IC (I^2C) 210
Ion-channel 223, 245, 246, 278
Ion-exchange 144
Ionic activity 151, 156, 157
Ionization constant 143
Ion-selective electrode (ISE) 149
Ion-selective field-effect transistor (ISFET) 149, 155
Ion-selective membrane 155

L

Lab-on-chip 267
Laser induced fluorescence (LIF) 208
Light addressable potentiometric sensor (LAPS) 40, 159
Linear expansion coefficient 51, 52
Linearity 15, 30, 62, 65, 68, 72, 82, 128, 157
Linearization technology 30

Liquid film 157
Liquid junction potential 145-147, 150
Logarithmic and exponential circuit 22

M

Magnetoelectric sensor 106
Management node 216
Mass transfer rate 164
Mass transport 166
Measurement 1-4, 6, 9, 13, 14, 16
Measurement range 62
Membrane potential 151, 156, 254, 257
Metabolism 157, 158, 223, 231, 248, 266
Metal oxide semiconductor (MOS) 191
Field-effect transistor (MOSFET) 157, 191
Michigan parallel standard (MPS) 210
Micro total analysis systems (μTAS) 201
Microarray 223, 244, 259, 260, 261-267
Microelectrode 5, 11, 14, 154, 164, 208, 283
Microelectrode array (MEA) 38, 42
Micro-electro-mechanical systems (MEMS) 201, 218
Microfluidic chip 137, 201, 202, 204, 205, 207-209, 266
Microfluidics 201, 245, 248, 266, 268
Microorganism 2, 8, 223, 226, 227, 231-233, 248, 259
Microorganism based biosensor 231
Microphysiometer 248-251
Molecular imprinting 138

N

Nanobelt 269, 278
Nano-biosensor 11, 268, 269
Nanomaterial 223, 268-271, 275, 278, 280, 281
Nanoparticle 269-274
Nanopore 269, 270, 274-277
Nanorod 181-183, 278, 283, 284
Nanostructure 269, 270, 274, 275, 279, 283
Nano-technology 10, 11, 270
Nanotube 278, 280, 281
Nanowires 269, 270, 278-282
Negative temperature coefficient (NTC) 290
Nernst equation 144, 145, 178
Non-invasive detection 6, 7

Nonlinear error 15, 54, 65, 83, 108

O

Ohmic drop 165, 166
Olfactory 8, 11, 12, 37, 38, 190, 256-258
Oligonucleotide 241, 242, 260
Oxidation reaction 168
Oxygen electrode 223, 229-232
Oxygen on demand (BOD) 232

P

Parameter converting circuit 13, 21, 22
Passive measurement system 20
Pattern recognition (PR) 138, 189, 190, 192, 193, 196, 197, 198, 258
pH glass electrode 153
Phase-frequency characteristic 18
Phase-sensitive detection 27-29, 77, 78, 80
Photocurrent 116-118, 122, 159, 162, 250
Photoelectric effect 115, 121
Photosensitive diode 119, 120
Photosensitive resistance 116, 117, 118
Photovoltaic 121, 122, 123
Piezoelectric effect 96, 97, 102, 181, 237
Piezoelectric materials 98
Piezoelectric sensor 2, 22, 47, 96, 99, 100-102, 237
Piezoresistive coefficient 50
Pm: phase modulation 24, 26
Polarization 123, 141
Polydimethylsiloxane (PDMS) 93
Polymerase chain reaction (PCR) 244
Polyvinyl chloride (PVC) 157
Positive temperature coefficient (PTC) 128
Potentiometry 150
Power-aware routing protocol 216
Precision 5, 8, 11, 16, 30, 48, 60, 122, 128, 131, 219, 240
Principal component analysis (PCA) 200
Proportional to absolute temperature (PTAT) 132
Protein chip 259
Pulsed laser deposition (PLD) 162
Pwm: pulse-width-modulation 26

Q

Quantum dots (QDs) 272

Quartz crystal microbalance (QCM) 98, 103, 191, 237, 242, 256, 273

R

Ratio computing 22
Received signal strength (RSS) 217
Reduction reaction 168
Reference electrode 144, 147, 148-152, 156, 168, 178, 198
Relative error 15, 30, 83, 84, 108
Relative humidity (RH) 14, 133, 183-188, 276
Relative location estimation 217
Repeatability 13, 15, 23, 183
Resistance strain sensor 2, 48, 58
Resistance temperature detector (RTD) 126
Resonant amplitude modulation circuit 68
Resonant frequency modulation circuit 68
Respiratory inductance plethysmography (RIP) 80
Response time 5, 19, 152, 157, 169, 170, 173, 185, 244
Reynolds number 201, 207
RS-232 Bus 211

S

Salt bridge 145-147, 150
Second-order sensor 18
Seebeck effect 129
Selectivity coefficient 151, 152
Self-assembly monolayer (SAM) 239
Self-organized network 210
Semiconductor gas sensors (SGS) 167, 170, 191
Sensitive materials 5, 11, 48, 157, 167, 186, 200, 256, 270
Sensitive membrane 5, 157, 161, 162, 250, 251, 273
Sensitivity 7, 11, 13-15, 17, 48, 49, 53
Sensor 1-11
Serial peripheral interface bus (SPI bus) 213
Signal demodulation 27
Signal detection 3-5, 202
Signal modulation 21, 23, 24
Soft lithography 202
Solid electrolyte (SE) 167, 177, 178

Specificity 5, 11, 225, 226, 228, 234, 235, 240, 242, 244, 256, 263, 273
Spontaneous potential 257, 258
Stabilization technology 30
Standard hydrogen electrode 145, 148
Static characteristic 13, 14
Steady state 14, 18, 165, 166, 182, 249
Strain effect 48, 58
Strain gauge 2
Strong electrolyte 142
Surface acoustic wave (SAW) 99, 180, 191
Surface plasmon resonance (SPR) 236, 242, 256

T

Temperature coefficient 50, 51, 60, 98, 99, 112, 113, 127, 133, 152
Temperature drift 23, 60, 122
Temperature error 51, 113
Temperature compensation 23, 54, 55, 108, 111, 122
Thermistor 127, 128, 180, 187
Thermocouple 2, 47, 126, 128, 129, 130
Thermoelectric sensor 47, 126
Thermoelectric effect 129
Time-of-arrival (TOA) 217
Transfer function 17, 18, 195, 196
Transient electrochemical methods 165
Transient response 16
Transition region 50
Triode 131, 132

V

Variable reluctance sensor 63
Volatile organic compounds (VOC) 191
Voltage sensitivity 52-55

W

Weak electrolyte 142, 143
Wheatstone bridge 60, 61, 141, 187
Wireless sensor network (WSN) 210
Working electrode 34, 147-149, 168, 169, 178, 233, 245, 276, 284

Z

Zero-order sensor 17